P9-CMW-391

TIME RESOURCES, SOCIETY AND ECOLOGY

TIME RESOURCES, SOCIETY AND ECOLOGY

On the capacity for human interaction in space and time

Volume 1: preindustrial societies

Tommy Carlstein
Department of Social and Economic Geography,
University of Lund, Sweden

London
GEORGE ALLEN & UNWIN
Boston   Sydney

First published in 1982.

This book was typeset, proofed and cleared for press by the author.

This book is copyright under the Berne Convention. All rights are reserved. Apart from any fair dealing for the purpose of private study, research, criticism or review, as permitted under the Copyright Act, 1956, no part of this publication may be reproduced, stored in a retrieval system, or transmitted, in any form or by any means, electronic, electrical, chemical, mechanical, optical, photocopying, recording or otherwise, without the prior permission of the copyright owner. Enquiries should be sent to the publishers at the undermentioned address:

GEORGE ALLEN & UNWIN LTD
40 Museum Street, London WC1A 1LU

© T. Carlstein, 1982.

British Library Cataloguing in Publication Data

Carlstein, Tommy U.
    Time Resources, society and ecology
    Vol. 1: Preindustrial societies
    1. Time allocation.
    I. Title
    306   HM131

    ISBN 0-04-300082-7
    ISBN 0-04-300083-5 Pbk

Library of Congress Cataloging in Publication Data

Carlstein, Tommy.
    Time resources, society and ecology.

    Contents: v. 1. Preindustrial societies.
    1. Time management. 2. Space and time. 3. Human
ecology. I. Title.
    H91.C28 1982   304.2   81-19118
    ISBN 0-04-300082-7 (v. 1)   AACR2
    ISBN 0-04-300083-5 (pbk. : v. 1)

Printed in Sweden by Infotryck AB, Malmö.

For Umbereen

PREFACE

Initially, this book was intended to deal with innovations and rural change in the Third World. This interest of mine in development problems dates back to the mid-1960s. In those days, the issues of development were dominated by economic theory and, being a Swede, the geographic 'law' of distance and interaction put me in contact with the writings of Professor Gunnar Myrdal, a leading economist. Myrdal, however, was hardly an economist of the orthodox kind and one of his many tenets was that economic theory (in those days especially) was too static and that the *temporal and dynamic* aspects had been drowned in all the equilibrium formulations. Apart from the many substantive issues in this stream of theory, I also found these time aspects very thought provoking.

My disciplinary affiliations were more in the field of Human Geography, however. I had entered the Department of Social and Economic Geography at Lund University and was fortunate enough to have a renowned scholar, Professor Torsten Hägerstrand, as my teacher (and later as my colleague in a long-term project on time, space, society and geography). In Geography, of course, the *spatial dimensions* have long been basic to our understanding of the earth.

So my idea was to use the time/space dimensions to design an analytical framework for dealing with problems of development. One day at an informal seminar, Torsten Hägerstrand presented some pieces of work he had previously done in population studies (Hägerstrand 1963). He drew some trajectories of individuals in a three-dimensional time-space on the board in his office, and I was taken aback by this pithy way of combining all the dimensions in one graph in which the population was related to its habitat. This *graphic projection* then became the synthetic core around which I organized my further studies and research.

On the advice of Torsten Hägerstrand I started to search for data on time use and daily life in the anthropological literature and found a very useful initial source in American Anthropologist. I was swiftly engrossed with all the materials of Social and Cultural Anthropology. A monograph by a countryman who had explored an island in the

Pacific (Bengt Danielsson: Work and Life on Raroia, 1955) gave me some initial data for the application of the time-space approach to the real world (Carlstein 1966). Later when I decided to take up Social Anthropology on a more formal basis, I was fortunate in finding as my teacher Professor Karl Gustav Izikowitz at Gothenburg University. Apart from his well-known contributions to the study of shifting cultivation societies (1951), he also had a profound interest in the temporal organization of societies.

As mentioned before, the original theme of my studies was innovations and their consequences for (regional) development. Innovations strongly affect and are affected by resource utilization in society (Carlstein 1978). A major resource involved, apart from land (space), energy, water and different forms of human-made capital, is *human time*. This seemed to be the most neglected of the resources dealt with in social science, although human time is a resource which cuts across absolutely all sectors of human life, and hence could serve very well in a more *integrative approach to society*. It would thus help to build bridges between society, ecology, habitat and economy.

The theme was found to be comprehensive indeed, so in this book on 'Time Resources, Society and Ecology', I will begin by dealing with preindustrial societies. This volume is an extension of a limited edition published in December 1980 as a Ph.D. thesis. (In Sweden, doctoral theses have to be published.) However, this main edition was reserved for Allen and Unwin, and it also includes the new chapters (8 and 9) on comparative time allocation in preindustrial societies. Due to some intervening tasks, these chapters were not completed until now. However, with the reader's and publisher's permission, I will in a year or two put forward my corresponding materials on urban-industrial societies (Vol. II).

Department of Social and Economic Geogrraphy,
Sölvegatan 13,
223 62 Lund,
Sweden

April 1982   Tommy Carlstein

ACKNOWLEDGEMENTS

I would like to express my great debt of gratitude and warm thanks to the following persons and institutions:

— The Swedish Humanities and Social Sciences Research Council for having funded my research for a good number of years and for having created a platform for my explorations into one of the lesser researched areas of the social sciences.

— Torsten Hägerstrand for having been my longtime mentor and for inviting me to join in on the time-geographic venture. This field was opened up by some powerful germs of thought of his in the 1940's and 50's, and to me it has proved to be one of genuine excitement.

— Nigel Thrift for taking on the job as informal copy editor of many chapters in this book. His embellishment of the language and question marks in the margins have lead to multitudinous improvements.

— David Seamon, Allan Pred, Kenneth Olwig, Jonathan Friedman, and Jaya Appalraju for ameliorating the language of various sections (when there was no time to send off the script to Nigel Thrift in Australia and back.)

— Douglas Johnson at Clark University for helpful comments on my initial outline of the book.

— Professor Fredrik Barth and Dr. Ethel Wester for allowing me to use some of their previously published graphs.

— Eva Särbring for redrawing most of my graphs and Rezsö Laszlo for doing the photographic work.

— All the field-working anthropologists, geographers and others who collected the data on which this study is based.

— My friends and colleagues in the Geography Department at Lund, especially the members of the research group on time geography.

Last but far from least, I want to thank,

— My wife Umbereen for her support and willingness to put up with the unbalanced division of labour at home so that her husband could allocate more than full time to his work, and for helping with various other tasks in conjunction with the book; and my daughter Arjumand for quietly putting up with my being either absent-minded or "absent-bodied".

Again, thank you all.   T.C.

CONTENTS

PREFACE

ACKNOWLEDGEMENTS

1. ECOTECHNOLOGY, CARRYING CAPACITY, AND
   TIME-SPACE RESOURCES   1
   Introduction   1
   Biological ecology and human socio-cultural ecology   1
   Intensification theory   10
   The carrying capacity of terrestrial space-time   19
   Time resources and the carrying capacity of a population
       time-budget   22
   Objectives of this study   33

2. LIFE PATHS AND LIVING POSSIBILITY BOUNDARIES:
   ELEMENTS OF THE HÄGERSTRAND TIME-
   GEOGRAPHIC MODEL   38
   The time-space fragmentation of resources   38
   The Hägerstrand time-geographic model   39
   Living and activity possibility boundaries   52
   The Lund school of time-geography as a time-space
       structuralist approach   56
   A note on spatial structure and structuralism in human
       geography   60

3. HUNTING-GATHERING   65
   The hunting-gathering ecotechnology   65
   The Bushman hunting-gathering society   78
   Advanced hunter-gatherers: Relaxing the constraint on
       mobility   94

4. NOMADIC PASTORALISM   103
   Some characteristics of nomadic pastoralism   103
   The time-space structure of the nomad pastoralist habitat 106
   Aspects of short-term and long-term carrying capacity   111

Divisibilities and coupling constraints affecting capacity
    utilization   121
Agro-pastoralism as a mixed ecotechnology   126
Water scarcity versus snow abundance: The polar conditions
    of pastoralist ecotechnology   136
Intensification, expansion and sedentarization   141
Nomadism, multiple residence and daily commuting: Some
    concluding remarks on pastoralism and mobility   144

5.   SHIFTING CULTIVATION   147
    The shifting cultivation ecotechnology   147
    Rethinking rural settlement and land use   149
    Land cropping cycles and the intensity of space-time
        occupation   155
    The carrying capacity of the local prism habitat   162
    Village size, fusion and fission   175
    Shifting cultivation systems: Some tentative conclusions   185

6.   SHORT FALLOW CULTIVATION: REINTERPRETING
    THE STRUCTURE OF LOCAL AND REGIONAL
    INTENSIFICATION   187
    On the ecotechnology of short fallow (medium intensive)
        cultivation systems   187
    Intensification at the regional level   190
    Intensification at the local level   199
    Spatial zones of intensification   211
    Fertility improvement as time demanding activity and
        interaction   221
    Intensification at the domestic unit/farmstead level   228
    Population, activity and settlement as a composite system:
        A time-space structuralist perspective   244
    Structure, contradiction and intensification: Villagization
        in Tanzania revisited   250

7.   IRRIGATION AGRICULTURE   257
    The ecotechnology of irrigated cultivation   257
    Local irrigation systems   260
    The temporal allocation of water in space   269
    Regional irrigation systems and large scale coordination   287
    Irrigation and intensification   289
    Towards a composite space-time model of rural land
        resource mobilization   292

8. TIME ALLOCATION AND THE CARRYING CAPACITY
   OF A POPULATION TIME-BUDGET   301
   Time allocation and packing at the aggregate level   302
   Time allocation at the individual level   321
   A remark on time, money and economic anthropology   327

9. TIME RESOURCES IN PREINDUSTRIAL SOCIETIES   331
   Methodology and dimensions of time mobilization   333
   Generalizing the 'labour utilization identity'   344
   Time resources and alternative concepts of intensification   348
   Hunter-gatherers   360
   Shifting cultivators   364
   Short fallow cultivators   372
   Irrigation agriculturalists   379
   Time use intensity at the household level   386
   Population dynamics and the household level   393
   Time demand and sex role asymmetries   403
   A note on pastoralists   405
   Specialization, caste and other forms of vertical segmenta-
       tion in a population time-budget   407
   Class, productivity and factors of intensification   410
   Intensification theory: Some findings and complications   415

AFTERWORD   419

BIBLIOGRAPHY   423

INDEX   438

1 ECOTECHNOLOGY, CARRYING CAPACITY, AND TIME-SPACE RESOURCES

INTRODUCTION

In this chapter we will present the 'problematique' of this study and define its general objectives. Since we will look into issues that concern several disciplines, in particular human geography and social anthropology but also ecology, economics and sociology, the total composition of the material is likely to make the reader feel that important insights from his particular discipline have been left out or given very cursory treatment. And so they have, but this is the price to be paid for not operating within established fields. We can rest assured, however, that the reader will be sympathetic to the particular aims we have chosen, and that critical and constructive responses will be forthcoming.

BIOLOGICAL ECOLOGY AND HUMAN SOCIO-CULTURAL ECOLOGY

In plant and animal ecology, analysis is conventionally set within terms of the food-chain and food-web relations between various species as well as with respect to other input-output or interaction relations. Recently ecology has striven towards generalization of trophic relations in the form of energy flows, where a system overview is a achieved at a general quantitative level before the qualitative differences of energy are taken into account, as for example, in the differences between carbohydrates and proteins (H.T. Odum 1971). The importance of energy as a paramount resource is thereby stressed, and many crucial dependencies are highlighted, ranging from the autotrophic as in plants absorbing energy directly from the sun, via higher trophic levels of animals, to humans, who are high level consumers in the ecosystems found on earth.

Many of the basic concepts, models and theories of plant and animal ecology (biological ecology or bio-ecology for short) are also applicable to human socio-cultural ecology. The field of human population dynamics and the carrying capacity of human habitats is a case in point. Particularly for pre-industrial societies where the dominant modes of production are based on a biotic technology (i.e. hunting-fishing-gathering, pastoralism, or agriculture), ecological concepts are of major relevance. Even within the wider resource spectrum of industrial society, many general concepts and models may be useful, such as that of ecological balance (Wilkinson 1973).

The anthropologist Geertz (1963) adopted an ecological stance in his study of regional and economic development in Indonesia, but he nevertheless expressed scepticism about the broader application of conventional ecology to human society:

> The adaptation of the principles of ecological analysis and the concepts in terms of which they are expressed (niche, succession, climax, food chain, commensality, trophic level, productivity and so on) to the study of man can be constructed in a variety of manners, not all of which are equally useful... some... amount to hardly more than sloganeering. (Geertz 1963:5)

He was also critical of narrowing down the scope of ecological analysis to explanations of the territorial arrangements that social activities assume, i.e. to man's adaptation to space. In his eyes such 'locational theory' may be useful but hardly exhausts the subject.

By contrast, another anthropologist, Redfield, had virtually taken the opposite stand in line with the spatial location and organization paradigm and with the ecological approach of Park (1934) and Hawley (1950). Redfield states that,

> for the study of urban communities, the ecological system is quite inadequate, so that the conception becomes here a very different one in human ecology of American sociology: it becomes a study of spatial and temporal orders of settlement and of institutions without reference to animals, plants or the weather. (Redfield 1960:29)

Conventional ecology thus has to be considerably extended and revised in order to come to terms with industrial modes of production, complex technology and urban communities. (And, in fact, the energy theoretic models of H.T. Odum (1971) and many other ecologists is an important step in this direction.)

The solution, as Geertz sees it, is that of Steward's cultural ecology, rather than locational analysis. Ecological principles and concepts should be confined to 'explicitly delimited aspects of human social and cultural life for which they are particularly appropriate rather than extending them, broadly and grandly, to the whole of it' (Geertz 1963:6). The cultural ecological format implied the existence of a core of culture which was adapted to the natural habitat and its organismic relations, but that the rest of culture could not be understood simply by referring to space and ecology.

During the 1960's, the cultural ecological paradigm was further extended by American anthropologists by an ecological systems theoretic framework being applied to show how cultural institutions were useful in adaptation to the natural habitat (cf Vayda 1969 and Rappaport 1968 for a number of succinct statements of this case). Other anthropologists felt that this was a transgression, since it largely replaced environmental determinism by a bio-ecological variant. For all its numerous other merits, this form of 'vulgar materialism' (Friedman 1974) undoubtedly suffers from a reductionist bias which Steward (1955) with his culture core concept was able to avoid. But similar interesting experiments in extending and broadening the frontiers of human ecology are still essential contributions.

An increasing number of ecologists have recognized that when dealing with human populations which have culture and society, it is grossly inadequate to look into population dynamics alone, as a function of birth, migration, death and who eats who. Although a strategic form of interaction in nature is eating or being eaten, human society contains numerous other forms of interaction, for instance within the human population, which ultimately reflect on the way humans occupy their natural habitat and interact with plant and animal populations there. Rappaport (1969:184) is well aware of this.

> The concept of ecosystem, though it provides a convenient frame for the analysis of interspecific trophic exchanges /i.e. how different species consume one another/, does not comfortably accomodate *intra*specific exchanges taking place over wider geographic areas... Some sort of geographic population model would be more useful for the analysis of the relationship of the *local* ecological population to the larger *regional* population of which it is a part, but we lack even a set of appropriate terms for such a model.

If this form of geographic model is to be useful, it must no doubt also include time if it is to live up to expectations, since the choice is not between spatial or ecological analysis but rather how they can become compatible and how they can be expanded, reformed and/or revolutionized to cope with new and urgent problems.

The geographer Clarkson (1968) saw the relationship between

spatial and ecological analysis as one in which the former starts where the latter leaves off:

> Spatial analysis is concerned with factors affecting the location of specific activities... In agricultural location some of these factors will be features of the natural environment: soil, temperature, slope, hydrology, etc... For shifting cultivators, for example, the distance from dwelling, even a temporary dwelling, is much more heavily weighed that the distance from market (which has almost no weight at all) and the distance from water may be of more importance that all the rest combined.
> Ecologic analysis is concerned with the interaction of the factors which define the activity itself, rather than with how the factors affect the location of activity. Ecologic analysis concerns itself with the emergent system formed by the factors interacting and with analyzing how the system functions. Locational analysis, in a sense, begins where ecologic analysis leaves off — it takes the system investigated in ecological analysis as given and goes on to relate it to location, albeit without specifically stating this aim.

A major reason for the incompatibility of locational analysis with human ecology arises when space is primarily regarded as a locational matrix and not also as a *resource,* both in the concrete sense of land and in the abstract sense of space or room to accomodate populations and resources. Many of the classic theories in human geography such as von Thünens theory of agricultural location (cf Chisholm 1962) or Christaller's central place theory (1933), have largely neglected space as land and room (cf Hägerstrand 1973b). On the other hand, if the process of interaction between species and within the human population are to be analysed in any detail, the locational dimension of interaction is essential, as will be indicated later in this book. The perspective must then be enlarged to include time-space location. The dimension of time, both as a locational dimension and as a resource dimension, has received scant attention in geographic enquiry and social scientific enquiry in general (cf Carlstein, Parkes and Thrift, 1978, vols. 1-3 for a statement of the case.)

In fact, even the more detailed study of food-chains and trophic relationships requires an explicit time-space approach. As Parrack (1969) notes on the flow of energy through ecosystems:

> Much of the early work was confined to describing what eats what in food chains... Such work... is useful in that it indicates energy relationships, but it rarely includes quantification of these relationships in much detail. It is rather easy to make diagrams showing that mice eat grass seeds and that owls eat mice. It is something else to determine how much energy is available to a

mouse in a gram of seeds and how the mouse uses that energy, or how much energy an owl derives from a mouse and how much it spends in hunting mice.

What is interesting to observe from this statement is that in order to get at detailed energy relations, the *activities* of a species must be studied, because different activities demand varying amounts of energy as well as give (by eating) certain amounts of energy. To get at these facts one must study the *time use of individuals in space*, and *time-energy budgets* for different types of individuals are becoming increasingly important in bio-ecology. This situation also applies in modern energy-oriented studies in ecological anthropology, such as those by Kemp (1971) on the Eskimos, Lee (1968, 1969) on the Bushmen, Rappaport (1968) on the Tsembaga swidden cultivators, and a number of other studies in similar vein.

At this stage we can therefore summarize the discussion by observing that once we get down to the nuts and bolts of human ecology, a combined resource and locational framework is in all likelihood the most powerful approach, and that of the resources worthy of consideration, terrestrial and territorial space and population time are of as much importance as the traditional focus of attention, energy.

A definition of human ecology and ecotechnology

Although human ecology can be roughly defined as the relation between a human population and its habitat-environment, this definition is dangerously crude, and a number of additional aspects must be explicitly incorporated, as in Figure 1:1. Biologists working their way towards application of ecology to human populations and society, tend to see the latter as a black box and speak of 'man' and nature. This is detrimental to any deeper analysis of human ecology, since most of the ways in which human individuals and populations intervene with 'nature' are mediated by all sorts of socio-cultural mechanisms. Even nature itself is partly shaped by human socio-cultural activity so that, for example, to explain the cultural landscape grafted onto a 'natural region' requires understanding of technology. Or when we want to define the carrying capacity of a region for a human population, we again must specify the technology or technological culture of this population and the social interests which lead to the exploitation of some resources and organisms in the environment rather than others.

Human beings therefore have a culture and a culture based society. The sector of culture called technology defines the materials in nature that people are able to use for sociocultural ends and projects.

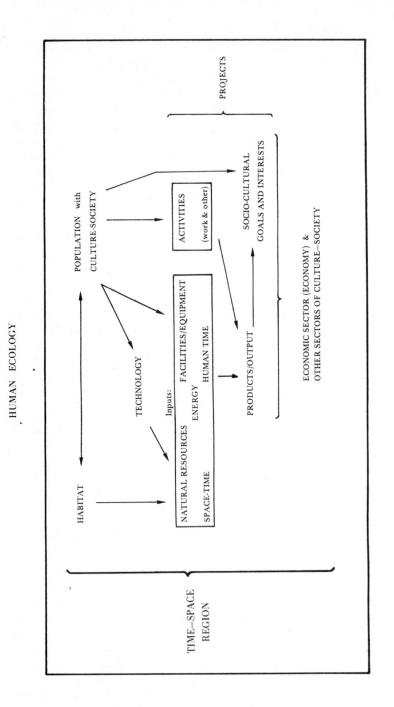

HUMAN ECOLOGY

FIGURE 1:1   A definition of human ecology as relations within society-habitat.

It is culture which defines what is a natural *resource.* Iron or uranium, which are resources to an industrial society, may merely be an obstacle in the landscape to a group of stone age hunters. Culture-society also defines many of the activities for which resources are used. As pointed out earlier, human ecology cannot confine itself to a theory of interaction by eating or trophic exchange between species. It must rest on a much more general theory of interaction. A new time-space or time-geographic interaction theory to serve this purpose is outlined below.

All activity and interaction is a resource demanding process. It cannot thrive on thin air. *Energy* is obviously one major resource in all forms of ecology, while living space is another. Both are interrelated in many ways: for instance plants require terrestrial space to collect solar radiation and convert this into other forms of energy. The field of energy analysis is a good example of how conventional bio-ecological theory can be generalized and extended into human ecology, by considering how energy is also used in society and in various abiotic technical processes. The total energy conversion system thus goes beyond its biological base but still remains in contact with it (cf Odum 1971). The recent literature in this field has grown at an almost explosive rate.

Embedded in culture and society, human individual and collective activities are directed towards many forms of pursuits and production other than the getting of a full belly. Major inputs into these production processes are natural resources, land as space for given periods of time (space-time), facilities and equipment, and human time. The activities which we normally associate with work and associated inputs and outputs fall within the economic sector of society-culture, while non-work activities cover other sectors. However, all the sectors contain projects using various inputs to reach individually and collectively defined products, and not just the economy.

When societies-cultures are classified into such categories as food-collecting (hunting-gathering-fishing), agricultural, and/or industrial, this is a classification according to *technology*. It is not a matter of economy, since this has to do with the social form of organization of production, distribution and consumption. Economic forms relate to reciprocity, redistribution, or market exchange, to follow Polanyi's scheme (1968), or to mercantilism, capitalism or socialism, to mention a few other forms of economic system.

All these elements, subsystems and systems are located in terrestrial space and have temporal extent, which is why the broad subject of human socio-cultural ecology must pay attention to the underlying existential dimensions of space and time.

A definition of ecotechnology

The term *ecotechnology* employed in the present study pays tribute to the fact that technology is an intermediary factor between ecology, on the one hand, and economy, on the other. The kind of technology of a given society has implications both for its ecological situation and its economic system, without it being necessary at this stage to decide what determines what.

But the term ecotechnology also points to another set of systemic properties, to which Hägerstrand (1974a, 1974b) called attention by trying to place 'ecology under one perspective'. One of the core issues in the time-geographic approach to human ecology is that people also interact with 'populations' of non-living things:

> In all kinds of ecology the concept of 'population' is central... each separate organism in a population has a limited life-time between a moment of birth and a moment of death... Man clearly forms a biological population among others... But these facts are only a part of the total picture where man is concerned. Between man and other populations and between man and man lies a world of symbols and artifacts of his own creation... One possible way to bridge the gap between biological and human ecology would perhaps be to view symbols and artifacts in the same way that we view populations in general... note that fabricated entities have certain formal population characteristics... /and/ typical 'life-times'. They have birth rates controlled by socio-economic forces. They own different degrees of mobility and they have age-distributions... The similarity between the man-made world and living communities goes further. Technology exhibits its own 'food-chains' in the sense that more complicated artifacts must be composed of less complicated parts in long chains... both living and non-living things have space filling properties... The trajectories of both living and non-living entities can be seen as being occupied with 'tasks' /or activities/, having various duration and various location in space-time. (Hägerstrand 1974b:272-5)

In other words, by analogy, technology has its own kind of ecology, and if we are to get a better understanding of both pre-industrial and industrial technology, we must deal with its concrete components and their relations. This is an area ignored by biological and social science alike. Social scientists have commonly refused to see 'dead things' as social or have left them aside for the natural scientists. Social scientists have also commonly refused to look upon artifacts as social in the sense that they impinge on how individuals interact with one another. These 'dead things' are, at best, seen as symbols and are not considered to be genuine ingredients in social situations and processes. No wonder many social scientists — anthropologists, human ecologists, social philosophers, etc. — have concluded that we live in a technologically runaway world (cf Mumford 1966 for an early and comprehensive treatise on the topic). The concept of ecotechnology is intended to cover these broader aspects of technology in relation to ecology, economy and society, although only a few aspects of this concept will be dealt with in this study.

Classification of societies by ecotechnology

The different societies presented in these chapters have been arranged in order of increasing complexity of ecotechnology. As we move from hunter-gatherers via pastoralists to shifting and sedentary cultivators, and so to urban-industrial and post-industrial societies, the focus will shift from interaction within the habitat to interaction within the population. In following this focus we do not pretend that post-industrial society is less dependent on nature or that it does not have habitats of its own. We merely take into consideration the fact that advanced societies are dominated to a lesser extent by coordination with the rhythms of the natural environment and that to a greater extent they are attuned to the man-made and social environment.

By presenting the material in this order, we are able to wrap up a number of dimensions in the same 'package deal':
— the level of techno-organizational complexity and ecotechnology,
— the approximate stage of human cultural evolution,
— the level of population density,
— the level of spatial mobility,
— the degree of activity coordination with nature versus society,
— the intensity of utilization of land (space) resources,
— the intensity of utilization of human time resources, and
— the intensification process itself.
It is necessary to stress at the outset that these dimensions do not go hand in hand in making up some unilineal pattern of cultural evolution, although the above types of ecotechnology have appeared historically in the rough order of chapterwise presentation here (except that agriculture preceded nomadic pastoralism). Rather, our aim is to discuss these forms of society and ecotechnology mainly with respect to the intensity with which space and time resources are utilized and the underlying mechanisms of intensification and interaction, leaving aside macro-historical and evolutionary processes. To begin with then, we must look into some of the leading issues of intensification theory.

INTENSIFICATION THEORY

Ideas of carrying capacity have a long lineage, stemming from practical notions of how many people and animals can live off a given piece of land and what land can produce. Ecology and economics are both derived from the same Greek word, *oikos*, referring to house, household and husbandry. Man-land relations were an established focus of interest in classical economics which, like its modern version, explored the various relationships between the conventional factors of production — land, labour and capital.

Economics was also closely associated with demography, and one exponent of this combination was Thomas Malthus. He carried out a lot of empirical work to back up his thinking of what happened when the population changed numerically and how this related to the carrying capacity of land. Given the state of the art in his days, his accomplishments were impressive although his emphasis on temperate climates and therefore on rainfall agriculture proved to be a theoretical stumbling block fostering a rather *static conception* of land use and carrying capacity.

For economics and allied sciences this became a handicap for years, and the whole battery of aggregate land-labour and capital-labour ratios rather enforced its own static underpinnings. In the later debates on the economic development of the newly independent Third World nations, many of the doctrines which saw daylight proved detrimental to development policy. They were based on statistically abstract, crude ideas of agricultural systems and operations, such as the zero marginal utility of labour doctrine. Inputs of more labour into agriculture were seen as futile, so the 'excess' population had to be skimmed off and shifted onto urban areas and sectors. Empirically minded agricultural economists, such as Clark and Haswell (1964), took great pains to rectify this type of theory, but it took a long time for similar efforts to influence the debate.

It was in this atmosphere that the agricultural economist and historian Ester Boserup developed her theory on population growth, technical change and land and labour use intensification. Her theory was carelessly labeled Neo-malthusian (like any work on population growth), although she never regarded birth and population control as a panacea or even discussed the matter.

Her main contribution perhaps was to give land use in agriculture a more dynamic interpretation. She did not consider carrying capacity to be a constant but a function of technological level and change. This had also occurred to Malthus, but he thought that population growth would outstrip productivity increase and technical improvements anyway. Malthus theory takes population growth to be the dependent variable in this two variable drama, for Boserup it was the reverse. Population grew for other reasons than increased food supply. In fact, population growth generated more food by making more labour available and by using land more intensively. It further triggered off other adaptive changes in technology and food production, some small and others more substantial.

In the tropics, expansion could also take place along the temporal margin by using land more intensively over time and reducing the perennial and even seasonal fallows, the latter by means of irrigation. Classifying successive land use intensities (increased 'frequency of cropping') was a major step forward in dissipating the static notion of land use which had dominated economics, and also, to a considerable extent, anthropology and geography. For geographers Boserup's dynamic land use concept was particularly inspiring, and being well aware of the operations of tropical agriculture, she was able to detail the mechanisms of agricultural transformation which led towards more complex and intensive forms.

Coincidentally, the anthropologist Geertz (1963) had been working in parallel with Boserup on a specific region, and his analysis of population growth, colonialism and agricultural 'involution' in Indonesia turned out to be a good example of Boserup's thesis. However, there were certain important differences. Geertz did not describe the stages of intermediate intensities between shifting cultivation and irrigated rice cultivation. But he did illustrate how the driving force behind intensification was not only population growth but also colonial exploitation. The latter system in effect constitutes a withdrawal of land, as well as labour, resources, leaving less land for the Javanese to share, the effect being equivalent to additional population growth, i.e. land use intensification. Labour tax had the same effect: more human time had to be mobilized for cultivation so that subsistence requirements could also be satisfied. Colonial exploitation therefore has the same consequences as population growth albeit the cause is socio-political rather than bio-social.

The next important contribution to intensification theory was by another economic historian, Wilkinson (1973), who generalized the intensification theme both 'backwards' to include hunter-gatherers and 'forward' to cover industrial societies. Rejecting the various idealist interpretations of economic development as a result of progressive human nature, he formulated an 'ecological model of economic development'. From this standpoint, societies were of two kinds, those in ecological equilibrium and those where this balance was upset by population growth. The former lived within the carrying capacity associated with a given technology and had 'settled down into a known and proven way of life which allows it to deal with all eventualities without innovation...' When population grew, society could no longer cope without resorting to innovative ways of intensifying land use and improving technology.

In this way, Wilkinson explains how the industrial revolution in England was induced by an increasing shortage of land and land based resources such as wood (for ironmaking, housing, and fuel). This in turn led to the exploitation of coal and provided incentives for mining equipment among which we find forerunners of the steam engine, and so on. One can only admire the consistency with which the puzzle is fitted together and his explanations do to a great extent tally with what is historically known.

Wilkinson's materialist counterpoint to Victorian ideas of human progress may be a healthy purification rite, but automatically one wonders what the contribution of socio-economic factors such as England's market and trading system was. Although his technological and ecological analysis is fruitful and cannot be dismissed simply because it does not account for the whole truth, the hazards of relying on a single prime mover behind social change, such as population growth, are evident.

Independent of Boserup and Wilkinson, Sahlins (1972) put forward a more idealistic theory of intensification. His model covers a less broad spectrum of ecotechnologies, from hunter-gatherers to agricultural systems slightly more intensive than swidden agriculture. For Sahlins, intensification of the 'domestic mode of production' makes people work more intensively and for longer hours, so that the 'original affluent society' is replaced by a less leisurely form. Sahlins' study was one of the few comparative time use studies in social anthropology. Although carrying capacity is discussed, ecological dynamics is not, and Sahlins (1972:49) repudiates the idea that 'demographic pressure' has anything to do with the intensification of production, which is seen as wholly attributable to political mobilization within the framework of kinship. The state as a force of mobilization is not considered (as Geertz (1963) and Wittvogel (1957) did), since Sahlins's essay deals only with pre-state societies. Again, while many of his arguments are no doubt valid, his version of intensification still leaves one discomfited by incompleteness as well as by contradictions in the explanation of driving forces behind intensification in production, time use and land use.

Recently another American anthropologist, Harris (1977) has tackled the problems of human ecology, intensification and cultural evolution on a big scale. He explains the transition from food collecting all the way up to industrialism and the associated emergence of a number of institutions — from cannibalism to democracy — as a consequence of intensification in the use of land and labour. Contrary to Sahlins and in line with Boserup and Wilkinson, population growth is the prime mover. Reproductive pressure and the concomitant build up of population density paves the way for a number of institutional responses and developments, which have ecologically adaptive functions. So they have at least up to a certain point, where environmental depletion and crisis lead to more technological and institutional innovation, later permitting a new round of population growth. One typical consequence is the increase in work load and working hours, as already explained by Boserup, Wilkinson and Geertz, and Sahlins too for that matter.

Harris like his predecessors outlines a number of important mechanisms of human interaction, ecology and society. But his general synthesis is a bold overstatement of the case. His critics claim that he makes cultural evolution a simple matter of 'protein and profit' (Sahlins 1978), but this too is an exaggeration. Nevertheless, one must agree with Godelier when he makes the general point that:

> ... what is called the ecological approach, as soon as it tries to become a *general* theory of social life and history, comes to grief in so far as, on such essential points as the causality of the economy and/or the environment, the nature of the functional relations between social structures, and the driving-forces of the evolution of systems, it relies upon the dogma of vulgar materialism ... /which/ reduces all social structures to be nothing but epiphenomena of the economy which is itself reduced, through technique, to a function of adaptation to the environment. (Godelier 1972:xxxiii-xxxiv)

Godelier's characterization, congruent with that of Friedman (1974), is largely applicable to Harris, Boserup and Wilkinson.

However, in rejecting a number of ultimate aspects of these studies, there is no reason to throw out the baby with the bathwater by ignoring either relevant component mechanisms and sub-models, or even the whole theme of intensification. For instance, Boserup's model can be of considerable use in planning for agricultural intensification, so that land and labour resources can be mobilized for increased food production. And even those who are not 'vulgar materialists' have to find ways of incorporating the role and consequences of population growth.

Anyway, as soon as one cycle of theorizing is over, another cycle of theoretical bodybuilding usually begins, and we are left with the dismembered theories of yesterday to use as raw materials in new attempts at synthesis (hopefully in combination with other materials as well). In this study, we do not address the overwhelming drama of cultural evolution. Nor is the task to find some ultimate cause, prime mover or form of determinism, be it population growth, ecological adaptation, minimization of effort, progressive human nature, prestige seeking, economic profit, or political dominance. We shall not even attempt to find some optimal balance of such causes, when it is hard to sympathize with one.

Instead the task is a modest attempt to contribute to a new cycle of Progress in Social Science, or with respect to Geography in particular, to a new PIG-cycle. Structural kinds of causality (cf Chapters 2 and 6, cf also Godelier's variant, 1975), can seldom be articulated in the weak conceptual models we have almost grown accustomed to in social science. We need to refine and renew our conceptual tools and build up better platforms for exploration. In the attempt, we shall also stretch our models to cover a few of the salient features of so-called post-industrial societies.

The need for models and methodological stamina

Much of the creative and provocative theory on intensification presented above relies on an unnecessarily weak methodological base. This makes it genuinely difficult to resolve some of the important theoretical issues. Without the development of better models and methods to remove these handicaps, the emergence of good second generation theory in the field will be hampered. Not that *models* can (or are intended to) replace *theory*, but they can help to articulate many of the mechanisms and socio-environmental processes involved. They can also clear up muddles and eliminate pseudo-theoretical issues.

It is tempting to undertake this task and also some reformulation of theory, although shortage of space in this volume dictates that many admittedly essential systems and processes will have to be sacrificed. But it is hoped that the results presented here can be applied in combination with already existing theory rather than as a bad substitute for it.

A major area in which intensification theory particularly leaves much to be desired is that of human time resources and activity-cum-interaction theory. Much of this present volume is written with a view to rectifying this omission. If the venture is successful is a matter left to the reader.

But before we look into time resources problems, the topic of carrying capacity and population density will be discussed, especially with respect to geography and intensification theory.

Carrying capacity and population density

The traditional concern of geographers dealing with human ecology has hovered around population density and carrying capacity, although some — especially physical geographers — have adopted a more explicit bioenergetic point of departure. A few geographers have frankly dismissed ecosystems and bioenergetic analysis as irrelevant, for instance, Stoddart in his review of organism and eco-system models in human geography:

> While the ecosystem concept has proved useful in several branches of geographical work, it has become apparent that its influence is seminal rather than definitive and lies not in the ecosystem con-cept as such, but in its general system properties... the biological emphasis on energetic and trophic structure of ecosystems, for example, is clearly of peripheral geographic significance, but the fundamental concept of *system* in geography is fundamental. (Stoddart 1967:537)

From a pure theoretical standpoint, it is unlikely that single concepts such as 'systems' or 'ecosystems' in themselves solve any problems. Only when mutually consistent and interrelated clusters of concepts are articulated in models, can something substantial be accomplished. To dismiss trophic and energetic analysis as Stoddart does, when many regions of the world suffer from overexploitation and acute or incipient food and fuel shortage seems more than bold. Moreover, energy systems have numerous spatial geographic implications. This is all the more obvious today when entire countries (such as Sweden) are considering reorganizing their energy systems so that a large por-tion of total energy is derived from biomass ('energy forests' which form a renewable resource. This will surely be a space-demanding innovation with numerous repercussions. Indeed, the industrial revo-lution in England was partly triggered off by an energy crisis and a shortage of land based energy resources, as Wilkinson (1973) demon-strates. Of course, when Stoddart made his statement in the mid-1960's spatial organization thinking and the quantitative revolution in human geography had only just gained momentum, so that resource management and human ecology figured less prominently on the research agenda.

The substantive links between bioenergetics and land-cum-popula-tion dynamics are quite clear. Carrying capacity is a central concept in the ecology of population dynamics, as it describes some of the possible and impossible relations between a population and its habi-tat. More specifically, it denotes the maximum population number which can in terms of food (energy) be supported in a region of given size and quality (e.g. varying quality of land). It is often measured in terms of aggregate population density per unit of space.

When measuring carrying capacity, land is usually assessed simply in terms of spatial area, but in a strict sense this is only appropri-ate for vegetation, where there is a close correlation between catch-ment area and solar energy influx (given other variables such as moisture and differences in soil quality). The gross solar inflow in mega-watt-hours per square kilometer and year, for example, is only an ultimate limiting factor for vegetative biomass production. It is only when the food-web structure of autotrophic and heterotrophic species in the area is specified, that the carrying capacity of land for the human population can be determined.

In practice, what is often done is that carrying capacity is assessed in simple population density terms under a good many *ceteris paribus* assumptions with respect to bioenergetics. Land may be classified according to known empirical population densities in similar regions, and areas where production can be intensified under certain conditions can be identified in a similar fashion.

In the present study, when classifying societies and ecotechnologies according to population density, we are simply considering the fact that hunting-gathering and nomadic pastoralism are empirically associated with low regional population densities, shifting cultivation with higher densities, and cultivators with more fixed forms of cultivation with yet higher densities. Still, both the latter fall within the medium density range compared to societies with hydraulic agriculture. The real high densities or even hyper densities are found in urban-industrial societies, at least locally for city habitats. But since cities are supported by broad and often overlapping hinterlands, the local densities of cities do not reflect local carrying capacities of land in the conventional sense. Imports to cities may come from very distant places and cover all sorts of commodities, not just food. The concept of carrying capacity proposed further on is a more general but different notion which also applies to cities.

In other words, population density is at best an *indicator*, and at worst a cover-up for an assortment of phenomena, similar to the GNP-index in economics. For one thing, it must be specified whether one is dealing with local, regional or global densities and how the supporting area of the population is defined if certain major ambiguities in the analysis of various places are to be removed. One must also avoid regarding the human population as being fixed in space like vegetation, an impression often conveyed by population dot maps (cf Figure 1:2, top right). Conditions such as the interaction of the population, their travel and movement patterns, the in- and out-migration and relative durations of visits outside the region, the extent to which local-self sufficiency has broken down and is compensated for by interlocal, interregional and even international trade (flows of goods, services and information); all of these are conditions which must enter the picture and which are not readily summarized in simple man-land ratios or population densities.

For agricultural societies, some students have used more refined methods of assessing carrying capacity. For shifting cultivation systems, for instance, it was recognized early that simple population density models were too crude. Allan (1949) and many after him, for instance Peters (1950) and Carneiro (1960), had to take the land use system into account, particularly cultivation periods in relation to fallow periods. Potentially arable land also had to be distinguished from non-arable space. Further the average food requirements per capita had to be estimated in relation to population age and sex structure, since food requirements vary with age, sex, external temperature and the structure of activities over time.

These and other complications were thoroughly considered by the anthropologist Rappaport (1968) *   and more recently in the deep probing study by the geographer Christiansen (1975). Both considered the issues of time-energy budgeting. For Rappaport, dealing with the Tsembaga swidden agriculturalists of New Guinea, the carrying capacity for the pig population was also an important factor, which in turn varied with the cyclical pig population dynamics (the so called 'pig cycle'). The issue of carrying capacity for domesticated animal populations is also relevant for other eco-technologies, e.g. pastoral nomadism, agro-pastoralism and mixed farming. For instance, Brookfield and Brown (1963) elaborated on Allan's initial index of carrying capacity by further taking into account the territorial structure of adjacent sub-populations within the same tribe and various classes of land use. Christiansen (1975) investigated the subsistence syndrome of a Melanesian island, Bellona in the British Solomons. He is one of the very few scholars who has emphasized the four basic resources of water, land, energy and human time within the same unified context. Like Brookfield (1973) and Brown (1973), Christiansen restudied the same area at several yearly intervals, which allows a vary valuable insight into the process of intensification as well.

In sum, carrying capacity must not be a static index, but rather a composite measure of fluctuating populations in shifting environments, specific numerical values for population size only being given when many other relations have been made explicit: There is every reason to support Lea's statement that:

---

*   Rappaport conducts his analysis in the energy domain rather than in the land domain, so to speak. In so doing, he can assess the energy output from agriculture, but also the human energy inputs can be calculated by means of time-energy budgets. The latter entails classifying human activities according to their energy requirements per time unit and then measuring the time spent on various activities by different categories of the population. In this way, commensurability between land and human time resources can be obtained in the energy domain. This is an interesting and useful approach for other purposes, but in the present study, the objectives are different and we will look into land and human time resources in their own right.

There is no theoretical reason why the concept of carrying capacity should not be viewed more dynamically than it has been to date. Allowance could be made for changing crops, distance to settlement sites, increasing intensity of land use, changing perception and evaluation of the environment, technological innovation, changing land-tenure systems and socio-economic change. (Lea 1973:59)

This brings us straight back to the topic of intensification again.

Geographers and intensification theory

Several geographers had noted at an early stage the correlation between population density and the kind of agriculture practised. Grove (1961), for example, working in Nigeria, wrote that, 'nearly everywhere, there is a broad connection between population density and intensity of agriculture'. Still there was little theory on this topic until Boserup's dynamic model of population and land use and her comparative approach provided further stimulus to geographers dealing with agricultural change. Her ideas fell on fertile soil, not so much because geographers agreed with all her points, but because her general model of intensification was fruitful and important.

The leading geographer in this field is Brookfield whose work on Melanesia and on general regional development has been outstanding in terms of theory, method and empirical relevance. Melanesia has become and interesting area of convergence of anthropology and geography, and the micro-geographic study of the Chimbu in the New Guinea Highlands showed how well the disciplines could fit together (Brookfield and Brown 1963, the latter being an anthropologist). Brookfield also emphasized the necessity for 'Local study and comparative method' (1962), something which comes naturally to geographers wanting to connect local studies to regional wholes. In his work with Hart (Brookfield and Hart 1971), forty-four Melanesian local societies were compared with respect to population density, crops, intensity, and other cultivation system variables. The intensification theme was also central in the work by Brookfield (ed. 1973) and associates on 'The Pacific in transition'.

Waddell (1972), Clarke (1966) and Maude (1973) have also made separate contributions. Clarke applied the intensification model very early to a series of neighbouring New Guinea communities, and his findings were very much in line with those of Boserup, while Waddell (1972), who was among the first geographers to employ a more consistent space-cum-time approach, concluded that although 'high densities can only be supported by intensive agriculture', the productivity of labour does not decline with land use intensification in such a straight forward manner as Boserup proposed. 'The Central Highlands evidence suggests that intensification is a more complex process than that'(Waddell 1972:217-8). Some of the complications are caused by the coexistence of parallel types of cultivation systems at different levels of intensity, for instance as typical of the Kapau-ku, so carefully described by the anthropologist Pospisil (1963). The Chimbu evidence mobilized by Brookfield (1973) also indicates that agricultural evolution or involution is not a simple unilinear pro-cess, and that it takes refinement in method to sort out the trends from the long term cycles. For some of the islands in the Tonga Kingdom, Maude points out that the tuber based cultivation system,

> appears much better able to withstand population growth and intensification than the shifting cultivation model described by Geertz... On *prima facie* evidence, Ester Boserup's thesis more closely fits the Tongan situation, and her contention that differences in agricultural intensity are the consequences of differences in population density is supported by the material... However, the intensification of Tongan agriculture in response to population growth has not stimulated any basic improvements in the bush fallow technique of maintaining soil fertility except amongst strongly commercialized farmers ... The adoption of cassava and dryland varieties of taro... has helped maintain yields per man-hour as cropping has been intensified, while the adoption of the metal push hoe has enabled farmers to cope with changes in fallow vegetation... /leaving/ Boserup's thesis in the category of 'not proven' but probable... (Maude 1973:181-84)

Although no geographers have so far (to the knowledge of the present author) tried to present a broad spectrum model of intensifi-cation that also covers industrial society (as did Wilkinson and Harris), in the present study we shall try to probe into this field by furnishing some additional material and ideas, but we shall do so on the basis of a rather different conceptualization of carrying capacity.

THE CARRYING CAPACITY OF TERRESTRIAL SPACE-TIME

For our present purpose, we can put forward a very general concept of carrying capacity that applies to a set of basic resources and populations (the latter defined in the ecotechnological sense of page 8 above), each resource or population having their time-budget.

The carrying capacity of *terrestrial space-time* can thus be defined as the limited ability of a given area to accomodate space-demanding people, organisms, artifacts, materials and the activities associated with them. Since space can be more or less continuously occupied over time, we must actually deal with terrestrial space-time (which may also be called settlement space-time for an area settled by people). The carrying capacity of space-time is thus a function of the spatial area and the observation period, as in Figure 1:2 left. This can be quantified in terms of square-meter-days for a building or hectare-years for the lands of an agricultural village.

For any space-time region, there are limits to the packing of entities demanding space-time, as all living and dead things do. Unlike the carrying capacity limits of bio-ecology, these limits cannot be exceeded even temporarily, although a given regional space-time budget can be more or less densely or intensively packed.

The space-time concept of carrying capacity can be applied at all scales from that of a matchbox to that of a building, a region or the entire globe. It is thus pertinent to both rural and urban forms of land use, and covers all activities, not only those associated with food production, which is all that the traditional carrying capacity concept is designed for.

One can, however, readily discern the close connection between the two different perspectives. The gross energy influx from the sun, for instance, is a function of both the temporal and the spatial catchment area. And plants are space-time demanding entities which can not be packed to any level of density. Since all organisms require life-space, competing individuals and populations displace each other in space-time, as do dead objects, houses and cars, for instance. In this way, both organic and inorganic systems can be accomodated within the same conceptual framework. In the early chapters of this book, the conventional and the space-time concept of carrying capacity are so close that further specification is superfluous, while in the later chapters on industrial societies, we are indeed referring to the more general notion of space-time carrying capacity.

While the space-time volume of a regional territory has a fixed limit, carrying capacity is more relative at the local population level, since the number of individuals whose space-time requirements are to be satisfied depends on the space-time area which they can reach from given spatial locations, and this in turn is a function of the time people can afford to spend on travel. With respect to travel time we come up against another resource limit and capacity constraint, that on human time. In this way — and many others — there is a link between the space-time budget and the population time-budget.

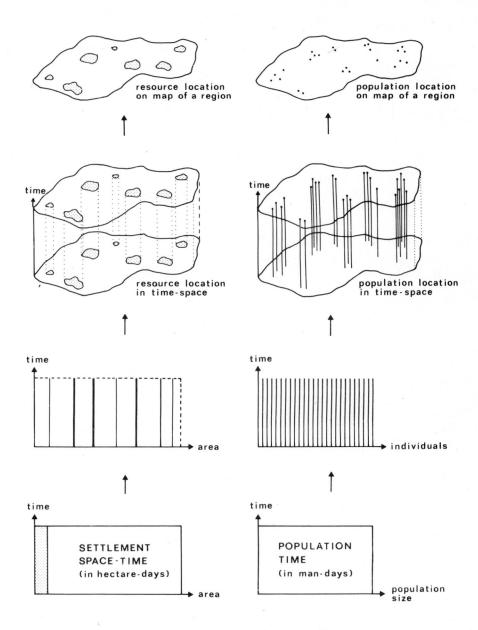

resource location
on map of a region

population location
on map of a region

time

resource location
in time-space

time

population location
in time-space

time

area

time

individuals

time

SETTLEMENT
SPACE-TIME
(in hectare-days)

area

time

POPULATION
TIME
(in man-days)

population
size

FIGURE 1:2   The relationship between spatial population distribution and the population time-budget (right), and the corresponding relation between the distribution of a given category of land (e.g. arable land) and the space-time budget of a region (left). Both resources can be said to have a limited carrying capacity to accomodate time demanding human activities or space-time demanding entities respectively. Note that the individuals dipicted to the right are not as stationary as they seem to be in the graph, and this is rectified by the path concepts presented in Chapter 2. (Cf i.a. Figure 2:2.)

TIME RESOURCES AND THE CARRYING CAPACITY OF A POPULATION
TIME-BUDGET

Time as a dimension of activity

Time, like space, can serve as a descriptive framework for human
activity and interaction. Activities are observable but multifarious
and the social scientist has to sort them out into clear categories
before they can be analytically useful. Activities carried out by
humans both occupy their time and have spatial location. They commonly also involve other inputs such as materials, tools, equipment,
domestic animals, buildings, and so on. In all societies, individuals are
constantly doing one thing or another: they talk, sit, sleep, work, eat,
fight, pray, play and so on. If one interviews people and asks them
what their social structure is, the question will probably be answered
by a smile. But if one asks who does what, when, where and to
whom and for what reason, the query is more likely to be understood.

Social scientists are prone to categorize activities in very broad
and superficial classes, such as cultural, social, economic, political,
religious or military. While this may be helpful for purposes of initial
orientation, it is otherwise a very coarse way of dissecting reality
so as to make sense of the myriad of processes concurrently taking
place in human communities. In the micro-worlds of ordinary life,
we are always faced with a mixture and flux of activities generated
simultaneously and sequentially in time. We may choose to ignore
this texture and lump elements together into crude classes and then
try to rearticulate them at an aggregate level, showing how one
abstract cluster relates to another. But unless we have been very
careful in how we sort out the activities constituting the processual
components of social and environmental systems, how can we be
confident that we have not really lost touch with the dynamics of
these systems? Perhaps we have created a caricature of reality
which is more lifeless than Frankenstein's monster ever was?

To some this is an issue of diachronic and synchronic study:

> The use of the terms 'synchronic' and 'diachronic' has the advant-
> age of putting to the forefront the *fact* of time *(chronos)* and
> avoiding the impression that a structure can really be analysed
> without analysing its evolution. In this way one gets rid of the
> old ambiguous manner of talking that contrasted a 'structural
> analysis' with a 'dynamic analysis', as though one could exist
> without the other; as if time was a variable external to the func-
> tioning of a system which could be introduced into this function-
> ing 'after the event' (Godelier 1972:260)

Although we are told that structural and dynamic study are really one and the same animal, this attractive cross-breed is seldom consumed by us hunting-gathering scientists. Some economists and structuralist-marxists claim to eat it all the time, but when their meals are deciphered we find that the macro-perspectives of both have already disregarded the unity of structure and process at the micro-level. They are instead dynamizing their structural abstractions.

If one says that evolution is something that also takes place in the daily round of activities, one can more readily subscribe to Godelier's statement and critique of introducing time *ex post facto*. But this is seldom the case, and micro-time scales are usually passed over. More common are illustrations of how otherwise timeless entities and structures undergo transformation, quantum jumps at perennial intervals. This is also how many geographers have depicted changes in spatial patterns and structures of activity, where distributions on maps are shown for a sequence of strata in time between which certain jumps of lumps have taken place. Although on many scores diachronic and historic-chronological study certainly is useful, it still does not make analysis holochronic, and perhaps we have lost many fundamental aspects of real dynamics because our concepts were too static in themselves.

It is seldom that anthropologists who have ventured out into the villages of the world and recorded how people act over time have been accused of conducting dynamic analysis. Studying such activities as they occur or may occur in time and space is more associated with 'boring repetition' than dynamics or evolution. Yet few would maintain that social dynamics are not a result of action. We need not pretend that all activities are of equal importance and that some are not more critical than others. But the literature is full of diachronic studies in which activities in general do not consume time and where human time is not a resource although time is supposedly incorporated. One must conclude that the entrenched distinction between synchronic and diachronic enquiry is not so clever as conventional wisdom has it. It is actually a rather shallow one in the light of more recent philosophy on the role of time in social and environmental science. And quite often we find that synchronic analysis is a fancy name for what amounts to a study of patterns and structures without any real temporal dimensionality.

All too few social scientists practice what one could call *holochronic analysis* comprehending activities, time resources, synchronization, sequential ordering, temporal organization, process-cum-interaction, structural transformation and evolution. One of the few exponents of such an orientation is W.E. Moore in his book on 'Man, Time and Society' (1963), a study rarely quoted outside a narrow circle of specialists on time and society.

It is also not without justification to place some of the work by Marx in the holochronic category. Unfortunately Marx moved rather too swiftly from time utilization to social labour time, surplus value, capital accumulation and other macro-structures, and there are many features of both primitive and advanced industrial societies that his theories and models do not cover with the kind of temporal (and for that matter spatial) precision that we are advocating here. Moreover, while appreciating the contributions of Marx, there are good reasons to doubt that his conceptual categories from the last century are diachronic enough to do all the jobs expected of social science in this century.

Even though time may have been adopted as existential, process-ual and resource dimensions, it does not follow that the spatial dimensions of existence are given corresponding attention, this in spite of the availability of theory on the spatial structure of society and habitat. To omit spatial organization in social science may be as fruitless as playing chess without the board. But environmental science that treats society and habitat as if it was void of temporal structure and dimensionality is similarly counterproductive. Several geographers and social ecologists, among them Hawley (1950), Meier (1959, 1962), Hägerstrand (1963, 1969, etc.), and Chapin (1965, 1974) have looked more consistently into the time-space organiza-tion and processes of society and habitat, but they are in a minority. (For a review and presentation of time-space approaches in social science with special emphasis on social and economic geography, cf Carlstein, Parkes and Thrift, eds. 1978).

There is a growing literature on time use and activity analysis in anthropology, geography and allied disciplines, and many examples will be referred to further below. Connell and Lipton (1977, a geo-grapher and an economist), have recently reviewed the local commu-nity (village) studies on Africa, Asia, Oceania and Latin America. Most of this literature is written by social anthropologists.

In geography, Kay (1964) was among the first to carry out a time budget survey of an entire village for a year, and a recent full fledged time and land use study is that by Waddell (1972). The Hägerstrand group started their work on time allocation in relation to spatial organization in 1966 (cf Hägerstrand 1963, 1966, and Carlstein, Lenntorp, Mårtensson 1968.) Brookfield commented in an essay on environmental perception with reference to New Guinea that:

> the need for data on the perceived environment is only one part of a larger agenda of data-collection facing human geographers now emancipated from the essentially descriptive task of studying /spatial-regional/ differentiation. Work-organization, *time alloca-tion,* /financial/ budget-allocation studies are desperately needed

to illuminate what actually goes on in society and enterprises. (Brookfield 1969:75) /my italics/

Of course, for the New Guinea area several good anthropological time and land use studies were already available, such as those by Salisbury (1962) and especially Pospisil (1963).

Yet it is curious how studies in social anthropology with an explicit time use dimension, such as the comparative essay by Sahlins (1972), have managed to remain totally uninfluenced by the time-budget tradition in general and rural sociology (cf i.a. Sorokin and Berger 1939, Szalai et al., eds. 1972). The comparative survey of Connell and Lipton (1977) reveals with utmost clarity the need for methodological stamina in future anthropological time use studies.

Individual capability constraints

At the individual level it is the experience of every practical person that in order to reach some goal, produce something or get somewhere, she or he has to act. Action as a sequence of activities consists of meeting other people, giving and taking services, using tools and materials, moving around, and so on. But individuals have limited capabilities to act. These *capability constraints* as Häger-strand (1970a) has aptly called them, vary with age, sex, cultural conditioning, experience and skills. One such capability constraint is the fact that a human being is indivisible and can only appear in one place at one time. (Many others are discussed in chapter 2.)

Because of capability constraints, most activities are *mutually exclusive* in the sense that an individual cannot carry them out *at the same time.* For one thing, a person seldom has the mental ability to concentrate on more than one activity in one go, or is unable to mobilize enough energy to do so. Likewise, limited abilities to use hands, feet and body, to listen, speak, see, smell, feel, lift, carry or locomote also mean that most activities exclude one another and can only be done sequentially.

An individual's activities also involve the input of other persons, tools, materials, buildings and facilities. These entities are spatially localized, are often in themselves indivisible and unable to appear in more than one place at a time, and have to be met or picked up where they were left. While some persons can call upon others to bring things to them or otherwise assist, only a limited set of individuals can be spoonfed in this fashion. In practice, people must move around themselves and only when the necessary inputs are within reach can the activity be started. Many activities require special sites and settings and cannot be performed elsewhere. In short, for individuals to carry out a host of activities, they become subject to a set of *coupling constraints* (cf Hägerstrand 1970a) as a result of the necessity to get into contact with various localized inputs found in their environment. This need makes activities space consuming and that is another reason why they are mutually exclusive and can only be done in both spatial and temporal separation. The moral of the story is that a good many activities are physically incompatible and cannot therefore be carried out simultaneously by one person but must be done one after the other.

Is allocation always the consequence of scarcity?

For those deeply committed to economistic models of scarcity and choice, it may seem surprising that choice must be exercised more as a function of physical incompatibility and capability constraints than as a result of sheer quantitative scarcity of some resource. The latter explanation is in many cases a piece of academic sophistry, and though it is not without foundation, it is — like a good deal of our economistic legacy — not all that basic either. The fact that activities displace one another is sufficient reason for having to select among them, and often enough this choice is not in terms of relinquishing certain activities all together but is more a matter of arranging them in a temporal order.

What is really fundamental to all action of individuals is what we might call place allocation: the fact that individuals have to decide where they want to be next, knowing very well that they can only be in one place at a time (and that some activities are only possible in certain places). It is only prisoners who remain in one place because of a scarcity of other places to go to. Normally place allocation has little to do with scarcity of places: it is rather a function of the indivisibility constraint that we have learnt to live with. Since we shift places over time (within our capability constraints on speed of travel), we can actually speak of *time-place allocation*. (This is what we in the next chapter define as path allocation.) Time-place allocation underlies all processes of human time allocation to activities, a topic to which we shall now turn.

Is human time a resource?

Although time may not be a culturally recognized concept and resource in some societies, all human activities occupy what people of contemporary industrial society call time, whether we are interacting with other things or organisms or with other people. Few persons even in clockless cultures believe that activities get done in a jiffy, being surrounded by other processes which indicate what we call time. (For an early study of time reckoning and time indication, cf Nilsson 1920.)

Judging from the general interest in human time in social science discourse, most scholars simply ignore it except in its most blatant manifestations, some paraphrase it and call it by other names (cf the substitute concepts below), yet others disregard it because they do not believe time to be a scarce resource in the particular societies they are dealing with. A few even question the idea that time resources have any relevance. All sorts of gut reactions can be encountered.

To depreciate time resources and time allocation because time as such is not a defined or unified category in some cultures is tantamount to advocating that social scientists should stick to the folk model level of reality and never go beyond it. It is an overtly empiricist reaction and a poor exercise of the sociological imagination. Others cannot reconcile themselves to a resource being so intangible as human time, when land, water, vegetation, food and even energy are so concrete. That a human being can be a resource to others (and even herself) may not seem too objectionable, but that a *temporal portion* of a person can be is to some an appallingly abstract idea.

To compound the picture, the existential dimension of time — time as flow, relative location, chronology and history — is often confused with *human time*, which is a much more specific thing. The general concept of time has many forms and attributes, and although these forms are interrelated and are in many respects commensurable, they are by no means one and the same phenomenon (cf i.a. Carlstein, Parkes and Thrift, eds. 1978).

Human time is a resource since *all* activities necessarily require it as an input and since we have limited capacity to act in relation to time. In an ultimate sense this is because we all consist of atoms, molecules and bodies in relative motion within the physical universe. However, this does not imply that explanations of society and habitat should be reduced to physics. It only means that the capability and capacity constraints on how humans occupy time and space are rooted in certain physical and biotic realities which cannot be ignored in social enquiry just as they cannot be the sole focus.

One can admit that for societies unregimented by clocks and calendars, human time may seem a somewhat reified resource. But there is a close correlation between physical amount of an activity and temporal duration, not least because of the various socio-cultural norms and standards on performance. Given these capability and other constraints on output per time unit, human time also becomes a *finite* resource, at the individual and the population system level. Both the life spans of individuals and the population are limited (cf Figure 1:2, right side). But even for an arbitrary observation period as in the figure, human time for the population becomes a finite resource: only a limited amount of activities can be accomodated within a population time-budget. In short, human time is a resource allocated to activities even where such time is not necessarily 'scarce'. Underlying all time allocation are the more fundamental processes of time-place allocation mentioned earlier, but the major consequences of this perspective will become more evident below.

Convenient ways of obfuscating human time

But why then has human time been so neglected if it is really that strategic to social science? There are several reasons why the study of human time has been a field of conceptual obfuscation and intellectual fragmentation. Although many innovative scholars have worked on the topic, their ideas have somehow been nipped in the bud and have never entered the mainstreams of their disciplines.

This is largely because time in general and human time and temporal organization of society in particular have already been partially incorporated into various inferior substitutes. Scholars have been too often caught in their own categories. In economics, economic anthropology and economic geography, for example, *labour* is such a substitute. Labour is time spent on work. But the idea that all activities, not just work and production, take time has diffused very slowly. Only a few economists have made an active effort to include time as a resource, for example, in consumption (Soule 1955, Becker 1965, Linder 1970). Godelier (1969) makes a brief remark in one of his economic anthropological studies that work is not in short supply in primitive societies, since 'productive activities', at least for the men, only occupy part of their available time. This may be true, but the formulation reveals the underlying production bias in most economic thinking, as if non-work activities were entirely residual. This bias is shared by liberal and marxist economists, and goes back to classical economics. For Adam Smith, David Ricardo and Karl Marx labour is of prime importance, while human time as a unified resource is misconceived. Soule (1955) was among the first economists to protest about this. *

* An anthropologist who has looked into questions of consumption and included a number of time aspects is Mary Douglas (Douglas and Isherwood 1979). For a discussion of time and consumption theory, cf Carlstein and Thrift (1978). Consumption and 'living and action possibility boundaries' are taken up in Chapter 2 below.

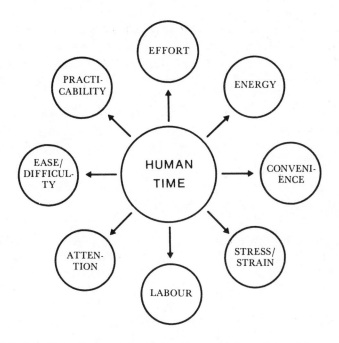

FIGURE 1:3   Substitute concepts for time which tend to obfuscate the unified perspective on human time resources.

Not only human time is obfuscated by concepts like labour and work. The list of substitutes is long, as evident from Figure 1:3. Human time is referred to as energy, effort, attention or bother, or is paraphrased as strain, stress, ease, difficulty, and not least by the magic word 'convenience'. Time is only invoked when some of the more roundabout concepts fail to fill the gap. To some, time may even be money, but we shall leave this to Chapter 8.

Examples of 'fragmentation of time resources' in the literature are legion. Scudder, for instance, writes that:

... much *energy* is expended in weeding during the rains /32/...
... when larger harvests are reaped, less *effort* is put into nursing grain crops /48/... Instead he spent most of his gardening *time* in his large UNDA /a kind of garden /56/.... the main drawback is the *energy* expended in getting from one garden to another. When the task of cultivating becomes to *arduous* /and time demanding/, the owner usually allows kin with closer access to take over cultivation /69/... Now many growers can take their own tobacco to the line of rail without *neglecting* /i.e. spending less time on/ their agricultural tasks./92/ ... If the crop is left until the rainy season...

most kin will be *too busy* preparing their own land to help out./111/ ... the occupants of most bush hamlets would have to expend considerable *energy* /and time/ driving their stock daily to the Zambesi and returning home with water for household use. /146/ (Scudder 1962) /my italics/

This way of referring to basic time allocation mechanisms as they impinge on ordinary life mystifies the role of human time as a resource. Sometimes the lack of time to do too many activities is explained as a 'difficulty'.

For marriages within a three-mile radius... a man can, without much *difficulty*, work land simultaneously both in his own natal village and in that of his wife. (Leach 1961:86)/my italics/

A scrutiny of the anthropological literature reveals another mysterious force in human affairs, namely maximization of *convenience*. It seems that this is a variant of Zipf's (1949) general principle of effort minimization, a kind of ahistoric law of behaviour. The doctrine of convenience is especially applied to the time costs of traversing space but also more generally, and most often it refers to a mechanism for saving time:

... someone contemplating a visit to the Lalbari barber had to weigh the (possibly) superior quality of the service against the *convenience* of receiving a shave and haircut during a leisure moment back in the comfort of one's own courtyard. (Elder 1970:122)

There are usually only one Imam and one barber within *convenient* reach of any given customer... (Barth 1959:47)

Indeed, the traditional range of behaviour and allocation in a Fur village indicates that the Fur do not subscribe to any kind of prohibition in joint conjugal households — such arrangements are just not very *convenient*. (Barth 1967b:677)

When Douglas in her excellent essay on 'primitive rationing' takes up the role of money, she remarks:

Money may sometimes have emerged from the barter situation... On this familiar argument, the *inconvenience* of barter and the *difficulty* in arranging credit lead to the adoption of a medium of exchange. (Douglas 1967:121)

One might interpret this as barter being too time consuming in a structural situation of commercial expansion. Barter presupposes double coincidence of wants so that you can sell what your trading partner wants, and so that he can supply the very good or service you in turn want. To find a trading partner with these qualifications may be a time consuming affair. Many people may also be producers and have to spend time in this role. They are unlikely to be able to fit too much bartering into their time budget. The introduction of money may be 'convenient' and time saving by relaxing the double coincidence of wants constraint and thus permitting a greater number of transactions per time unit within the broader population.

However, the convenience theme is most persistent with respect to time, distance and settlement:

> After harvesting, the grain is stored in grain bins... built close to the ... gardens for *convenience*. (Long 1968:83)

> In a relatively short time the soils and woodlands within *convenient* walking distance of the village were in short supply. (Long 1968:83)

> These *conveniently* situated plots may be cultivated for longer periods than they are rested... (Netting 1968:86)

> A modest number of people usually sooner than later reduce the food resources within *convenient* range ... (Sahlins 1972:33)

> /pigs houses/... are the same size as women's houses but are built close to the gardens so that the women who live in them can *conveniently* tend the crops and look after the pigs... the competing ends of *convenience* in cultivating nearby gardens and maintaining the fertility of the soil are explicitly weighed against one another. (Salisbury 1962:12, 83)

> Landlords who are resident outside the village in question usually lease their land to a local resident. It is *inconvenient* to work land more than two miles from the cultivator's home. (Leach 1961)

Even so, it is clear that convenience as measured in miles varies a good deal from one society and place to another and with the kind of activities, and much of this variation can certainly be attributed to time allocation.

The objective of this discussion is not, of course, to run down the studies quoted, which are all of fine quality. Besides, Leach (among others) has genuinley looked into human time allocation processes at the village level in his essay on 'The economy of time in dry zone rural Ceylon' (Leach 1967) as well as having explored the more symbolic aspects of time (Leach 1961b). The same pertains to Barth, Sahlins and Salisbury. The point is to show how certain concepts cloud matters rather than illuminate them, but once we can see through these categories, we can restructure our ideas about society, resources and time.

Apart from this conceptual aspect of time resources, it is strange that time has been virtually undiscovered in social and political anthropology, while the theme has been debated in economic anthropology for years. One would have thought that human time offered an attractive framework for integrating different 'sectors' or spheres of activity such as production, education, ritual, recreation, decision-making, military activities, arts, and so on. All processes of social interaction involve time consuming contacts among people. When some individuals participate in one collective activity, they cannot also participate in other group activities elsewhere, and when new groups are introduced, others may be reduced in size, frequency and duration. All groups and institutions must be maintained by time demanding interaction (matters to which we shall return in later chapters).

One may assume that here again the scarcity-and-choice paradigm looms large. If human time is not 'scarce', it is not used, it seems. Attention is again turned away from activity and process and focused on the institutional structures and contexts of activity, rather than on how the latter are generated by and articulated through activities. The main result is an unnecessarily static and morphological conceptualization of society.

The same reasoning applies to social and economic geography. The structure of activities in space is regarded as more interesting than the processual and time resource aspect of activities. Only travel has been explicitly analysed in terms of time costs. The more recent interest in space-time geography is a trend which will perhaps remedy this situation.

The carrying capacity of a population time-budget

Finite resources and other constraints place upper limits to what is humanly possible under various forms of ecotechnology and social organization. Much has been written about the carrying capacity of land in this spirit. But there is also an analogous kind of carrying capacity which is less acknowledged — that of a given population system for time demanding activities. This is a capacity dimension, essentially internal to the population, rather than a relation between the population and an external resource, land. Roughly speaking, if a population has a given size and composition and this factor is multiplied by the observation period, a population time-budget is arrived at which constitutes the aggregate population time supply (cf Figure 1:2 right). Only a limited volume of time demanding individual and collective activities can be packed into this time-budget.

In some ways, population growth (through natural increase or net immigration) might imply increased carrying capacity of the population time-budget in absolute terms. But when we said that human time is a resource internal to the population, the objective was to emphasize that individuals demand one another's time, and if the population grows, so do the time demands to which it is exposed. In relative terms, this need not imply increased carrying capacity. For instance, if more babies are born, more of the time of the adult generation may have to be spent on them. Population increase is thus a double-edged sword, especially in a context that leads to agricultural intensification and increasing working hours, but more about this below.

Any socially organized and culturally conditioned human population has limited capacity of performance within its existing institutions and resource endowments, as well as a limited capacity to transform their institutions and habitat. In fact, a population has a limited capacity to carry culture itself, since all culture is produced, reproduced, communicated, and even stored through time demanding activities, a topic we will discuss at some length in Chapter 12. These limitations and potentialities can perhaps be better appreciated within a general model on how human populations are able to utilize their time and space resources for various human projects.

But the constraints on carrying capacity are just one form of capability constraint under which human societies operate. In any social, economic, technical, political, or ecological system or subsystem — however we prefer to label similar phenomena by fuzzy macro-terms — one finds a series of capacity-capability constraints which make some manifestations possible and others impossible. In Chapter 2, we will outline a number of basic capability constraints that set up production, consumption and general *activity possibility boundaries,* limits within and with which human populations must live.

OBJECTIVES OF THIS STUDY

This piece of basic research is directed towards several objectives of theoretical, methodological and substantive nature. Above all, however, an attempt is made to construct a useful kind of general interaction theory which has potential for extension and can serve to explore a new range of topics as well as reinterpreting some of the existing work in social science. This interaction theory employs the dimensions of space and time to help us structure our conception of society-habitat and to handle substance with which we are already familiar in new and perhaps more productive ways. Space and time are thereby a means rather than an end in themselves.

The other objectives can be listed as follows:

1) Furnishing interrelated sub-models of interaction

Since it is impossible to give theoretical coverage to all the kinds of society that have existed, having ecotechnologies such as hunting-gathering, nomadic pastoralism, agriculture or industrialism, we will instead try to present a set of coherent sub-models. A good motive for building up a family of such interrelated sub-models — describing anything from the annual cycle of a pastoralist group through the dynamics of a conference, to the effect of specialization in a Japanese house or the daily round of a household in New Guinea — is that such models in themselves constitute a kind of *pre-theory* which facilitates theoretical synthesis of substantively different phenomena.

2) Integrating the subject matter of anthropology and geography

Traditionally, the contact surface between anthropology and geography has been considerable. The literature is full of examples of mutually supporting enquiries in the field of tropical agrarian systems, human settlement, land tenure, market systems, population analysis, and human ecology in general. Recently in anthropology, Carol Smith (ed. 1976) assembled an impressive amount of anthropological studies with explicit spatial dimensionality covering a wide range of socio-economic and ecological systems. These indicate very clearly the potential scope for cooperation, and one would only wish that human time and the temporal structure of society and habitat had been similarly incorporated. We shall try to do so here instead.

The empirical and some of the theoretical material in the first half of this book are largely anthropological as are some of the ideas on intensification, while the theoretical approach is of geographic origin but of a new and rather unestablished format, viz. that of 'Time Geography'. This designation, however, is more telling of the conceptual approach than the disciplinary pigeon-hole, and it is designed to forge links between disciplines rather than sorting them out.

3) Developing human time allocation theory

A more substantive target is to promote the understanding of human time resources and their use and application within social systems. In neglecting this field, we have missed an important integrative device with respect to the multitude of substantively different activities we see in societies around us. Not that time can serve as a vehicle for holism, because time itself and the capability constraints on our use of time make holism and total coverage-cum-integration of knowledge impossible by definition (cf the chapter on the carrying capacity for culture below). What we can do is make different sub-systems of knowledge more additive by making them interlinkable without excessive amounts of time-consuming reformulations. By increasing the speed with which the pieces in the puzzle can be fitted together, we may be able to attain better temporal economy of thought and grasp larger portions of reality within one theoretical system.

In dealing with time allocation, we also further the development of economic science in the direction of economizing in real terms rather than in financial terms. Economics tends to assume that all use and economy of resources is adequately reflected in monetary terms, and yet the pecuniary framework has given rise to very contradictory principles of how to best reach various goals. In fact, traditional as well as modern societies, have huge pockets of activities that are not mediated by means of money, the so called subsistence and informal sectors, and it is important to deal with resource allocation in real terms as an ongoing process. A direct focus on human time, settlement space-time, energy and water, for instance, paves the way for this.

4) Promoting the comparative study of time resource utilization

A major objective in this study is to encourage comparison of scattered studies. Regional geographers were often inclined to stress the unique features of their regions and this turned out to be detrimental to general theory. In current anthropology and geography, there is a wealth of local case studies with various aims and conceptual formats that make them difficult to compare, as evidenced by Sahlins's (1972) struggle to extract comparable data on time use, labour intensity, kinship and exhange. Some of this unnecessary idiosyncracy of each study is due to lack of conceptual standardization rather than lack of interest to include certain kinds of factors and variables.

Perhaps the integrative notation system of 'Time Geography' (cf Chapter 2) may enlarge the scope for comparative studies of activity systems, settlement systems and the use of time and land resources. If, for example, certain basic data of a demographic kind had been more uniformly presented, it would have been much easier to compare local societies with respect to anything from village size to potential labour supply. Although one should not go afield with too many preconceived categories but form these according to their local relevance, it is still annoying that presentation of data is so patchy. It would be useful, for instance, to achieve a measure of standardization of activity, time and labour utilization, so that the specific features of different places emerge as real social facts rather than as a function of unique terminologies and systems of description.

Some brave attempts have already been made in this direction. For Europe and North America, the Multi-national Comparative Time-Budget Project used a standard classification of activities to get at time use variations in a dozen countries (cf Szalai et.al. eds. 1972, and Carlstein and Thrift 1978 for a comment). A diligent team at Sussex University also made a laudable effort to register all village studies available on developing countries in order to extract useful information for the World Employment Programme sponsored by the ILO. They stress the need for a more general and comparative data collecting framework but one which also does justice to local variation (cf Connell and Lipton 1977).

One problem in this context is that the categories used are either free from theoretical connections alltogether or else they are so laden with them that many things go unnoticed. The latter is often the case with the transplantation of categories from modern economics to settings where these are less relevant. What we will do here is to link the comparative analysis of time use, time allocation and the intensification in time use to time geographic theory, which may not be the best body of theory, but which at least is not so culture and sector specific as economics tends to be.

5) *Integrating some separate fields within human geography*

The final objective is to integrate some of the separate approaches and bodies of theory in human geography and to develop this discipline into a more dynamic socio-economic and ecological science. This will be attempted by means of a more pervasive and consistent incorporation of the temporal dimension, which in turn implies a reappraisal of the relative role of space.

While geographers have contributed much fruitful material on the spatial organization of society-habitat, they have with few exceptions (for a survey, cf Carlstein, Parkes and Thrift, eds. 1978) contributed significantly less to our understanding of temporal organization. Since one approach supports the other and increases the joint powers of explanation, it is difficult to defend a continued separation. In a time-space framework some of the seeming contraditions of interest between what we may call the spatial organization school and the human ecological school in geography may be resolved. The emphasis on locational analysis associated with the former is all the more powerful in a time-space framework, and the resource-cum-environment bias associated with the latter becomes more meaningful when resources are looked upon as localized and allocated in space and time. Locational analysis and spatial organization were more connected with the quantitative technical endeavours of the 1960's while human ecology and resource management were associated largely with the growing concern with environment, energy and development of the 1970's. However, this is a very crude perspective and there are many bridges between these coexisting streams.

What we can do here is to probe into this complex a bit further from certain new vantage points and see where the journey takes us. We will use a time-space framework to rearticulate much of the substance with which we are already familiar. In this experiment to interrelate habitat, ecotechnology and society, one's thoughts often turn to Forde's classic study 'Habitat, Economy and Society' (1934), which was an early and very inspiring enquiry into the theme. Whatever may be accomplished in the present study, an endeavour is made here to work in the same spirit.

2 LIFE PATHS AND LIVING POSSIBILITY BOUNDARIES
Elements of the Hägerstrand Time-geographic Model

THE TIME-SPACE FRAGMENTATION OF RESOURCES

When aggregating population, land and other resources for an entire region, a process inherent in the assessment of carrying capacity, it must not be forgotten that one ends with a statistical abstraction. Neither the population nor the resources people utilize are all concentrated on one place. They are scattered — and often mobile — over the surface of the region. Resources are fragmented in space and time. Even at the local level, all the materials, equipment, constructions, animals and plants used are not concentrated in one spot, since they occupy space, displace one another, and impose limits on spatial packing. Dispersal means that in order to use resources or interact with other members of the population, space has to be overcome through movement.

Different units of action in society (individuals, groups and organizations) are thus placed in an environment where resources are *accessible* only in certain spatial and temporal situations. General *availability* in the purely quantitative sense for an overall region is not enough. Moreover, the problem is not simply one of gaining spatial access to resources through movement but is further complicated by getting a hold of them as they become available in time, for instance at certain seasons or times of the day. Time and space are the main dimensions of practical action and adaptation to the natural and social environment, since any use of resources and people in human projects entails arranging activities in space and time so that the right inputs are combined at the right times and places.

It is therefore essential prior to tackling the issues of human ecology to have an analytical method which can cope with resource allocation and utilization in its spatio-temporal matrix. The time-geographic model provides one such vehicle for getting closer to some of the essential features of resource utilization by human populations in social and natural environments.

THE HÄGERSTRAND TIME-GEOGRAPHIC MODEL

Many theories of resource utilization, social interaction and economizing have paid little explicit attention to the human predicament of existing in space and time or to how the dynamics of action are directly associated with these dimensions. At the root of this state of affairs has been a *language problem*, where the categories used allow one to perceive some things but not others.

A classic example would be that of the economic disciplines. If we are in the habit of categorizing basic resources as land, labour and capital, then we are likely to ignore that human time or energy may be equally important factors of production. Neither Adam Smith, Karl Marx, or Maynard Keynes defined energy as a factor of production, nor did any other leading economist prior to the 1970's. It was not defined as important by conventional wisdom and it took a rather startling event — an 'energy crisis' — for the significance of this factor to become evident.

There have been few successful languages conveying the message of how relative location in space and time of humans and resources affects *capacity and performance* of socio-environmental systems. The model and notation system of time geography offer one such possibility, since they can handle — descriptively or deductively — features such as resource utilization, settlement, interaction in and between groups, organization of domestic units, and time allocation. The graphic format of the time geographic notation system places it in a useful midway position between verbal description and mathematical formulation with the effect of building a bridge between the two.

Hägerstrand devised the elements of the model when he was working with Swedish population problems in the 1940's, mainly to allow classification of different forms of spatial mobility over time (Hägerstrand 1947, 1963). In 1966 a special research group was formed to explore the topic of *'time use and ecological organization'*. In 1969 Hägerstrand published an article giving a basic outline.

Population as a web of individual-paths in time-space

In ecology, food-energy relations are conventionally modelled as food chains between different species and the total system of chains in an ecosystem is referred to as a *food-web*. But in any population system one finds also another web which is composed of the paths of the individuals describing their movements in space over time.

The time geographic model of society and habitat designed by Hägerstrand (1963, 1969b, 1970a) is what he called a socio-environmental web model. A cornerstone in this model is that the human population is looked upon as a *web of paths* which flow through a set of time-space locations.

> Fortunately geographic space can be represented on a plane. A human population living in the area which the plane represents, is supposed to have arranged several sets of 'stations' serving its needs. The most important of these sets of stations are *dwellings* and *places of work* ... The stations can be represented on the plane as points ... Now if we add time (t) as a third dimension we get a picture of a simplified 'time-space'... (Hägerstrand 1973) /Figure 2:1/.

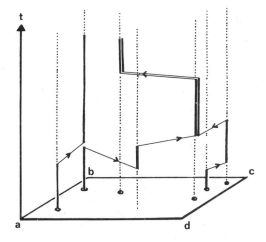

FIGURE 2:1   Three individuals in time-space. (Source: Hägerstrand 1963)

This time-space region contains the social system and is the setting of every-day life. As time flows, organisms and objects of different life-span describe paths which together form a large and complex web, where paths are born, move around (some more, some less) and die, combining all the time into different constellations, depending on the associated divisibility and mobility constraints.

The various paths can be roughly classified according to their divisibility properties. While human individuals, animals and other organisms, and units of equipment and tools are basically *indivisible*; materials, energy (in some forms) and information is largely *divisible* during their 'life span' (cf Hägerstrand 1970a, 1974a, 1974b). As for the mobility criterion, a population of plants would be largely stationary in space during the life span of the component individuals, while a population of humans by contrast would move and not be spatially fixed all the time (Figure 2:2).

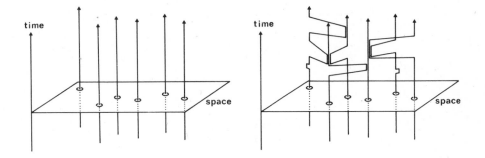

FIGURE 2:2   Stationary paths of plants (left), mobile paths of humans (right).

The map which has been the traditional instrument of expresssion in geography has the disadvantage of treating all organisms and objects as if they were more or less stationary. Though spatial flows may be mapped, it is still useful to have a mode of expression which is more attuned to a dynamic view of nature and culture. A major advantage with the time geographic notation system is that movements and changes in location can be registered in the *paths* or *trajectories* in a time-space map, just as the path of a jet-plane can be seen in the sky a few minutes after it has passed. A sequence of events and activities for individuals and objects thereby becomes frozen into a kind of historic-geographic matrix. This is in contrast to the traditional overcommittment to the two dimensional map as an analytical tool.

In a time-space region, each individual can be visualized as a continuous path starting in a point of birth and ending in a point of death. Depending on the observation period, individual-paths can be referred to as day-paths, year-paths or *life-paths*. This corresponds to the concept of *life-line* in demography, an idea initially conceived by the demographers Becker and Lexis (Lexis 1875), mainly as a temporal concept. Hägerstrand generalized this idea to a time-space concept in his population mobility studies, but he also looked into time perspectives shorter than the year, which is the conventional unit of time in demography and demometrics.

The indivisibility of people means that one individual cannot exist in two places at one time and therefore has to allocate his path in time-space. Recombinations of individuals into groups is thus done under the constraint of indivisibility. Since it is a time-consuming process to move around in space, the individual-paths are always tilted upward in relation to the geographic plane as in Figure 2:3. A

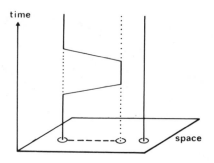

FIGURE 2:3   Two individual-paths in time-space, the left person moving twice.

path can never be parallel to the plane since that implies travelling at infinite speed and at no expenditure of time.

The paths of individuals are not isolated from one another, of course. On the contrary, they come into contact with one another and are coupled into *bundles*. These have typical geometric structures for various types of groups found in society. Households, schools and shops, for instance, can all be graphically depicted. In a daily perspective, a household is characterized by the individuals being together during the dark hours of the day, later radiating outward when members go to work, school, the shop or market place, and elsewhere, and later getting together again in the evening, as in Figure 2:4. In the generalized picture of a school as a bundle, the individual pupils leave their homes in the morning, attend school classes during the day, and later return home (Figure 2:5). Since

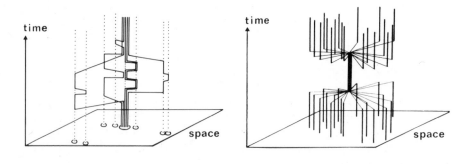

FIGURE 2:4   A household.   FIGURE 2:5   A school.

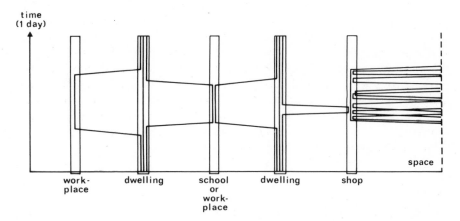

FIGURE 2:6   Couplings between individual-paths at dwellings, work places, schools and shops. (The entire paths of individuals visiting the shop in the period are not shown.)

individuals are part of a larger social web, bundles can be regarded as linked together 'sidewise' in space, as in Figure 2:6. The picture is broadened here and depicts the couplings between households, shops and school units in a day, using a simplified two dimensional time-space.

It is clear that a large population looked upon in this way forms an intricate web, and this is amplified by the fact that other entities besides the human population can be represented as paths. Humans form bundles with materials, equipment, domestic and non-domestic organisms, vehicles, and so on, which can also be regarded as forming populations (living or non-living) but which are a necessary part of human (socio-technical) ecology, as Hägerstrand (1974a and b) has pointed out. Moreover, there are also various elements found in the settlement system of a region:

> The physical environment in which individuals act may be said to consist of *channels of transport and communication* and *stations*, such as dwellings, work places, establishments for education and welfare, shops, and arrangements for recreation. Individuals reside in these stations, move about or send messages between them in order to collaborate economically and socially, make use of services, or otherwise having their needs and wishes met. These stations appear in the country /or region/ in different numbers and form geographic patterns with varying relative distances, from the high degree of packing and differentiation in city areas to the low degree of packing and lesser differentiation in the sparsely settled regions. (Author's translation of Hägerstrand 1966).

In the settlement system, stations such as buildings or fields can be depicted as pillars in time-space. In a short time period such as a day (Figure 2:7), these stations appear as unchanged and stable although the activities going on within them generally alter. But if the time perspective is extended to a yearly scale, elements in the settlement system would change in size, some would cease to exist while others would be added to the landscape. These stations form the spatial bases (points of origin and destination) for activity and interaction, but, of course, the structure of this system varies considerably among types of societies and regions.

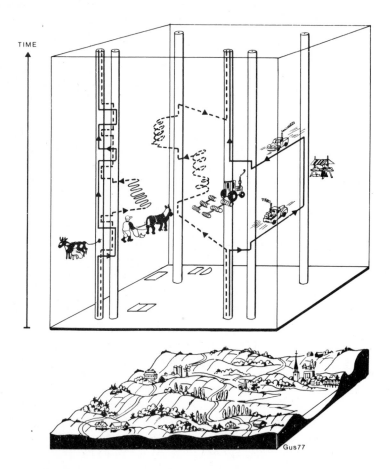

FIGURE 2:7   A day in a rural area. The figure shows four individual-paths, two for males (dashed) and two for females (solid). Stations are indicated as vertical pillars. (Drawn by G. Gustavsson 1977)

Not only do tangible biophysical bodies have location and structure in space and time, but so do many legal rules and other symbolic or normative elements in society (cf Hägerstrand 1970a, 1973b, and Carlstein, Parkes and Thrift 1978, vols. 1 and 2). An obvious example are the control areas or *domains*, which are the physical manifestations of norms, laws and contracts regulating the access to different time-space locations. Hägerstrand (1970a) has defined a domain as,

> ... a time-space entity within which things and events are under the control of a given individual or group... Some smaller domains are protected only through immediate power or custom, e.g. a favourite chair ... or a place in a queue. Others, of varying size, have a strong legal status: the home, land, property, the premises of a firm or institute, the township, country, state, and nation... Thus there exists a hierarchy of domains... (Figure 2:8, cf also Hägerstrand 1973b).

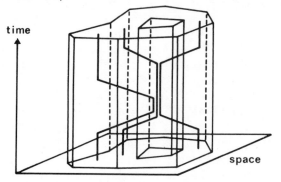

FIGURE 2:8   Two domains and three individual-paths. Domains of different kinds have specific structures and extent in time-space.

The role of the graphic system of description

There is no doubt that graphic complexity would be enormous if large populations in sizeable regions for a long time period were to be drawn as paths; the picture of merely one day in a small village is quite complicated even if computer plotters were programmed to do the actual drawing. But the important task of the graphic notation system is not to thrive on visual complexity, but to reveal the underlying logic of human society and ecology in space and time. Simple graphs can be strategically employed to pinpoint the elements, relations and mechanisms which have principal importance for social-environmental structure and operation. Complexity can be broken down into analytically manageable portions, not least because there are many regular, repetitive and highly structured components and sub-systems in the web. Moreover, the graphic format facilitates the design of proxies and methods for numerical calculation and assessment.

The constraints on path allocation in time-space

Human interaction with the environment can be looked upon as a *path allocation* problem in space and time, both when it comes to interaction amongst persons and between them and objects and organisms or places. Each individual is acting under constraints where many options are excluded. The fact that humans are indivisible is one example of a *capability constraint* (Hägerstrand 1969b, 1970a), the major impact of which is that a person cannot be in two places or more at the same time. It is this constant choice between where to be at what time which is referred to as path allocation, and it is a process which goes on whether the individual himself is aware of it or not. When a person shifts location, another capability constraint enters the picture, namely that he or she can only travel at a certain maximum speed. A simple consequence of this is that there are a limited number of places that a person can visit in life or some shorter time unit. Since all individuals are located in space (and moreover since interaction is time consuming), path allocation also entails *coupling constraints* among individuals (Hägerstrand 1969b, 1970a). In other words, if a person wants to interact with others, his path must form a bundle with theirs. This generally implies that he cannot allocate his path to those of yet another set of individuals at the same time. An individual can only participate in a limited number of bundles within any given time span.

In fact, each newly born individual is so constrained in his capabilities that he cannot survive alone; human life itself is a collective project with certain minimum doses of coupling constraints among persons. The constraints on the abilities of individuals can partly be relaxed by making activities collective and joint, but this in turn increases dependence on other individuals in the environment, it imposes additional coupling constraints. A similar relation of dependence exists between humans and their artifacts or domestic organisms. A person can travel faster on horseback than on foot only by being coupled to the horse, which itself is indivisible and can only be in one place at a time. Tools and machines allow people to do things beyond their capability as unequipped individuals, but only if one accepts the bond between human and facility. It is thus a true paradox of culture that individuals can only reduce the impact of their initial capability constraints and gain freedom of action by accepting the coupling constraints associated with these facilities.

Human projects and goal constraints

The concept of project is fundamental in time-geographic analysis since it incorporates the goal dimension in human affairs (cf Häger-strand 1973b). In a sense, the future is always project while the past is always product or output. A human individual starts off as a product, but from his subjective point of view life is a project (with a trajectory), although her or his time may also be an input in the projects of others. A characteristic trait of human societies is that many projects pursued are collective and interlocking with consequent demands for syn*chron*ization and syn*chor*ization of component individuals (*chronos* is time and *choros* is space in Greek).

Many of the projects carried out by individuals, groups and organizations are founded on culturally transmitted goals and interests, and are part and parcel of the material and immaterial culture and human made environment which is passed on from one generation to the next. Cosmologies, ideologies and religious principles furnish particular streams of interest, aspirations and goals in life, while institutions and material facilities are the socio-cultural instruments passed over as means for the new generation to make their goals attainable. The goals of individuals and groups can be classified in a spectrum ranging from subsistence goals, the making or earning of a living, to the life-goals of reproduction and propagation and the goals based on aspiration, virtue, enthusiasm and pleasure; the kind of goals pursued when most of the tasks of maintenance and subsistence have been completed.

Projects are vehicles of goal attainment, but they are also in themselves constraining. They channel human action in certain directions rather than others and therefore work as *allocative mechanisms*. Once the goals are set and decided upon, it inevitably means that some paths of action have been chosen in the set of alternative ones. Take the farmer, for instance, who has just planted his annual crop, this being the first stage in one of his projects for subsistence. He is in fact constrained *by his own choice* and determination to go on cultivating until the harvest is completed, because if he does not, his whole cultivation project will fail and his subsistence goal will never be reached. Projects act as volitional constraints on the individuals and organizations who act them out. This element of constraint by choice has given rise to the term positive constraint in certain operations research and planning terminologies.

But projects are also affected by *negative constraints*, those which are instrumental in character and related to man's dependence on the natural order of the universe and the socio-cultural order of human society. These constraints, which cover the broad classes of capability and coupling constraints, are given particular attention and content in this study.

Projects are realized through human action. There are also other ingredients, but human action is a necessary albeit not always sufficient input in projects. Action is composed of discrete processual units, packages or quanta, which are termed *activities* in the present study. Since activities are more or less organized and interdependent for a variety of reasons, they can analytically be said to form a sub-system in society, referred to as the *activity system* further below.

Projects can thus be defined as sequences of future activities of individuals and groups designed to materialize in a predefined result, which upon completion becomes a *product* or some other socio-cultural *output*. As activities, projects involve and demand resources: people's time, space-time in the settlement system, energy, and in many instances materials, tools, constructions, other organisms, and various other facilities.

Human intention, volition and regulatory constraints on action

Time itself would not be an interesting resource if it were possible for individuals to achieve anything by the wave of a magic wand. It takes time to carry out human projects and see them materialize into results and forms of output. Moreover, human time allocation has to be compatible with human path allocation, since time cannot be allocated in directions which the paths cannot, but more about this below.

A third set of constraints are those associated with how human activities are channelled with respect to human intentions. Hägerstrand (1970b) calls them by 'styrningsrestriktioner' (i.e. steering constraints) but the English translation calls them by the somewhat unfortunate term 'authority constraints' (1970a). These constraints are both negative and positive and have to do with the volitional, normative and institutional channelling and regulation of activities. The whole set of mainly invisible regulatory devices in human society reflecting goal-seeking, volition, intention, decision, norms, rights, duties, contracts, and agreements which regulate the more or less goal directed behaviour of human beings is endorsed in this category apart from the exercise of authority and various forms of control. For want of a better term they will be referred to at this high level of generality as *regulatory constraints*, wherever a more specific term is not used. Regulatory constraints are also associated with tenure, ownership or holding, resource control, domains and so on when these constraints are viewed negatively. By contrast, goal constraints are usually regarded as positive constraints in operations research and systems theory, although goals still have constraining effects (cf section below).

Space, time and human conditions

We can summarize some of the important conditions and facts of life which circumscribe what can be accomplished in human societies operating in different environments and under various cultural and social orders. In an essay on 'Space, time and human conditions', Hägerstrand (1975b) called attention to a set of basic conditions which gave the limits to possible structural and organizational forms:

1) the indivisibility of the human being (and many other entities, living and non-living),
2) the limited length of each human life (and many other entities, living and non-living),
3) the limited ability of the human being (and many other indivisible entities) to take part in more than one task /or activity/ at a time,
4) the fact that every activity (and project) has a duration,
5) the fact that movements between points in space consumes time (cf Figure 2:9 for some graphic interpretations),
6) the limited packing capacity of space,
7) the limited outer size of terrestrial and territorial space (whether we look at a farm, a city, a country, or the Earth as a whole), and
8) the fact that every situation is inevitably rooted in past situations. *

The point Hägerstrand (1975b) makes here is that,

the interaction between these fundamental conditions could be and ought to be the object of precise theoretical research. I feel

* The latter is because the trajectories or paths of people, objects and organisms must come from somewhere and go somewhere. The path conceptualization thus allows us to understand many of the continuity constraints of material existence (cf p. 55-6 below).

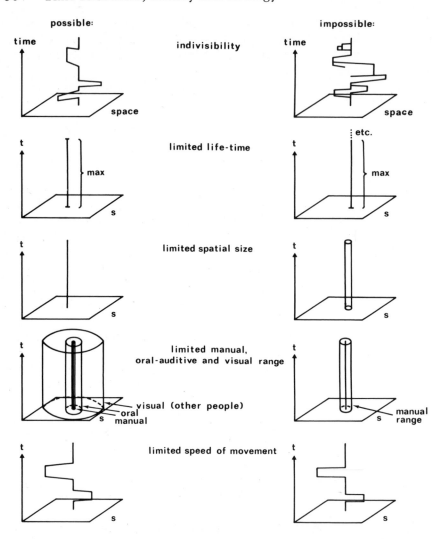

FIGURE 2:9   Basic individual capability constraints, which have implications for aggregate structure of society. Left: possible outcomes. Right: Impossible outcomes. Note that the limits of oral-auditive and visual range may vary between individuals. The visual range drawn pertains to a given level of optic resolution, such as that for recognizing a particular individual.

that this research is the starting point from which more practical consideration should develop concerning better or worse institutions (capitalism versus socialism, bureaucracy versus participation), concerning better or worse technologies (private transport versus public, videophones versus letters) and concerning better or worse cities (circular, multi-nuclear, bandlike, or no cities at all).

What this approach essentially signifies is an attempt to commence with the more general parameters of human ecological and socio-cultural existence first and then move on towards more specified forms of analysis. The fact of human indivisibility is a biotic constraint, the impact of which advanced technology has only partly reduced (e.g. by telephones allowing transmission of messages without physical presence in the same location as the counterpart), but which is still with us. We do not remove these constraints because we have to travel less to communicate in some situations. But it is the aggregate effects of all the constraints in relation to the various goals and projects endorsed by the human population which really counts. Within this general perspective, the various idiosyncracies of specific cultures, societies and regions can be placed.

The anthropologist Barth (1966) is thinking along similar lines in his essay 'Models of Social Organization':

I believe that the study of social anthropology cannot today be advanced much by sophistication and refinement of its current total stock of concepts and ideas. Rather, we should make a careful selection among them, and among concepts available in related fields, to isolate *a minimal set which is logically necessary and empirically defensible.* The implication of any such set should then be *explored and exhausted before further complexity is added.* /My italics/

In a comment on the role of the spatial dimensions in sociology, the present author made a similar statement:

For a student of change and development, there may well exist a *primary level of analysis* of human activity and interaction, group formation, function and participation, studied from the angle of time allocation in space. It may not be really necessary to cram into sociological analysis such vast amounts of *specification of substance* about people and the sectors to which they belong. Many major constraints on social institutions and processes lie at a deeper level than this and it would be good strategy to concentrate on them first... (Carlstein 1972)

How this strategy will be employed is amply illustrated below where the deductive power of very simple realistic assumptions will be illustrated, not least with respect to interaction within a socially organized human population.

LIVING AND ACTIVITY POSSIBILITY BOUNDARIES

There are good reasons for calling attention to the previously mentioned set of basic human conditions and constraints, since they are
strategic in understanding the limits to possible structural and organizational forms of society-habitat. Without this understanding based
on the theoretical demarcation of boundaries between possibilities
and impossibilities, self-deception lies close at hand. We must, in
other words, be able to see the core and structure behind the manifestations and appearances that are more directly observable. To
do this is no simple task; much of the success of coping with this
problem depends on the *efficacy of basic concepts* and how these
are interrelated and articulated in models. The time-geographic
approach is one attempt to select key concepts, which correspond
to various fundamental conditions in society-habitat, and by em-
bedding them in the rigorous existential framework of time and
space, an effort is made to evaluate them in terms of their impact
on living and existence, production and reproduction at all scales.

An important method of achieving this is by delineating the
production possibility boundaries as an aggregate expression of a
specific set of constraints that intersect and jointly circumscribe
the possible forms, thereby separating feasible from infeasible
solutions. Production in this context refers to the generation of any
kind of socio-cultural output within the total formation of society-
habitat, not just production in the narrow sense of generating 'goods
and services' or production as a preliminary stage to 'consumption'.
This restricted view of production occurs both in the conventional
marxist format where 'production is the pre-condition of consumption, and consumption reproduces the conditions for production',
as Godelier (1972:159) puts it, and in the neoclassical economic
framework, where the production concept is distorted in favour of
the output of material goods, rather than the sum total of output
from all sorts of human activity. In a time-geographic framework
with all its emphasis on the use of universal resources such as human
time, terrestrial space and energy for *all kinds* of activity, the classic
notions of production and consumption can stifle a deeper understanding of the workings of society-habitat.

Observing the world around us, we find that many processes
habitually classified as production simultaneously involve some form
of consumption, for instance, the consumption of materials extracted from nature, of space and of human time. On the other hand, a
good many household activities subsumed under consumption in
conventional economic accounting are actually a form of production,
not least of services. Conventional wisdom has it that consumption
is the use (or acquisition through purchase) of goods and services for

want satisfaction, but we 'consume' many other things which do not fit very neatly into the crude categories of 'goods and services', such as various composite socio-environmental situations. We not only consume the food at a dinner, but in many respects we also consume the whole dining situation by participating in it, just as we may consume a situation such as a forest hike without eating the trees. In all these different situations we consume space (space-time) and we also consume time the instant it is generated, as it were, since we cannot store human time. In other words, the categories that may be useful in the economic accounting framework of neo-classical economics, may not be equally enlightening for other purposes and may even be of limited value for the purposes to which they are traditionally put.

Irrespective of all this, if we want to be attuned to the physical realities of ecology and material existence, neither 'consumption' nor 'production' activities can occur unless they are associated with transport or movement of some form. People in urban-industrial society, for instance, do not get all commodities delivered to them at the instant they are wanted, but spend a considerable amount of time travelling and shopping. Similarly, primitive populations had to collect and gather. So which *movement and transport possibility boundaries* can be identified * and why are we never told of them? Or why is there so little discussion of *consumption possibility boundaries* in economics? And why are there factors of production but seldom if ever factors of consumption, distribution or transport? (Cf Carlstein and Thrift 1978). The time-geographic conceptualization tends to fit edgewise into the categories and models handed down by tradition since the days of Adam (Smith), Karl Marx and even Maynard Keynes. One might enquire if there are valid reasons for persevering with so many of the categories which have held economic and social science in a spell for so long, or whether it is time to *try out different concepts* which *may eventually* do a better job.

But suppose we accept 'consumption' as a fruitful category; time-geographically and in its *concrete* manifestations, consumption then becomes a process of resource utilization in which human individuals must be in *physical contact* and *proximity* to various inputs used in different time-space situations. The human individual is a localized input-output entity who requires and desires inputs such as food, water, air, information, equipment, shelter, settlement space and human time (in the form of the time of others as contact, care and

---

*   These movement and transport possibility boundaries are to a great extent expressed by the concept of 'prism' in the various chapters below (cf p. 73 and onwards).

assistance, as well as her or his own time. All these inputs and the situations (bundles of inputs) which they constitute, potentially compete with or displace other situations involving other participants. Much of the innate logic of 'consumption' (and 'production', for that matter) disappears when abstracted from the time-space context, as when consumption is regarded as a propensity of households to purchase commodities. Even when the volume of input and output per time period is specified, as is generally the case in economics, the underlying logistic aspects of assembly, disassembly, processing and transport still deserve attention. When these interaction aspects of consumption are investigated, it is obvious that there are numerous constraints that impinge on how consumption situations can be generated. If people are 'consuming' one another's personal services, for instance, different kinds of bundles of individual-paths must be formed before these services can be transferred (cf Chapter 11 on services). But a given population of individuals can only form and participate in a limited number of bundles in a given time span. So there must be *consumption possibility boundaries*, which derive from a set of factors that exist at a deeper level than disposable income and simple purchasing. (In saying this, we by no means deny the various institutional and regulatory constraints passed down through the monetary-financial system used in many societies, but these have already been subject to extensive research and teaching.)

Since the constraints underlying production, consumption, transport and other types of activity are essentially the same, the notion of 'production' possibility boundaries is narrow and misleading. It is therefore more appropriate to refer to the general concept of *living-and-activity possibility boundaries*. *

The carrying capacity of a population time-budget and that of a regional space-time budget are two special cases of the 'production' or living-and-activity possibility boundaries of any region, but they are very important cases. The application of these two carrying capacity concepts allows the construction of a very useful formal framework for sorting out possible and impossible modes of system performance. Using the notion of the population time-budget with limited time resources, for instance, we can go far beyond the

---

*   This term is admittedly a bit clumsy, so either the term *living possibility boundary* or *activity possibility boundary* is used in the text here. 'Living' can be regarded as a short form for 'conditions of life and living', but since the concept should also be applicable to non-living things (e.g. those associated with technology), the term activity possibility boundaries sounds better in relation to the use of machines or other facilities. To the extent that the word 'production' is used in this study, it either refers to the generation of something in the broadest possible sense, or else it covers a process of output in a narrow and specific sense such as 'food production.'

narrow concepts of production and labour and start to cope with the interdependence among the multifarious forms of socio-cultural output and use of the earthly habitat. Achieving this broad coverage is the objective behind the combined analysis of the population time-budget and the terrestrial space-time budget of a given region.

Situations generated and their roots: The logic and substance of allo-cation

From the general concept of socio-cultural output and an extended definition of production possibility boundaries we must now proceed to some aspects of the logic and substance of allocation. Allocative processes are not only *formal*, as some critics of the formalist school in economic anthropology have maintained (cf Le Clair and Schnei-der 1968 for some examples). Both the application and allocation of human time and the choice of time and place (path allocation) are *substantive* processes for arriving at some forms of result rather than others.

In 1916, when de Saussure made his classic distinction between *langue* and *parole*, between the syntactic and abstract whole of language itself, and the linguistic utterances in concrete situations, he looked upon *langue* as a generator and *parole* as that which was generated. But it is interesting to note that while individuals had such generators, they could only articulate a limited amount of speech in a given time period, although *potentially* and qualitatively, they had the ability to produce many additional grammatically consistent permutations of *parole*, of words. This limited *quantitative* capacity to generate speech implies an allocative process, in which some potential outcomes are never realized. Most people do not produce words at random but in fact *choose* them (not least when they are writing monographs). Like so many other every-day choice processes, we may not perceive it so consciously as to regard it as the choosing of words, but the allocative element is still there.

In this sense, allocation is a part of underlying structure, since not all intended (or unintended) actions can be put into practice as there are not resources and capacity to nourish them all. Taking the jig-saw puzzle analogy of Lévi-Strauss, in which structure lies not so much in the shape of the component pieces as in the design of the machine producing jig-saw puzzles, it is not insignificant to add that this machine could only produce a limited number of puzzles per time unit. There is, in other words, a capacity dimension to structure, which has perhaps received too little stress.

We can illustrate this point with respect to individual path alloca-tion and the generation of different social and other situations in which individuals participate. Hägerstrand's eighth point above, that 'every situation is inevitably rooted in past situations', is more than a way of saying that things have a history (a historical geography, to be exact). In Hägerstrand's words:

> It is evident that the huge amount of situations which constitute societal life mutually regulate one another through the fact that the component people or objects which at a certain time and place cease to constitute a situation will later become components in another situation at the same or in other places. Situations are thus linked to one another in a complicated web /of paths, human and non-human/, which is not without structure. (1974:226)

> The whole set of situations which constitutes the life of a society regulate each other first and foremost by the fact that we as individuals are indivisible ... /If a situation is defined as a specific configuration of indivisible and divisible entities/... it follows that there is an inevitable shortest time in which these entities can regroup themselves /in time and space/ into a new constellation, i.e. that there are definitive limits to the rate at which different situations can replace one another. (Hägerstrand 1974, translation of unpublished manuscript.)

There are, in other words, limits to the rate and quantity of situations which can be generated in a given system in a given time period and this is a property of the underlying structure which we can describe in terms of its associated living-and-activity possibility boundaries.

Aggregating the constraints which form the elements of a structure cannot be done simply by addition or as in a linear programing model. The process itself alters some of the constraints in the underlying structure, especially in the perspective of decades but even in shorter time perspectives (although some constraints will be more stable than others, of course). This means that living-and-activity possibility boundaries cannot be given a static formulation. In this sense, the structure is ultimately that of transformations. This is easy to say but difficult to apply, and finding ways of aggregating different constraints to discover structure is an equally difficult scientific art.

THE LUND SCHOOL OF TIME-GEOGRAPHY AS A TIME-SPACE STRUCTURALIST APPROACH

From the early stages of time-geographic developments by the Hägerstrand research group at Lund, behaviouristic approaches to the study of time use and human activity were rejected. Simple recording of what people did for how long, etc. might be indicative, but it could not furnish much explanation as to why such or such outcome was generated. From a policy viewpoint, such recording would be of little avail in evaluating how the structure and design of a particular habitat-environment influenced the range of activity choices open to its inhabitants, since everything appeared to be feasible (cf Lenntorp 1976, 1978 and Mårtensson 1978, 1980). Unless actual behaviour is viewed in relation to individual and environmental constraints on possible behaviour, it is difficult to determine in planning how the action possibility boundaries for a particular population can be broadened. To widen the range of choice was a major concern for those members of the group working with time-geography in the late 1960's, dealing as we were with issues of Swedish environmental policy, as well as basic time-geographic research (cf Pred 1973, and Carlstein, Parkes and Thrift 1978b:117-263). We were also unreceptive to economistic ideas that preferences and choices are revealed through actual behaviour (since it is hard to establish which alternatives were chosen away for a variety of reasons). Moreover, to adopt various statistical classification procedures for sorting out heaps of empirical time use data would be an empiricist approach leading to a number of valid generalizations, but still disclosing little of the underlying structure (whether individual, social or environmental).

One response to the environmental policy problems we worked with at that time was the development of an intricate computer simulation model, PESASP, i.e. Programme Evaluating the Set of Alternative Paths (cf Lenntorp 1976), which generated types of *possible* daily activity programmes on the basis of goal constraints and capability constraints of household members, coupling constraints of the urban environment and various legal and regulatory constraints imposed on that environment. Another part of the research programme explored a number of topics such as household daily life, public transport, innovations and their effects, local labour markets, and the temporal organization of industrial society (cf Hägerstrand, Ellegård, Lenntorp, Mårtensson, Wallin and Carlstein).

Although the time is not ripe to present a full-fledged time-space structuralist approach to society-habitat, the intention here is certainly to make a contribution in that direction. In format, such an approach would be closer to a materialist than an idealist perspective, and while there would be commonalities with several well-known streams of thought, philosophically a time-space structuralist interpretation of social-environmental formations would not fit comfortably into any of the established -isms.

At the present stage, it will suffice to posit a time-geographic view of *structure* as the result of the aggregation and interaction of constraints, whereby some solutions are feasible and others not. Such an interpretation of structure has several advantages:

1) It separates essence (deep structure or 'langue') from appearance (empirical manifestations or 'parole');
2) it considers that component sub-systems and sub-structures are parts of larger, hierarchically ordered systems and structures in which some constraints carry over from one structural level to another, from one sub-system to another, and from one system level to another;
3) it separates cultural and natural *substance* from basic *structural properties* (although the various conditions chosen as basic to time-geographic analysis are such that structural functions and consequences are better correlated to observables, i.e. it is easier to link structural models to reality); and
4) it regards constraints behind activity not only as negative, but also positively in terms of the goals behind various projects and activities pursued. The positive constraints of some sub-populations are also recognized to be the negative constraints of other sub-populations in many instances.

The motion of society-habitat itself through time means that all constraints are not invariable and fixed for all times. Some, of course, such as the indivisibility of the human individual, have been invariant throughout biological as well as socio-cultural evolution. But other constraints are much less stable, and are altered as a result of cultural evolution. Technology has been one key force in relaxing different constraints, although we must not forget all those constraints which are still with us and which shape the present structure of society-habitat, nor must we forget the new constraints which follow from technological innovations.

As a formal approach, time-geographic thinking shares several elements with other formal approaches, for example, some of the marxist-structuralist ideas of Godelier (1972) and Jonathan Friedman (1972, 1979), both of whom have also been influenced by systems theory and operational research. Godelier, for instance, speaks of 'the formal model of a possible economic system' in the following terms:

> What do we mean by a 'possible system'... Lévi-Strauss gives us the representation of the formal element common to every possible system of totemic thinking. A common formal element is an 'invariable' /i.e. invariant/ that persists all through every one of the possible varieties and variations of the system envisaged.

> Formalism is an ... approach, by which thought is detatched from every *actual* system so as to give us all the *possible* systems, and to rediscover the actual in them as a 'realized possibility'.
> (Godelier 1972:262)

Of course, we are not confining ourselves here to economic systems in the substantive sense, and although we do not employ the marxist model of infrastructure-superstructure (as Godelier does), we are, like Friedman, interested in looking into:

> ... the hierarchy of constraints which at each succeeding level determine the limiting conditions of internal variation and development. This ... implies nothing in the way of direct causality. Rather, it is a selective operation whereby the *existence* and *non-existence* of any particular structure is determined by its degree of compatibility with structures at the preceding level.
> (Friedman 1979:28)

Although the levels discussed by Friedman are different from those outlined in this book, many of the functions within the conceptual models employed are quite the same.

Friedman also points out that when applying a marxist model in a broad cross-cultural manner, one has to insist on a 'clear distinction between cultural form and material function.' Religion, being a relation to the supernatural, may in some places function as relations of production, for instance, although as a cultural category is is obviously more related to superstructure (ideology) than infrastructure (forces and relations of production). He similarly makes a formal interpretation of the forces of production by saying that:

> forces of production ... include the technology and the exploitable ecological niches, the totality of technical conditions of production and their possibilities for organization. (1979:25)

But he goes on to say that:

> The forces of production are not simply meant to represent an existent technology in operation, but rather, the *production possibilities* afforded by a given level of technological development. (Friedman 1979:29)

At this level of abstraction there is definite overlapping between the marxist-structuralist model employed by Friedman and that of time-geography. However, what we have defined here as living-and-activity possibility boundaries have broader coverage than the constraints subsumed under the notion of production possibility boundaries.

It is interesting to note that the conventional marxist distinction between infrastructure and superstructure is difficult to apply when time resources are examined on a broad scale, for although many aspects of ecotechnology discussed in this book might be placed in the realm of infrastructure (especially forces of production), time resources are just as involved in the articulation of ideology, politics and religion, normally taken to be part of superstructure. And relations of production are not simply structural but are in real terms mediated through time demanding activity. There are thus 'production possibility boundaries' that apply to the levels of relations of production and superstructure. In Chapter 11, for instance, we try to show some of the material conditions under which education and politics operate. However, for the purposes of this book, the place of time resources in the conventional marxist model is not a strategic issue. *

Moreover, it should be stressed that this study does not try to deal with total social formations and their evolution but only partial formations although a time-space structuralist approach can certainly shed a good deal of light on the boundary conditions affecting both the articulation and transformation of structure. In this study, we must first point to a number of essential aspects which, traditionally, have too often been neglected or misconceived and, in so doing, we are compelled to study what is perhaps a broader universe than that of the 'social formation', namely the formation of society-cum-habitat. We need to consider some of the biotic, ecological and environmental aspects, as well as those of ecotechnology, which have so far received less attention in social science theories.

A NOTE ON SPATIAL STRUCTURE AND STRUCTURALISM IN HUMAN GEOGRAPHY

An interesting and much needed explication of structuralism and the scope for structuralist approaches in human geography is that by Gregory (1978a, 1978b), who relates this approach not only to structuralism but also to several other current '-isms' and '-ologies',

---

*   In saying this, we are not denying the role of time in the labour theory of value propounded by Marx, but it is outside the scope of this study to present a critique of marxist conceptualizations of human time resources in general and labour-time in particular. When some consistent alternative conceptual system of time resource utilization, such as the time-geographic model and theory, is more fully developed in its implications, the present author hopes to be able to return to this topic in some later context.

such as empiricism, positivism, marxism and phenomenology. *
While he is appreciative of structuralism's search for 'deep structure'
and its generative orientations (i.e. how actual outcomes are based on
combinations and permutations of invariants, etc.), he is rightly
sceptical of the 'categorical' version of Lévi-Strauss' which, for all its
other merits, is ahistorical, static and idealistic, or what Godelier has
called 'structural morphological' (1972). One of the problems con-
fronting structural geography, as Gregory (1978b) sees it, is how to
reconcile *spatial structure* with *social structure*. To prevent structuralism in geography from becoming too morphological, Gregory tacitly

---

*   Apart from a misinterpretation of two of my figures (Gregory 1978b: Figs.
12i and 12ii), Gregory's whole explication of time-geography in his book 'Ideology, Science and Human Geography' is a misunderstanding. It is based on only
one or two of Hägerstrand's numerous time-geographic essays, supplemented by
an additional indirect source or two. This largely accounts for Gregory's treatment of time geography under the heading of 'reflexive explanation' rather than
'structural explanation', where it would have fitted in much better. 'Reflexive
explanation' covers phenomenology and behavioural geography, but our objective
was to incorporate certain essential biotic and ecological predicates, to cover
what Vidal de la Blache would have called 'genre de vie'. This does not mean we
subscribe to a phenomenological framework of the 'life-world'. Nor does it entail
resting content on an atomistic or behavioural level of society-habitat, although
the view one takes of the individual — psychologically, biotically and socially —
is quite important in the treatment of more aggregate levels, which is why we
placed new emphasis on the indivisibility and capability constraints of individuals,
for instance.
      A major aim in time-geographic discourse, however, was to build up the time-
geographic framework so that it could tackle the complexities of the structure of
society-habitat. Such a goal (or project) obviously entails more than incorporating intentionality at the individual level and juxtapose this onto resources in the
environment and individual meanings and perceptions of the environment. We
pointed to the kind of resources used at the individual level and the competitive
allocation of these, but 'competitive allocation' (to use Gregory's term for it),
has both counterparts and consequences at more aggregate levels. Moreover,
Hägerstrand (1973b) did not simply regard projects as 'empirically given, pre-
existing structures'. He rather emphasized the fact that instrumental action has
to be sequentially ordered in time and over space (and any one who dislikes this
notion is perfectly free to eat his food first and then cook it, or to go to office
before getting dressed). It is further clear that Hägerstrand's presentation of his
concept of *project* was intended as a very general concept. (If economic and
social science would have devoted more thought to projects rather than products,
it would be more future oriented rather than *ex post facto* oriented.) Rather
than implying an idealist framework of intentional action shaping the resulting
totality, it should be evident that Hägerstrand pointed to competitive allocation
and displacement effects which made the total outcome anything but the sum
total of intentions at the level of actors (be they organisms, human individuals,
groups, organizations or even states). Such a framework of struggle among and
between projects would to most people be highly reminiscent of a materialist
paradigm.

recommends development towards marxist-structuralist formulations. Geographers should make spatial structure something closer to the essence of things than manifest spatial pattern, while sociologists must also resolve the tension between social structure and spatial structure, by paying explicit attention to how human social beings shape or structure their habitat-environment (a process Gregory calls 'structuration') rather than dodging the issue.

From a time-geographic vantage point, however, spatial structure alone should not be the focal concern of human geography at all. Rather, human geography must consider the *time-space structure* of *society-habitat* as a composite formation. Spatial structure, even when elevated to a true structuralist status, would still be a temporally 'flat' version of what is existentially and practically a much more adequate conceptualization, that of time-space structure. In this latter format, many of the traditional incompatibilities between structure and process would be hammered out from the onset, and a *unified process-structure concept* is already built into the very premises and postulates of such a format.

To Gregory, reconciling social and spatial structure would require geographers to rethink their dominant geometric tradition (his underlying suggestion being that it should be abandoned):

> If geography aligns itself with this critical tradition /i.e. marxism-structuralism/, however, it will be obliged to confront some difficult questions about its current dominant geometric tradition. (Gregory 1978a:41)

Admittedly, much of the spatial organizational thinking in human geography would benefit a great deal from being rethought, but more in terms of what content is articulated and the interrelations between settlement system aspects and other aspects of society, than in the choice between geometric and non-geometric expression of different relations. The problem with the 'flat' spatial projections was that they were able to capture too few of the processes and activities with which the rest of social science showed concern, but the scope for including them is manifestly greater in a time-space geometric conceptual framework.

In such a perspective where time is also included as a dimension, the geometric tradition should not be abandoned but strengthened, since it gives more precision in the handling of a number of topics which have recently emerged as very important, such as problems of energy utilization, environment, quality of life, growth and development, and technology assessment. Geometrically structured models are of help both in formalization and in adding precision to analysis of society-habitat. However, a joint time-space locational and resource framework is necessary to avoid the trap of static morphology (cf

Carlstein, Parkes and Thrift 1978, vol. II, part II).

A holochronic framework, as argued in the preceding chapter, also allows a better treatment of motion, dynamics, change and dialectics, since the very logic of process is built into the 'elementary structures' themselves, so to speak. The conventional marxist format, for instance, incorporates process as dialectics, or in Gregory's words:

> Whereas the categorical paradigm /Lévi-Strauss' 'structural morphology'/ constitutes an invariant net to hold fast the eddies and flurries of change in a regular, unyielding mesh, the dialectical paradigm moves the net with the rhythms of this moving world. ... /The dialectical paradigm/... is a mode of reasoning founded on the perpetual resolution of oppositions. It is in these oppositions and their resolution that motion /sic/ is found. (Gregory 1978a)

It may not be necessary, however, to see motion as the perpetual resolution of 'oppositions', and the dialectical paradigm which has many virtues perhaps in the analysis of class struggle, is far from the only key to the resolution of structure versus process.

In a way, the problem both with conventional dynamics in social science (including dynamic spatial modelling in geography) and with dialectics is that time is only half-heartedly incorporated. Many dynamic models start from an essentially a-temporal conceptualization of reality and introduce time through the back-door by describing how stock variables make numerical quantum jumps between discrete and discontinuous points in time. This may sometimes be a useful procedure but is equally often inadequate. Likewise, dialectics are more frequently centered around the qualitative sequences of resolution of oppositions than the continuous flow of events in time, a continuity disrupted when certain effects accumulate into peaks triggering off a structural transformation. But in either case, time is more basic both to deep structure and appearances of phenomena. The case can be made that the perspective and role of time (and space) need to be improved in most conventional social science paradigms.

Let us finally add some temporal dimensionality to the concept of *invariant*. What is an invariant or not is related to the time scale chosen. In the daily round of activity, the settlement system may be looked upon as an invariant, and one can proceed from this point to assessing which kinds of activity are possible in such a system in an ordinary day (cf the PESASP-model devised by Lenntorp 1976, for instance). But in the longer time perspective of a year, it would be nonsensical to regard the settlement system as unaltered. By contrast, the indivisibility of the human individual is an invariant that has stood the test of millions of years. It is curious that it has been more opportune in social science to treat the *human mind* as an invariant, for instance Lévi-Strauss' 'savage mind' or 'homo economicus', our mythical friend in economics. It is thus in a somewhat hostile climate that time-geography pays tribute to the *human body* as physically indivisible. *   From the vantage point of 'deep structure', however, it would be a pity not to exploit the deductive power inherent in this basic premise. One can, of course, also look into the consequences of breaking the spell of indivisibility. This occurs frequently in human society, as when an indivisible domestic beast of burden is ritually slaughtered and divided up as meat (or perhaps as when the conceptual spark plugs are removed from a theoretical band waggon which has been taking us for a ride).

---

*   It may not be divisible, but it does not follow that it cannot be 'extended' in the sense often used when regarding technology as an 'extension of man'. The other capability and coupling constraints associated with such extensions of man are dealt with in subsequent chapters.

3 HUNTING-GATHERING

THE HUNTING-GATHERING ECOTECHNOLOGY

As empirical field studies of hunting-gathering societies become increasingly available in the anthropological disciplines, the foundation for more comparative and generalized insights is laid. In the late 1960's, a symposium was arranged on »Man the Hunter». In the proceedings from this meeting, Lee and Devore (eds. 1968) point to the fact that 'Cultural Man has been on earth for some 2,000,000 years, and for over 99 per cent of this time he has lived as a hunter-gatherer.' Although most social scientists have no personal experience of hunting-gathering, these societies are more than historical curiosities. If one is to understand contemporary and even future societies, it is necessary to stretch the scholarly imagination and look into all types of society from an evolutionary and human ecological perspective. Hence, there are sound reasons for starting off with a low density, low velocity of movement, and small-scale type of society and ecotechnology, where constraints on activity performance are more conspicuous and less complex. This is also an excellent context in which to introduce time-geographic concepts and analytical methods.

For the non-specialist, hunting-gathering societies are mainly of interest in comparison to other societies, for instance with respect to how resources are used, the intensity of social life, and so on. The potential width of comparison is great as indicated by Sahlins's challenging label on this form of society as the 'original affluent society' (Sahlins 1968, 1972). Affluence is relative in more than one respect, but a major point made by Sahlins when he summarizes his findings is that in many ways hunter-gatherers are not so badly off as many people of the present age complacently believe. The materials assembled by Lee and DeVore (1968) and others indicate that foraging populations did not enjoy a quality of life lower than that of agricultural society ushered in by the Neolithic Revolution, and that their necessary work load (measured as human time spent on work) was low by modern labour yardsticks.

With a renewed interest in human ecology by numerous scholars of the 1970's, many of our established images of progress as we move towards more dense, technified, and large scale societies have been contested, or at least placed in a more complete cultural evolutionary perspective (e.g. by Wilkinson 1973 and Harris 1977). There is thus a macro-temporal dimension to the optimist's cry that we now live in the best of all possible worlds, or to the pessimist's fear that this is true. The test must be made for different societies both along the historic and geographic dimensions.

Without giving a definite account of variations within hunting-gathering societies, a few common propensities found in them can be mentioned by way of introduction. These criteria are particularly relevant to the hunting-gathering populations of recent time who have been confined to marginal environments around the world as the more central areas have already been occupied by agricultural and nowadays also urban industrial societies.

Lee, in his discussions of Bushman society, has detailed the basic qualities of their economy as follows,

> In input-output terms, /such/ an economy exhibits an elementary form when the relation between the production and consumption of food is immediate in space and time... Such an economy would have the following properties: minimal surplus accumulation, minimal production of capital goods, an absence of agriculture and domestic animals, continuous food-getting activities by all able-bodied persons throughout the year, and self-sufficiency in foodstuffs and generalized reciprocity within local groups. (Lee 1969:49)

Roughly speaking, hunting-gatherers have very low population densities, they live in small groups of 25 - 50 persons and move around spatially a great deal. Ethno-linguistic communities can like-wise be comparatively small (a few hundred), and the sexual division of labour has a strong bias towards men hunting and females gathering. Moreover, a pattern of sharing collected food resources is the general rule. The amount of personal belongings is small, and property rights, especially in terms of land, are weakly developed. Larger concentrations of population occur mainly on a seasonal basis and the temporal variability in food resources is met by flexible organization and patterns of spatial mobility. Food storage is not strongly pronounced among the simpler forms of hunting-gathering ecotechnology, and the free movement of groups among local areas, combined with the patterns of inter-group visits, tends to resolve conflicts through fission, i.e. larger groups breaking up into smaller ones. Social control is thus weak as a function of the few platforms on which it can be based.

The time-space conflict between storage and mobility

If hunter-gatherers have survived for centuries at a simple level of ecotechnology, 'What then are the real handicaps of the hunting-gathering praxis?' This question posed by Sahlins (1972:33) suits the present context well when realizing that by *handicaps* we mean *constraints* and that praxis relates to practical action in space and time. A major dilemma of hunting-gathering praxis is the conflict between food *storage* and spatial *mobility*, implying that if this kind of local population is not mobile, it will exhaust the local territory and sooner or later reach its carrying capacity. Moreover, if the population has too much material wealth, and limited capacity for porterage, people become tied down to a given place, i.e. they become subject to a coupling constraint. They may then have to increase their work time, being compelled to travel beyond their depleted local habitat on a regular day-to-day basis.

Of course, if there were no handicaps on human praxis at all, no problems of this kind would arise. If we could carry any amount of objects like food items, storage would not render us immobile. If individuals could travel at high speed for sustained periods, storage would be compatible with spatial mobility, and if the carrying capacity of local land was unlimited, spatial mobility by reason of food quest would be unnecessary. But not even in post-industrial societies can people operate under such freedom from constraint, and many handicaps of pre-industrial life and culture (albeit modern technology has buttressed the constraints to a greater or lesser extent) remain with us. For this reason, the hunting-gathering epoch of human cultural evolution is far from irrelevant to contemporary societies.

Stations, storage and interaction with localized entities

People interact with a variety of material objects, plants, animals, and artifacts, which are used as inputs in human projects and activities (apart from people interacting with one another). As members of the physical universe, all materials are subject to the law of inertia, i.e. they are stationary and immobile in space until some force makes them move. Nature harbours many agents which reshuffle materials over the earth's surface, and where natural forces like wind, water or various organisms leave off, human intention and action may set in.

Whichever projects are pursued, each human individual is placed in an environmental situation where other individuals or objects are located at a greater or smaller time-space distance away. Distance must be overcome through time demanding movement. A hunter checking his traps would look something like Figure 3:1a while a gathering round would be depicted as in Figure 3:1b. All places where individuals stop to act or interact with some other individual, organism or object are *stations,* which form the elements in the settlement system and the spatial bases for activity and interaction.

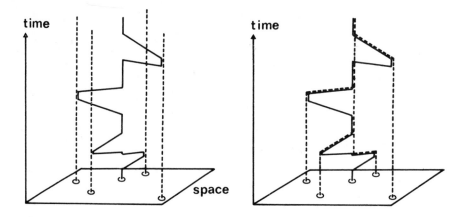

FIGURE 3:1   A hunter checking his traps (left) and a person gathering building materials in four different places.

In hunting-gathering societies, some stations in this sense have very short duration and are not revisited for years, while others are of longer duration and are more regularly visited or used. The very formation of stations in the settlement system is mainly attributable to the individual capability constraints on the portage and transportation of objects, organisms, and perhaps other individuals, for example children. It is impossible in any society for a person to carry along the whole quantity of food, water, raw materials, equipment, and other facilities used in the course of a few weeks. The volume of movable inputs must be reduced to the minimum compatible with the capability constraints on carriage. Since many materials used are *divisible*, this facilitates portability. The storability of objects themselves is often a precondition for whether they can be transported. Perishables are best consumed right away or else left behind. Interaction with materials and other localized objects depends on the factors of *divisibility, movability, storability, and portability*. While Figure 3:2a showed the interaction between a person and in-divisible, non-movable objects, Figure 3:2b illustrates the case of interaction with indivisible movables, where a person can carry all the objects he or she picks up.

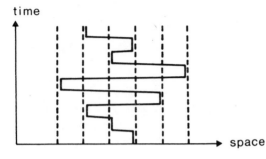

FIGURE 3:2   Interaction between an individual and objects.

a) interaction with stationary and non-movable objects.

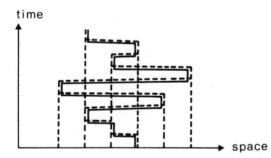

b) interaction with movables.

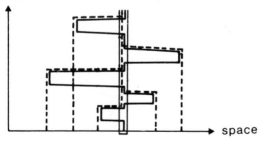

c) interaction with movables while subject to constraints on portability. This gives rise to a nodal pattern of interaction with stores. The objects are picked up one at a time and taken to a station or place of storage.

The principle of return in space

A different pattern of interaction in time-space arises when the materials used in a given period are too heavy or bulky for one individual to carry all at once, and yet these materials have to be collected for use in that period. In this case one or a few spatial locations must be selected as *stations of storage* to which the initially dispersed objects must be brought if they are to be consumed or combined into new forms. The relative locations of such stations of storage are a basic factor affecting time spent on travel between them.

But there is another factor which is strategic to the utilization of stores and that is the factor of returning to the same place. This is an absolutely fundamental principle in human and also in much of animal behaviour. The utilization of a person's dwelling or a bird's nest assumes that the place is revisited, while stores themselves assure people continuity in supply and accessibility to inputs at times required. The accumulation and use of a wide range of tools and facilities is based on the principle of return in space (Figure 3:2c) rather than random walks through space.

Returning to places is a way of providing feedback of information which boosts memory and identification with the environment as well as creating security of repetition in external conditions. Situations which are not in complete flux but are constant in a good many features require less time to be invested in learning and relearning and allow greater applicability of past knowledge and experience, hence boosting technical capability by reducing inherent capability constraints. The principle of return has many social and cognitive concomitants apart from physical ones relating to economies of time and travel in the utilization of material objects. Many sociologists enquiring into the issues of social order, continuity, turnover, and change have perhaps underestimated how continuity in the use of the material environment based on returning in space provides continuity in the social system.

The principle of reunion

If meetings between given sets of individuals occurred once but never again, no society would be possible. The condition that the same persons never meet again cannot be taken as the basic principle of social organization, as this would preclude the existence of socio-economic units formed among the population and prevent cultural continuity. Of course, in virtually all societies the whole spectrum of possibilities is there, between the extreme cases of each individual meeting the same persons every day throughout his life and that of interaction with another person just once in a life-time.

In practice, the overwhelming majority of individuals in all societies known have a minimum number of people whom they meet repeatedly over long periods of time, i.e. for years rather than days (cf Figure 3:3). Such interaction is a precondition for the pursuit of many human projects and is based on the *principle of reunion*, which is analogous and complementary to that of return in space. It is essentially a continuity principle of maintenance of both biotic and socio-cultural relations, such as that between adult and offspring as a minimum condition for human reproduction.

Again, the principle of reunion is a means of making repeated use of acquired social learning and of establishing and maintaining socio-cultural instruments such as language, norms, forms of cooperation, and other institutional capital facilitating human projects. Not only is reunion an economic principle for reducing overhead costs in cultural learning, it is also a way of keeping human resources and human time mobilized for various projects. For Homo Sapiens it is a principle both for generating and maintaining culture. The main point is not that culture is socially transmitted and shared through human interaction, which it is. Rather, recurring meetings make cumulative material and mental stores of culture possible and establish cumulative feedback in cultural process (feed-*forth* in time to be exact, as things can only be fed back in space). Were this not the case, common experiences and learning would get dissipated by too drastic a turnover of individuals in the various groups. It is only through repetitive couplings that culture can build up. For each population, sub-population or group, some balance must be struck between time allocated to external and internal participation, if their culture or sub-culture is to be maintained and reproduced.

Social units, domestic or otherwise, cannot simply exist in the minds of people but have to be maintained physically through meetings that demand time and the formation of bundles of individual paths. Most domestic units throughout the world consist of the same persons interacting with one another for hundreds or thousands of days, almost regardless of the kind of ecotechnology (hunter-gathering, agriculture or industrialism), while many other bundles are formed on a less repetitive basis. Industrial society is as characterized by coupling constraints as are pre-industrial societies. Technology only shifts and changes these constraints but never removes them all together.

Bundling of paths is thus imperative both for culture and collaboration by enabling individuals to carry out projects that are beyond their individual capabilities. The performance of collective activities assumes coordination, which generally means co-location in space

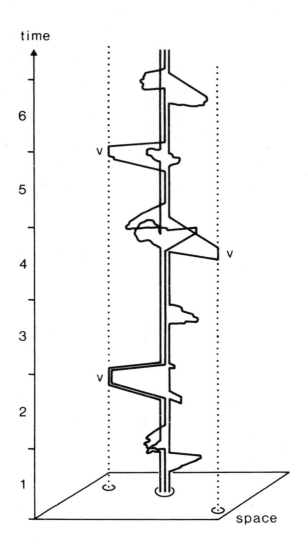

FIGURE 3:3   The principle of *reunion* is illustrated by means of recurring bundles among three individuals belonging to the same household/family. Although the period covered in the graph for three hypothetical hunter-gatherers is only *six* days, it gives an idea of a widespread phenomenon in all societies. Usually such recurring bundles among family members are formed for *thousands* of days rather than just a few days. (In this fictitious example, the loops indicated by V signify visits to neighbouring settlements or camps.)

and time. As figure 3:4 illustrates, co-location in time or space alone are both inadequate conditions for joint activities between individuals (although co-location in space only is the principle behind the famous silent trade case in economic anthropology). Co-location in both time and space is the general and basic precondition for collaboration and culture. In his work on socio-cultural causality, the sociologist Sorokin (1943:173) made exactly this point:

> Since human beings are destined to live and act collectively, one of the indispensable conditions for any possible collective action is a time synchronization or time co-ordination of the actions of the parties involved. If X agrees to meet Y, both must be at the agreed *place* at the agreed *time*. Otherwise, if X arrives five hours later than Y, neither the meeting nor collective action is possible. A mutual, coordinated timing of the behaviour of the members of a group has been the indispensable means for the adaptation of their behaviour to one another.

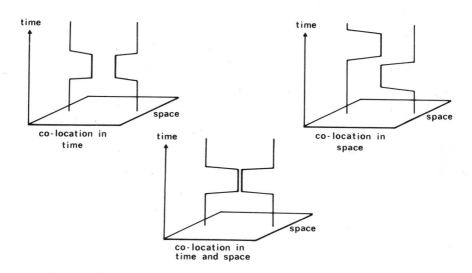

FIGURE 3:4   Coordination in space and time (two sample individuals).

Capability constraints on spatial mobility

In understanding hunting-gathering societies, attention must be given to spatial mobility, since both other individuals and stores in the environment impose coupling constraints. Whatever projects and activities an individual pursues at any given moment in time, his place or *station of departure* acts as a *situational determinant* on his immediate prospects for making use of other spatially localized inputs. The *accessibility* of these inputs is further determined by the individual's capability to travel, especially the constraints on his *speed*. The individual-path can only move outward within a time-space cone which is circumscribed by the maximum speed of movement from the point of departure. This potential action space looks graphically like a funnel, assuming that the individual can move in all directions at a uniform speed in a uniform landscape (in an 'homogenous plain' as geographers sometimes call it.) For simplicity, this is illustrated in a two-dimensional time-space graph (Figure 3:5a). For hunter-gatherers, for instance, speed on foot is slow, as is the case in any type of society. Assuming that an individual has to return to the same station at a fixed time, such as sunset, this coupling constraint circumscribes the area that can be covered, since the further out from this point one is, the sooner he must turn back to return in time. Hence one can speak of another funnel which is the

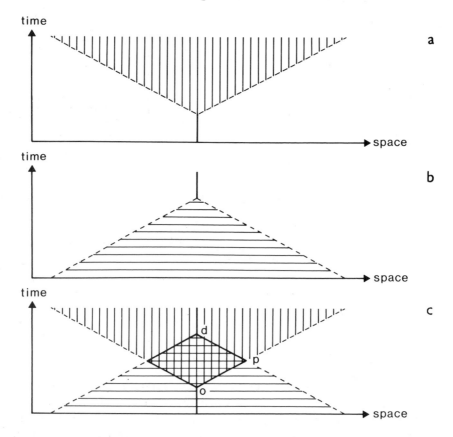

FIGURE 3:5   A day prism as a result of interacting constraints.

mirror image of the former (Figure 3:5b). The action space of the individual is thus doubly delimited, and the resultant portion of time-space can be called a *prism* (Figure 3:5c). On the assumption of homogenous speed of travel, a prism becomes symmetric if the station of origin and destination is the same. Otherwise it is asymmetric (Figure 3:6).

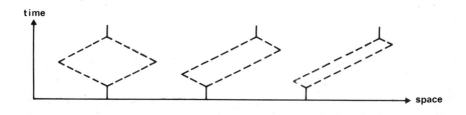

FIGURE 3:6   The effect of increasing distance between stations of departure and destination on prism size and shape.

It is obvious that an individual starting at point *O* (see Figure 3:5c) can move outward in any direction, but if he goes beyond point *P* he will not have time to reach point *D*. The volume of the prism is defined by all potential paths of an individual when traveling at maximum or less than maximum speed. Only the time-space locations inside this circumscribed area can be used. The individual is 'imprisoned', so to speak by his own capabilities of traversing space as well as by his own decisional constraint to be back at a given time. The potential action space is called a prism because of its shape, although this term may be geometrically incorrect when presented in three-dimensional graphs. There exist a variety of capability and coupling constraints which result in prisms restricting the use and occupation of time-space in the settlement system. In most societies, dwellings operate as the main station of return and reunion in the daily round, but other stations have similar functions. Speed is another factor influencing prism shape. For instance, if one rides instead of walks, the maximum speed is normally increased, paths slope more horizontally, and the potential time-space covered increases correspondingly (Figure 3:7). An increase in spatial mobility can have drastic social and technological consequences.

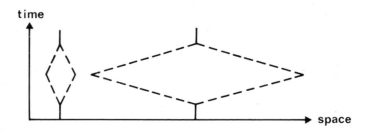

FIGURE 3:7   Prism size as a function of speed of travel (Left: walking. Right: horse-back riding.)

The introduction of horses among the American plains indians, for example, allowed them to catch up with the bison herds. Another determinant of prism size is the *time distance* between origin and destination, since a person covers a greater area in a longer time period than a shorter one, but the *spatial distance* between origin and destination also affects prism size and shape (cf Figure 3:6b. For a major treatise on prisms and transport, cf Lenntorp 1977.)

Mobilizing time supply within prisms

The fact that time supplied to any activity is mobilized within a prism and under coupling and capability constraints is commonly overlooked in time allocation theory. In analysing the daily life of a population, practically all individuals belong to some domestic group occupying a station of residence from which they depart in the pursuit of various projects and activities. The residential settlement in a region constitutes the set of stations from which time supply is spatially mobilized and on which the main day-prisms of individuals are focused.

Prisms, therefore, also describe the time-space area within which human time can be mobilized and hence the volume of time that can be spent on activities located at various distances away from the station of residence. The greater this distance, the greater the drain on time, and activities of certain durations are impossible to complete (Figure 3:8). Moreover, as travel time distance away from the station of return is increased, the remaining time for stationary activity is *reduced by a factor of two*, since time must also be supplied to the return journey.

The structural conflict between mobile and stationary activities is also important. The time supplied to movement is immediately

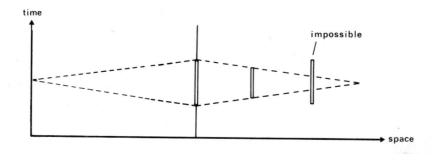

FIGURE 3:8   Activity duration and distance from day prism centre.

reflected in an individual's time budget by the balance between stationary and mobile activities (Figure 3:9). If a given amount of stationary activity is necessary in order to complete certain projects, gross inefficiency in the selection of routes, or the order in which stationary activities in different places are linked in space over time, upsets this balance. Too much of an individual's time supply is then lost in the sheer spatial mobilization of it to meet demand elsewhere.

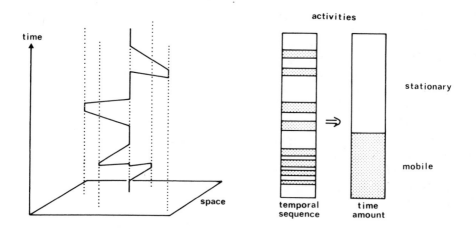

FIGURE 3:9   The balance between stationary and mobile activities in an individual's daily time-budget.

One can conceive of travel behaviour in time-space which is so inefficient that major projects are impaired. Since highly inoptimal time-space behaviour is an ecological threat to survival, the process of biological evolution of both animals and humans has built in mechanisms for economizing on time spent in movement. These mechanisms are so taken-for-granted that they often go unnoticed until their lacking application leads to stress and shortage of time for what is perceived as vital and highly desirable activities. Geographers have applied this tendency for humans to economize their time to both explanation and planning of spatial interaction. Much has been written on optimal spatial behaviour, 'minimization of movement', and calculation of shortest possible routes for transport in order to save time and energy. To economize human time spent on travel does not necessarily imply that people strive to eliminate movement all together, since travel is a means of reaching certain ends and products. Rather, the need is to keep time and energy expenditure on travel and transport from impairing the completion of various individual and collective projects. Thus, individuals and groups moderate their interaction in space.

THE BUSHMAN HUNTING-GATHERING SOCIETY

The prism concept is not only useful in sorting out the conflict between storage and mobility, but can also be employed to illustrate other features of hunting-gathering societies such as their foraging strategy and ecological adaptation. Adaptation itself is a dynamic relationship between population and environment. It has a pronounced time-space structure which is readily analysed in terms of time geography.

Consider, for example, the !Kung Bushmen * as described by Richard Lee (1968, 1969), which serves to give coherent illustration of a few basic mechanisms. The !Kung Bushmen, numbering today a few hundred, are one of the better known cases of a hunting-gathering society, and share with most twentieth century hunter-gatherers the predicament of inhabiting a marginal environment, in the case of the Bushmen the Kalahari desert of Botswana

* For convenience of exposition I will drop the !Kung specification and refer to this people as Bushmen, although this is ethnographically too vague. The exclamation mark in front of Kung simply refers to a click sound (!K) which is phonetically transcribed thus. Most of Lee's intensive survey refers to the Dobe group of the !Kung Bushmen, consisting of some 30-40 persons, but again this specification will not be repeated in the text here.

and Namibia.

In relating prisms to the carrying capacity of the local habitat and the seasonal adaptation of the Bushmen, two crucial resource inputs come to the fore: food and water. Given the desert environment, it is the latter which is the most scarce and hence is the more constraining in its effects on Bushman life. Water scarcity is structured in time and space rather than given at random. Thus, the time-space locations of water impose major coupling constraints on other Bushman activities: 'Since Bushman camps, of necessity, are anchored to water resources, they can exploit only those vegetable foods that lie within reasonable walking distance of water. Food resources that lie beyond a reasonable walking distance are rarely exploited,' (Lee 1969:56).

Food, whether animal or vegetable, is more abundant and more widely dispersed in time and space. Since both food and water are crucial to Bushman subsistence, it is the joint constraints associated with these resources which shape their adaptation and foraging strategy. 'The crucial factor in the annual subsistence cycle', says Lee (1969:59-61), 'is the distance between food and water... At any given moment, the members of a camp prefer to collect and eat the desirable foods that are the least distance from standing water...' The temporal stability of various constraints is an important factor to consider. The availability of resources in the habitat typically fluctuates over the annual cycle, which is most strongly felt in the changing relative time-space location of water and food resources. Other major constraints such as the mode of travel do not alter; the speed of walking is fairly constant and only varies somewhat between different age segments of the population. There are thus both fixed and variable constraints impinging on the Bushman daily round.

The day prism applied to Bushman camps and local groups

The camp station serves both as a place of return and a place of reunion although the spatial location of the camp itself shifts over the annual cycle. 'The camp serves as the home base for its members. Each morning some people move out to collect plant food or hunt game, and each evening the workers return to the camp and pool the collected resources with each other and with the members who stayed behind...' (Lee 1969:58). From a camp of thirty to forty people, individuals move out to collect plant food and hunt in small groups of two or three. If they leave in the morning, they have to come back before dark, providing they want to share the common evening meal. This leads to the typical day prism as described above (Figure 3:5).

Since most camps are located immediately by a water hole for most of the year, it can be assumed for the moment that water is available in the centre of the prism and is accessible at hardly any time cost. It is thus the maximum velocity of sustained movement which determines the volume of the prism, apart from the time that can be spent away from camp. Although Lee does not state the exact rate of movement, circumstantial data reveal speed clearly: 'By fanning out in all directions from their well, the members of a camp gain access to the food resources of well over 100 square miles of territory within a two hour hike.' This area having a radius of 10 km (6 miles) that is covered in two hours corresponds to a speed of 5 km/h. 'Since the average working day was about 6 hours long... /and/ the adults worked about two and a half days per week... !Kung Bushmen of Dobe... devote from twelve to nineteen hours a week getting food...' (Lee 1968:35-37). Consequently, on a working day two hours may be spent on travel up to 30 km away from camp if no stationary activities such as collecting plants or resting were included in the daily programme. But since the normative working day including travel is generally only 6 hours, the real day prism is much smaller in volume, which explains the 10 km (6 mile) radius as the standard range around a water supply point. It is henceforth this inner prism (Figure 3:10) which will be referred to in the Bushman case.

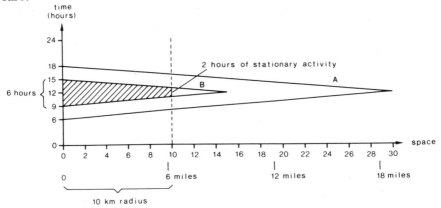

FIGURE 3:10   The relations between working hours, speed of travel and reach in a typical Bushman day prism. A: prism reach for a 12-hour work day. B: prism reach for a 6-hour work day.

To the extent that food quest activity involves motion slower than maximum walking speed (stalking animals can be a very slow form of movement) or if travel is done with frequent pauses for resting, the prism volume and hence the local territory covered is diminished (Figure 3:11). And when the social company of others in

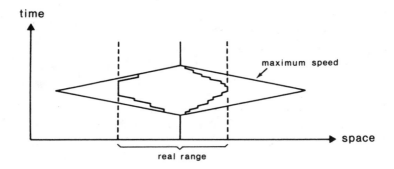

FIGURE 3:11   The people who leave the camp for 6 hours out of which 2 hours may be spent on stationary activities, for instance, would commonly intersperse stationary and mobile activities. Their real range would thus look somewhat different from that shown by the shaded area in Figure 3:10. The prism shape also depends on how the speed of travel is defined. Here it is assumed that people are walking at normal pace, neither running nor walking extremely slowly.

camp is given high priority in a society where the gathering and hunting expeditions themselves are carried out in extremely small groups as mentioned before, it follows that the actual local area used is much smaller than is indicated by the maximal 12-hour prism. This has implications, of course, for the time it takes to exhaust a local territory from food resources.

The carrying capacity of the local prism habitat

Bearing in mind that resources are generally located and fragmented in time-space, it is not really the average carrying capacity of the entire land area inhabited by Bushmen which structures the food gathering activities and foraging strategy. Rather, it is the carrying capacity of the *local* territory within prism range which is immediately decisive, and it is the size of the *local* (camp) population within the same prism range that matters first and foremost. By virtue of sharing the same station of origin and destination (station of return) and sharing one another's company (station of reunion), the Bushmen share the same general prism, perhaps with a minor allowance for age-sex differences in mobility. As Lee explains,

> The Bushmen typically occupy a camp for a period of weeks or months and eat their way out of it. For instance, at a camp in the nut forests (which form narrow belts along the crests of fixed

dunes /cf Figure 3:12/), the members will exhaust the nuts within a one mile radius during the first week of occupation, within a two-mile radius the second week, and within a three mile radius the third week. As time goes on, the members of the group must travel farther and farther to reach the nuts, and the round trip distance is a measurement of the /time and energy/ 'cost' of obtaining this desirable food. (Lee 1968:60)

The manner in which the Bushmen eat their way out of their local prism habitat can be illustrated as in Figure 3:12 covering three weeks with a typical day prism for each week. It is apparent that by the sixth week, they have eaten their way out to the crucial 10 km periphery. But what happens then when the local territory as defined by the six hour working day and 10 km radius has been covered and when food resources within time-space range have been largely tapped?

In this situation, the main option is to move camp to a new location where both food and water is within range. This is equivalent to shifting the base station for the day prisms, as in Figure 3:13. Due to the few stores and material possesstions, shifting residence for the Bushmen takes only a minimum of time: 'Although the Bushmen move their camps frequently (five or six times a year) they do not move them very far. A rainy season camp in the nut forests is rarely more than ten to twelve miles /16-20 km/ from the home water hole /used in the dry season/...' (Lee 1968:35). However, since water is comparatively plentiful in the rainy season due to temporary pools, the location of water sources imposes weak coupling constraints then, and little time in the daily round is spent on fetching water.

Annual cyclic variations in food-water distance and prism content

The time-space structure of the environment changes with season, and the Bushman foraging strategy correspondingly shifts. From November till April, food is abundant and there are many water holes. With the termination of the rainy season in April, a three-month dry season begins. Food is still abundant but there are no more than eight water holes in the entire regional territory (one in each local territory). In the second half of the dry season, from August to October, there are still eight water points but now food gets more scarce (Lee 1968:61-2). The structure of mobility and the rate of exhausting the local prism habitat previously described largely pertains to the six-month rainy season (rainy by Kalahari standards). The hunting-gathering strategy of this season is that which is least costly in travel time and thereby also in total work

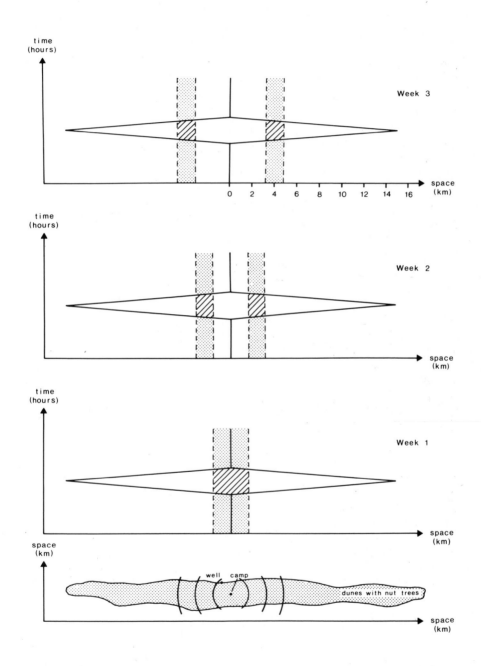

FIGURE 3:12   The sequence in which a Bushman population exploits its local prism habitat.

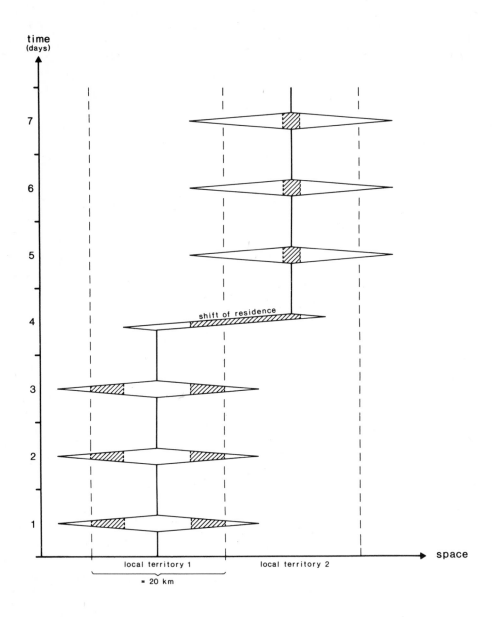

FIGURE 3:13   The shift of residence and prism habitat by a local Bushman population. Note how distant lands are used before the shift, and nearby lands after (hatched area).

time. But even then, satisfying both food and water requirements assumes the flexibility of adaptation that shifting residence gives.

> Towards the end of the rainy season, a temporary pool may dry up before the nuts in its immediate vicinity become exhausted. In this case the residents move camp to one of the larger summer /dry season/ water holes which usually persists until the early autumn (April or May). (Lee 1968:61)

In this case, it is not the food-water distance relationship which wields influence but the temporal availability of water in a given place. The Bushmen have no technology for storing water over periods longer than a few weeks and consequently *storage* cannot be an alternative to *movement*.

In May the dry season commences, and during the first half of the dry season, all the Kung Bushman groups are living by the permanent water holes, from which they eat out an increasing radius of desirable foods (Lee 1969:62). Now the option of shifting camp site is no longer open, since there is not water at other sites as in the rainy season. Still, however, the staple food — the mongongo nuts — are within reach, but more time has to be spent on travel, up to three hours in one direction from camp. But food is still relatively abundant, and Bushman subsistence time or time spent on food quest activities is still relatively small.

By the second half of the dry season, from August to October, the distance between food and water reaches its annual peak.

> People must either walk long distances to reach the nuts, or to be content to eat the less and less desirable foods, such as bitter melons, roots, Acacia gum, and the heart of the vegetable ivory palm... A round trip of up to twelve miles /19 km/ can be accomplished in a single day, but for trips to more distant points an overnight hike must be organized, involving the packing of drinking water and the carrying of heavy loads over long distances. (Lee 1969:61-2)

We are now dealing with a two-day prism, in which a much larger area is potentially within reach (Figure 3:14a and b). The shape of the prism is influenced by the fact that little movement is done by night. A corresponding three-day prism is also conceivable although three-day trips are not mentioned by Lee, as the potential surface covered increases exponentially with the distance from prism centre. This is not to say that all of this surface can in practice be covered in a short time in terms of actually hunting and gathering in all places. However, the main point is that it takes longer to eat one's way out of a local prism habitat if the prism is larger and more is contained within it (given that other competing groups are not encountered within this broadened range). In other words, mobility with respect to shifting residence can always be traded off against daily movement, the latter making subsistence effort and working hours increase. What is gained in settlement space(-time) by increasing range is lost in human time.

> The alternative tactic to the longer trips is to stay at the home base, and to exploit foods of lesser desirability in terms of taste, ease of collecting /time and effort collecting/, and abundance. At a given dry season camp one sees both alternatives in evidence. The older, less mobile members stay close to home and collect the less desirable foods, while the younger, more active members make the longer trips to the nut forests. As the water-nut distance increases, more and more attention is given to the lesser foods. (Lee 1969:61)

Again, this alternative is a good example of the time-space structured mechanisms in operation. One is that by lowering the quality of life by accepting to eat less desired — but still edible — food, some segments of the population do not have to go farther and farther afield to collect this food. Hence the carrying capacity of a local prism habitat is not simply naturally determined (although there are ultimate natural capacity constraints involved also), since the residential mobility pattern in the relatively plentiful rainy season is also a function of a normative constraint as to what is wanted and edible food. Prism volume and content is also normatively defined by what is regarded as an individually and socially acceptable length of the work day on those days when the Bushmen actually work. The rate by which the local population eats its way out of a local prism habitat is thus a compound result of constraints rooted in nature as well as society.

The prism model thus captures several substantively different matters such as the coupling constraints associated with places of return and reunion (water holes, camps), the capability constraints on how fast individuals are able to move for various distances, and the normative constraints of how fast they want to move and what activities they want (and in a subsistence sense have) to perform in their environment. The usefulness of prisms and indivdual-paths as descriptive devices is that all these aspects can be captured in one sub-model, and the consequences of a given situation can be derived for other related aspects of human interaction, one of which is the syndrome of how activities are allocated according to age, sex, and skill.

In the previous discussion of prisms, individuals were regarded as having equal travel capability, and so the individual prisms coincided with the collective or aggregate prism for the whole local population.

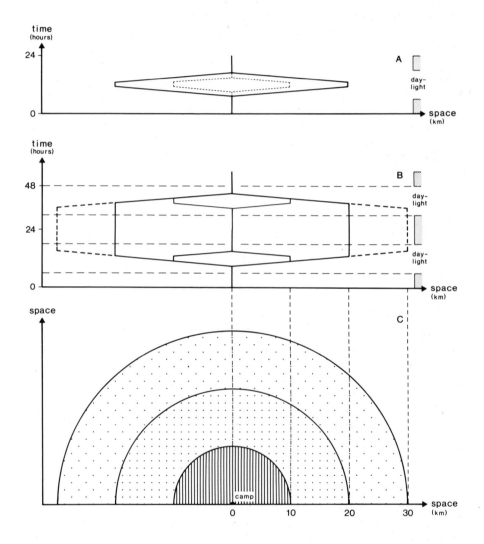

FIGURE 3:14   The accessible area within a two-day prism of the Bushmen.
A: The prism range for a hypothetical 12-hour work day (solid line) and a
6-hour work day as in the wet season (dashed line).
B: The two-day prism range assuming two 6-hour work days within a 20 km
radius (solid line), or two 8-hour work days within a 30 km radius (dashed line).
C: The spatial area covered in the above two cases.

This generalization is adequate for many analytical purposes but in other instances further precision as to how the local prism and content varies among population categories is necessary.

For example, consider age-sex categories. They are often spatially segregated. During the dry season when food is also getting scarce within day prism range, a more distinct spatial structure emerges in how various population categories mobilize their path and time for subsistence activities. It is typically the elderly and some children (the less mobile) who stay close to the prism centre (camp), while youth and the members of active age increase their range through overnight trips and by having two-day prisms (Figure 3:15). Working day norms are relaxed and modified to the exigencies of subsistence and environment during this season, only to be tightened again as the rainy season reappears.

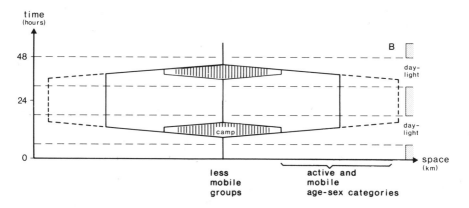

FIGURE 3:15   The spatial differentiation of activities by mobility category. The less mobile categories (children, elderly, incapacitated) stay close to the prism (camp) centre, as indicated by the lined surface, while the active and mobile categories further operate within a two-day prism range.

The annual cycle of residential mobility and the path of a local Bushman population can be summed up roughly as in Figure 3:16. It should be noted that the path of an entire local population is divisible unlike the paths of the individual members. The three seasons in the year are shown. In the rainy season they shift camp at five to six week intervals and then move to one of the permanent water holes at the beginning of the dry season. Towards the third period, the dry season from August to October, residence becomes compounded by an increase in temporary settlements outside the main camp by the water hole.

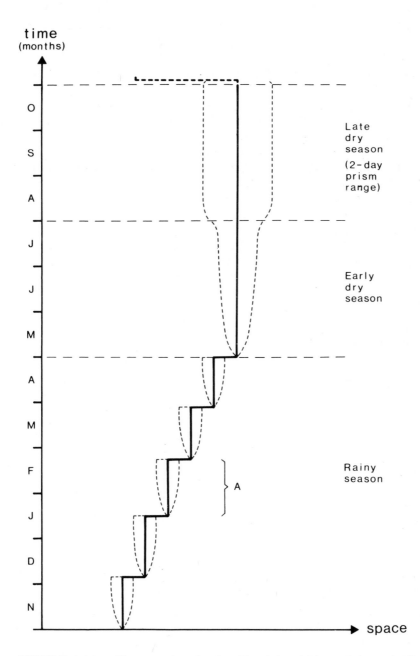

FIGURE 3:16   The annual cycle of residential mobility and the path of a local Bushman population, where the dashed lines circumscribe the area of exploitation. The period A covers five to six weeks, residential shifts being more frequent in the wet season when the water location constraint is less severe.

Sub-population size and the carrying capacity of the local prism habitat

Given the above premises it can be inferred that the larger the local population, the shorter the duration of residence in a certain locale will be, or alternatively expressed, the larger the number of individuals who share the same camp and basic day prism content, the greater the frequency of residential shifts, other things equal. Sahlins suggests exactly this when he says that 'a local group becomes vulnerable to diminishing returns — so to a greater velocity of movement /i.e. frequency of shifting residence/, or else to fission — in proportion to its size (other things equal).' (Sahlins 1972:34).

Fission is the anthropological term for reducing local population size by splitting up, some individuals going elsewhere, while *fusion* is the process by which it grows through groups merging and by 'in-migration' of individuals. Binford, among others, raises the topic of local population size (group size) by noting that,

> there may be size limitations to various organizational features of society... It seems that many of the people who have presented materials on composite bands /of hunter-gatherers/ have mentioned the size range of 15-25 people. What factors would operate to limit local group size? What factors can lead to an optimum group size of 15-25? (Binford in Lee and Devore 1968).

It is important to regard carrying capacity dynamically and not as a static and fixed ratio between sheer population size and a given territorial area. The population itself shifts and changes even over short time periods such as weeks and months (mainly through fusion and fission), and the habitat also fluctuates in content over the year, or from one year to another. Hence to conceive of the whole problem in terms of some maximum 'man-land ratio' is often not very enlightening. Moreover, the carrying capacity of the local habitat is partly a function of its size, which in turn depends on prism range and volume. The latter again is related to how far and for how long time people are able and willing to make their gross working day (i.e. a day including both work and travel). The amount of time that is spent on work and travel, finally, is part of the broader time allocation issue of how much other activity in camp such as socialising that they are willing to allocate away.

We saw before that the Bushmen do not shift residence from the water holes in the dry season, as water would then get out of daily range. Instead they spend more time on work and travel and broaden their local prism habitat (apart from exploiting inferior foods, which also amounts to increasing the carrying capacity of the habitat). If a local population adapts its behaviour and activities to the natural environment in order not to face a ceiling of carrying capacity too soon, it is precisely by extending prism range and enlarging the local prism habitat itself. At a general level, this may be the only alternative to moving or dispersing the local population and reducing its size through fission. This might also have been a feasible alternative for the Bushmen, were it not for the fact that the water holes are too few and impose the same coupling constraints as sharing camp with other individuals does. The Bushmen have a problem of prisms centered around water points which is not dissimilar to that of nomadic pastoralists discussed in the next chapter.

What happens then if group size, i.e. local population size, increases through *fusion* from say 15 to 30 people by another group of some 15 individuals joining the camp. This should, other things being equal, imply that the new local population would 'eat' their way through a given local habitat twice as fast. Or else they will, as Sahlins puts it, suffer from 'diminishing returns'.

That may be the solution adopted, but it is not the only necessary one because if they are only willing to walk a little bit farther out and spend a bit more time and effort on travel, the land within range will increase by the square of the radius increment. In other words, in the trade-off between human time and settlement space-time, as time for travel increases, the local prism habitat grows by much more. This is not to say that it can expand indefinitely, because then again we come up against limited time resources for travel in a given one or two day period. But the implication is that at the local population level (and given that the regional territory is not too densely packed with people), the population may increase *arithmetically* (through fusion), while by increasing travel, land and hence food resources will grow at a *geometric* progression. This is a curious inversion at the local level of Malthusian thinking (largely outmoded anyway) in which it is claimed that at the global or national level, population increases geometrically while food resources do so in arithmetic progression.

However, these are two very different issues, and it is in this context hard not to react to Sahlins's (1972:33-4) muddle of the two when he describes the 'cruelly consistent... Malthusian practices' of hunter-gatherers between how a local group may eliminate elders since they cannot effectively transport themselves, on the one hand, and how they must move faster through a territory, when the local group is larger, on the other. While admitting that both are related to the contradiction between storage and mobility-porterage, there is less consistency between the issues that it may seem. The 'draconic' population policies of senilicide and the propensity of hunter-gatherer women to space out births so that they only have one child to carry, have little direct impact on the fusion-fission processes that mainly regulate local population size. At least not so long as there is 'unoccupied' land for the local population to expand into. The fertility-mortality variables operate in an entirely different and longer time perspective than fusion-fission (in- and out-migration to and from a group), and the direct ecological effects of the former variables are less felt at the local level than as parameters of the regional level.

Carrying capacity of the settlement system versus population system

The main drawback with the population density concept is that it is too static and provides few clues as to underlying interaction. Population maps at least have the merit of showing how the population (or rather their residences) are spatially distributed, something which statistical measures of average population densities or man/ land ratios do not. But analysis must go beyond this, since the residential location of the population only indicates at best where the centres for their day prisms are (Figure 3:17). In societies where residences shift with the season or even more often as local territories

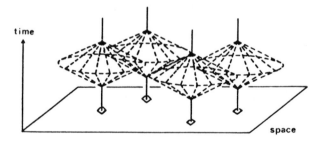

FIGURE 3:17   Stations in the settlement system as centres for day prisms.

get temporarily depleted, a more dynamic model is obviously preferable.

Realizing that   performance of joint activities by individuals presupposes physical contact and proximity, how does spatial population distribution affect interaction within the population? In human geography a considerable amount of theory and empirical study can be found on the relation between *distance and human interaction*, the most general finding being that the frequency of interaction tends to decrease with growing distance (cf i.a. Haggett 1965, Olsson 1965). There are many elaborations on this theme in locational analysis and several useful generalizations have been applied to planning problems. Yet these studies have often been clearer in terms of spatial organization than with respect to resource utilization in general. Let us confine ourselves here to two basic resources, human time and settlement space(-time) and describe a few basic mechanisms by which they are related to relative spatial location.

Firstly, spatial distance has a direct impact not noly on the frequency of interaction but also on the *duration* of interaction. Figure 3:18 shows how the latter is related to day prisms and the fact that people are not random walkers in space but are based at stations to which they normally return at some later time of the day or in some longer time perspective. In the first case (Figure 3:18a), the day prisms of the two sample individuals are widely separated so that no contact is possible in one day. Moving the base stations of the prisms closer by some two thirds of the way increases the scope for daily interaction to half a day's duration, and when the two individuals

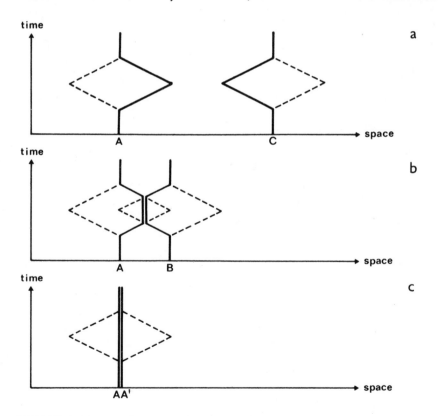

FIGURE 3:18   Distance and the duration of interaction (two sample individuals.)

share the same residential location, they can potentially interact over all 24 hours (Figure 3:18c). Although distance between prism centers for a whole population is one aspect of population density, spatial distribution (concentration versus dispersal) is also of great significance. On the crude assuption that each local population has a common prism, the extent to which prisms for the various local populations overlap largely determines the potential volume (frequency-cum-duration) of time consuming interaction within the regional population.

When prisms are set farther apart, as in Figure 3:19a, where population density and concentration is low, a minimum of time-space within the local prisms is shared. As population density increases, locally and regionally, so does the sharing of time-space. This indicates an important conflict in the synchronized and simultaneous use of human time and settlement space-time: The greater the population density/concentration, the less is the time cost for interaction within the overall population due to reduced travel time, but the smaller is the share of settlement space-time that is accessible to each individual and local population. What is a gain in human time is a loss in space-time. Population concentration can thus be adverse and beneficial at the same time, and some balance must be struck between boosting carrying capacity for time demanding activities in the population system and augmenting the carrying capacity of the local prism habitats for space demanding activities.

ADVANCED HUNTER-GATHERERS: RELAXING THE CONSTRAINT ON MOBILITY

The Bushmen so far described are an example of a very simple kind of food collecting society, which is not likely to be very representative of the advanced hunting-fishing-gathering societies that have existed in the less marginal environments of the world. Societies and ecotechnologies that have gone beyond the low velocity and 'high friction' modes of travel can increase their spatial range of interaction considerably, which has numerous consequences. Two obvious cases which come to the fore as having a more advanced ecotechnology are what Murdock (1968:15) calls sedentary fishermen and mounted hunters.

Sedentary fishermen

The most conspicuous and well-known cases of advanced hunter-fishermen-gatherers are the aboriginal peoples on the north-west

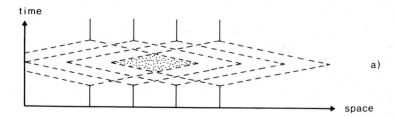

a)

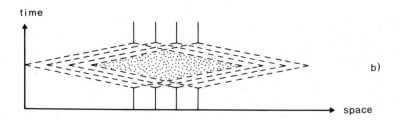

b)

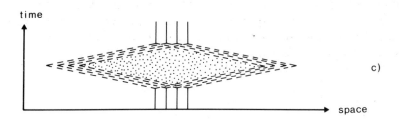

c)

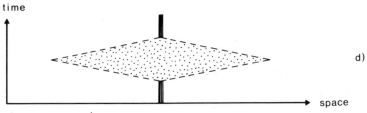

d)

(time = 24 hours)

FIGURE 3:19 Population density and the sharing of a local prism habitat: a) dispersed population and little overlapping; d) concentrated population and total overlapping of prisms.

coast of North America, peoples known as the Yurok, Tolowa, Chinook, Salish, Nootka, Kwakiutl, Haida, Tlingit, and others. As Suttles (1968:56) points out, they 'can offer some guidance in estimating the possibilities of cultural development under comparable conditions in prehistoric times.' The level of cultural complexity that this group of peoples have achieved is well on a par with those of of many sedentary cultivators. This complexity is facilitated by their plentyful and stable marine food resources in combination with their transportation technology.

It sounds like a paradox that increased capacity for spatial mobility leads to sedentarization rather than greater frequency of spatial shifting, but of course the logic of the situation is that water travel leads to a broadened prism range and increased carrying capacity of the local prism habitat in a region already rich in food resources. The conflict between storage and mobility is not only reduced for this reason, but also because of the relaxed constraint on *portability*, associated with canoe travel. Some peoples had large canoes which could take up to 60 persons, for example, although they had a whole range of sizes for varying purposes. Strategic spatial locations of settlements at the right times of the year also have the effect of letting the mobile food resources come to you rather than vice versa, as when the salmon go up the rivers. The broken and varied nature of the terrain and suitable coastline locations captured both marine and terrestrial food resources within local prism range (Figure 3:20).

Storage was an alternative to mobility of residence also in the sense that food supply was not continuous in space and time but

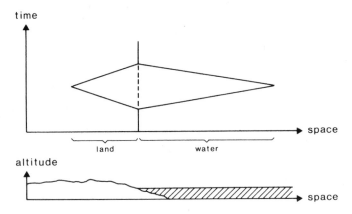

FIGURE 3:20   The differential shape of the prism depending on the travelling surface. It is assumed that travelling on foot in broken terrain is slower than using canoes at sea. The figure only gives an approximate idea, since the prism could also be shaped by the fact that people may move more slowly uphill than downhill, or that the wind may afford faster travel at sea in one direction than in the opposite direction.

there were great local and seasonal variations. To eke the most out of a habitat with this time-space structure, local carrying capacity could be used more fully with storage than without. Human time spent on food preparation was a viable alternative to spending it on travel. As Suttles (1968:64) reports in his essay on 'coping with abundance',

> A number of foods, such as dog salmon, must have required far more /time and/ effort in the storing than in the taking. Thus the limits in the exploitation of times of abundance may have been set less by people's capacity to *get* food than by their capacity to *store* it.

Another dimension of the carrying capacity of the local prism habitat is, of course, the number of individuals who share the same basic prism. Some of the strategically located settlements could reach population sizes of over a thousand individuals, a situation unheard of in more simple hunting-gathering societies, except possibly as an event lasting a few days in a season of abundance.

But relaxed constraints on portability and travel also allow the migration of large settlement units. The Nootka changed residences twice a year in whole local units to keep local population size compatible with the carrying capacities of their local prism habitats.

> Their winter villages were located near the heads of inlets for shelter, and their summer villages were near the shore. In each site they erected house frames... but each household owned only one set of planks to serve as walls and roof. Every spring and fall they carried their planks back and forth between their residences by canoe... The houses ranged in length from forty to one hundred feet and were between thirty and forty feet wide... (Coon 1976:55).

The case of the North-west Coast Indians of America is by no means unique. Hunters-fishermen-gatherers in other rich environments, both historical and recent, have reached similar levels of cultural complexity as the sedentary fishermen.

Mounted hunters

Mounted hunters are defined by Murdock as hunters,

> depend/ing/ largely on domestic animals which they ride in pursuing or surrounding their game... /They/ approximate pastoral nomads in a number of respects, the chief difference, perhaps,

being that their domestic animals do not provide a major part of their food supply. (Murdock 1968:13-15)

The best known mounted hunters are the North American Plains Indians. Eggan (1966) has classified them into two main groups, the *High Plains Indians* and their less nomadic neighbours, the *Prairie Plains Indians* Neither were hunter-gatherers in a strict sense. The Prarie Plains Indians also relied on agriculture, while the High Plains groups had reverted almost completely to hunting-gathering (the only plant cultivated by a tribe such as the Blackfoot was tobacco). From the time-geographic viewpoint taken here, however, the eco-technology of mounted hunters is particularly interesting to compare with that of other hunter-gatherers, but we shall concentrate on a few major aspects such as the role of the horse as a vehicle for spatial mobility.

The adoption of the horse among the Plains Indians is a fascinating innovation with far reaching consequences. It is a classic example of a transport revolution. Having been brought to the continent by the Spaniards landing in Mexico, the horse spread slowly northwards. Once adopted, its use meant that all previous constraints on speed of travel, portability, seasonal migration and settlement size were considerably relaxed. This opened up a new niche and the rich natural environment of the great plains could be exploited with much greater efficiency. The resultant impact on social structure was rapid, profound and traumatic and has been a favourite theme among many leading American anthropologists (i.a. Eggan 1966).

A cornerstone in the ecotechnological revolution affecting the Plains Indians was the increased *speed* of movement which the horse gave to its human riders. By virtue of the coupling constraint between man and mount, this new speed finally matched that of the quarry, the American bison or buffalo, which roamed the plains in millions. Previously these herds had moved much faster than the hunters ever could, and to get within bow and arrow range had been difficult and yielded few animals. Moreover, it was far from easy to kill a buffalo with one shot. So this technique had to be supplemented by both coordinated action and investments in physical infrastructure. For a tribe such as the Blackfoot, described very succinctly by Forde (1934:45-68) in his classic study of 'Habitat, Economy and Society', the only reliable method of securing their quarry in larger numbers was either to stampede them over a cliff or to impound them. The pounds were large permanent constructions used season after another close to the habitual bison trails. The dog, which was the main domestic animal prior to the horse, had been of little avail in this process. But with the horse (called 'big dog' for want of a precedent), radical change in the communal hunts was possible. The buffalo herds could be rounded up at many kilometers' distance and driven into pounds or over cliffs. Or else, small herds could be surrounded by riders and the animals shot one by one while running in a circle. Since the bison were migratory animals aggregating in large numbers on the succulent prarie pastures of the summer and dispersing at other seasons, the horse allowed a new form of time-space adaptation by the Indians. As herds split up maximally in the winter, for instance, so did the Blackfoot following them northward. As the buffalo could provide all the basic necessities, such as food, clothing, shelter and a variety of artifacts, there was little need to get tied down by agriculture, and for the High Plains tribes, the reversion from agriculture to hunting was nearly complete.

The neighbouring Prairie Plains Indians around the Missouri River area maintained more of their life style from before the equestrian era. These tribes had semi-permanent residence in earth lodge villages but participated in large communal buffalo hunts in the summer season. The womenfolk practiced the more stationary occupation — agriculture — while the men hunted buffalo and put up the military protection:

> The earth lodges were large ..., lasting for several generations, and the associated gardens were cultivated in common by the women of the household. After the fields had been planted, the able-bodied men and their wives prepared to go on the summer communal hunt, leaving the older people and children to look after the village and crops. During the summer hunt the travelled with tipis and camped in a camp circle, like their Plains neighbours. (Eggan 1966:59-60)

We can see this solution in Figure 3:21 for a seven day prism of a group of mounted Indians. Women, children and elderly who remain behind without horses have very limited prisms and ranges of operation indeed. Even if they were away travelling on foot for more than three days (before returning), their seven day range is rather limited compared to the corresponding range of mounted hunters. If this seven day prism was enlarged to cover a month or two, we would get a fair idea of the seasonal range covered by the Prairie Plains Indians in their summer hunts.

We can now return to the contradiction between storage and mobility which formed our starting point in this chapter. The High Plains and the Prairie Plains Indians solved this problem differently, but on a much higher technological level than simple hunter-gatherers did, such as the Bushmen.

The use of horses made spatial mobility compatible with storage, both in a daily and a seasonal perspective. For the High Plains groups it facilitated the combination of a more nomadic life style with

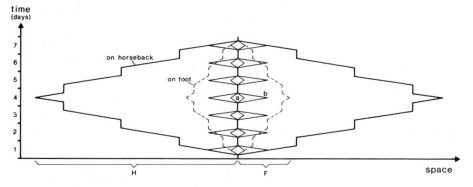

FIGURE 3:21   Differences in range and reach of local prism habitat when the period varies from one day to seven. The small day-prism (a) is that applying to women, children and aged who stay behind in the main camp/village during the summer hunt and who do not have access to horses. Dashed lines indicate the size and shape of their potential seven day prism. For males remaining behind but have access to horses, the day prism is broader (b). The main hunting party of males on horsback during a seven day hunt is illustrated by the largest prism. The seven day prisms are partly shaped by the fact that people and horses are assumed to rest during the night for some hours.

periods of fixed residence in sizeable camps without rapid depletion of the local prism habitat, simply because this habitat was so large. It was particularly large if hunting expeditions of more than a day were undertaken by selected active males, while the remaining age-sex categories, such as children, women, and aged, plus some active males could remain behind. (This is again the situation shown in Figure 3:21.)

It would have been of much less utility for the Plains Indians to extend their hunting ranges, if the summer glut of meat could not be of benefit throughout the entire year. But two conditions made this possible: Firstly, the meat could be made storable by drying it in the sun, smoking it and mixing it with fat, a process resulting in the nutritious food known as pemmican. Secondly, the meat could be transferred on horses to the winter areas of human consumption. Although dogs had formerly carried burdens up to 20 or 25 kilos on a *travois*, the horse had still relaxed the portability constraint by leaps and bounds.

Horses also increased spatial mobility by letting the High Plains Indians follow the buffalo herds on their seasonal migrations, and this too paved the way for a more even exploitation of buffalo throughout the year. Horses also allowed people to live in larger and more comfortable dwellings, the tipis, than in pre-equestrian times. Last but not least, the scale of hunting activities and social life could be altered, and Indians could aggregate into camps which at some seasons had populations numbering thousands rather than hundreds.

Apart from the altered storage and mobility conditions, the Plains Indians illustrate another typical syndrome, that of mobile individuals in relation to stationary domains or territories. The extreme mobility afforded to peoples with traditionally rather stable territories, induced a good many conflicts, not least since the very vehicle of transport was the major bone of contention. With the new mobility, the local prism ranges got out of step with the prevailing system of territoriality, as shown in Figure 3:22. And since horses from the very outset were in short supply and the Indians were poor at breeding horses, raiding and horse-stealing quickly developed into a major sport. The horse itself thus became both an end and a means. But increased mobility was also the germ of the *lingua franca* sign language which gained a common coinage on the plains enabling more peaceful interaction between groups of different linguistic stock.

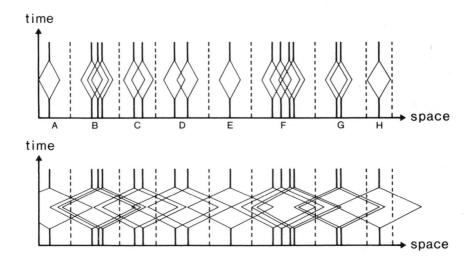

FIGURE 3:22   Relations between prisms and domains for ethnic groups A-H before the introduction of the horse (top). The daily prism habitat was narrow and overlapping prisms were more common within the same ethnic group. After the introduction of horses, prisms became broader (bottom) and different ethnic groups came into much more frequent contact with one another. The figure illustrates some principal relationships without being a description of an exact historical situation.

In terms of time use, the impact and secondary effects of the horse were no less spectacular. Horses were both a time-saving and a capacity increasing innovation, or as Forde puts it,

Between periods of intensive hunting /the Blackfoot/ had considerable leisure, which on account of their nomadism could be applied only to a limited extent to arts and crafts, and appears to have been devoted largely to the elaboration of ceremonial which affected every aspect of their life... (Forde 1934:68)

As for their fishermen counterparts on the Northwest Coast, the habitat of these hunters was anything but unproductive and marginal and their material, social and spiritual life was rich by pre-industrial standards, and their contacts with other groups were on a very wide front for a hunting-gathering people.

It is both tragic and ironic that the splendid, albeit sometimes cruel, life style of the Plains Indians of America, which was made possible by the white man of the 16th century, was to be undermined by the superior hunting techniques of the white man a few centuries later. Then the rifle in the white man's hand reduced what was perhaps sixty million buffalo to a few hundred. The same instrument was used to destroy the traditional hunters of buffalo. With the white man's expansion into the plains, the area again 'reverted' to agriculture.

In the next chapter we turn to nomadic pastoralists, whose spatial mobility is again sometimes enhanced by eqestrian or similar techniques. But, unlike the hunters of the plains, pastoral nomads use their domestic animals as a main source of food and by contrast occupy typically marginal habitats. For them, living possibilities, prisms and constraints appear in yet more different configurations.

4 NOMADIC PASTORALISM

SOME CHARACTERISTICS OF NOMADIC PASTORALISM

Nomadic pastoralists, or pastoral nomads, like many present day hunter-gatherers occupy marginal habitats. Their form of spatial mobility which has rendered them the epithet of 'nomads' is highly related to the marginality of their resource base, while their dependence on domestic livestock, on the other hand, sets them clearly apart from hunters dependent on wild food. The criterion of low regional population density and a less than 'sedentary' form of livelihood associated with the occupation of marginal habitats warrants their place next in this book following the hunter-gatherers.

The objective of this chapter is not to produce a concise theory of nomadic and other types of pastoralism, but to furnish *models* which epitomize *salient features* and *key mechanisms* found among nomadic and transitional types of pastoralists. The present chapter draws on the materials assembled in Johnson's comparative survey on 'The Nature of Nomadism' (1969) in conjunction with various case studies.

The term 'nomadic pastoralist' used here evaluates pastoralism as an ecotechnology first and mobility second. If the choice of technology is such that animals are tended and the main source of livelihood is not agriculture, then the fact of spatial mobility over time is largely induced by the livestock herds in relation to the resources in the habitat.

According to cultural evolutionary theory, nomadic pastoralism has evolved from agriculture mixed with animal husbandry, rather than vice versa. A different train of events has most likely been the case with the evolution of the reindeer ecotechnology in northern areas, where pastoral nomadism could have developed from hunting.*

Generally speaking, however, in the temperate and tropical regions far south of the reindeer belt, a close contact between pastoralism and agriculture persists; if not through the direct participation of nomads in cultivation activities, then through raiding, trading, or other forms of interaction.

A secondary reason for including pastoral nomads in this study is their structures of mobility. Mobility itself is a frail basis for classifying societies, since most of the conventional taxonomies of the sedentary to nomadic continuum tend to overlook more than they include. The element of mobility captured by the concept of nomadism has been applied to hunter-gatherers, ethnic groups like the Gypsies, sections of the population of modern U.S.A., and others, apart from what is generally considered to be nomads proper, i.e. pastoral nomads. Any simple verbal classification is bound to be problematic, since terms such as 'full nomad', 'semi-nomads', 'semi-sedentary nomads' and the like are only vaguely correlated to how people actually move around spatially in the annual cycle. The exercise is even more doubtful, if one wants to pin a general label onto the various time-space structures of how the stationary and mobile activities are intertwined. Taken in a broader perspective, the general point can be made that many of the traditional categories for dealing with spatial mobility both in pre-industrial and industrial societies are too crude. This has caused enough problems when simply describing nomadic pastoralists, as Johnson (1969) has lucidly shown, to make it apparent that these traditional categories are inadequate to describe societies in general.

It would be an improvement to employ *geometric* schemes of classifying mobility rather than verbal categories. Johnson has taken a step in this direction by analysing the spatial geometry of different forms of pastoral nomadism using all three spatial dimensions, including altitude. Horizontal nomadism covers the two sub-categories of pulsatory and elliptical nomadism, while vertical nomadism is divided into the three sub-forms constricted-oscilliatory, limited amplitude and complex nomadism. Although this perspective is more enlighten-

---

*   The domestication of reindeer may have been for their use as decoys to attract wild reindeer and also for riding or pulling sleds. Forde points out that 'the use of tame animals as decoys in hunting their wild fellows /is/ a practice which is almost universal in the reindeer area, especially during the rutting season ... So long as the herder is willing to remain almost as compliant with the natural movements of the herd as the hunter, the reindeer has characteristics which facilitates its domestication...' (Forde 1934:368) The practice of milking reindeer, however, is one which has diffused from other agricultural or pastoral nomadic peoples in adjacent territories. Reindeer herding is also quite compatible with both hunting of reindeer and other animals as well as with fishing.

ing than many of its preceding schemes, it is still defective with respect to the time dimension. What is needed is a time-space approach to classifying structures of stationary-cum-mobile activities. Such structures have a direct qualitative dimension as well as a quantitative one. Only this will allow a more precise understanding of the relation between society, habitat and ecotechnology.

On the continuum between a purely sedentary society (from the viewpoint of residence) and a hypothetical pure nomadism or completely migratory existence, there are many possible combinations, depending on the following main factors mentioned by Johnson (1969:12):

1) the combination of animals being herded;
2) the role that agriculture assumes in the group's economy;
3) the amplitude of the yearly displacements;
4) the seasonality of the natural regime;
5) the physiography of the tribal area;
6) the quantity and quality, viewed both spatially and temporally, of grazing and water resources.' (Johnson 1969:12)

In this chapter, we will endeavour to make explicit some of the main dimensions and mechanisms of pastoral ecotechnology, society and habitat by looking into the interaction of constraints that circumscribe the living possibilities of nomadic pastoralism. To round off this picture, a few cases of agro-pastoralism and hunting-cum-pastoralism will be included. To this end, the concepts of carrying capacity, prisms, capability and coupling constraints, fusion-fission and other concepts already presented will be applied so that differences and similarities to other ecotechnologies discussed in this book will emerge.

Capability constraints and coupling constraints on pastoral activities

While the Neolithic Revolution and the introduction of agriculture (the tending of domestic plants) was based on the principle that humans controlled the reproduction of plants, pastoralism is founded on a corresponding control of animal reproduction. Instead of using plants as main energy converters, pastoralists rely on the bioenergetic detour of animal products.

Which are the constraints and possibilities of this kind of ecotechnology? No doubt, the herding of domestic livestock imposes coupling constraints on the human population. So do plants, but animals are mobile, generally speaking, rather than stationary (cf Figure 2:2). If animals are to be harnessed as energy converters to heterotrophic humans, the individual capability constraints of the animals and their coupling constraints to the environment (what they must consume where and when) are more or less transferred to their human caretakers. The animals set the pace of movement (both as herds and as individual packing and riding animals) and their food and water requirements circumscribe the parts of the overall region which can be used. However, this particular form of animal eco-technology also relaxes many human capability constraints. Humans can carry more goods on the back of a camel, horse or donkey than when walking on foot and carrying freight themselves, for instance, and animals can be used to carry a supply of water. But the cost for all this in terms of coupling constraints is that of having to tend the animals even when they are not of immediate use (though they may be milked, ridden or slaughtered on some later occasion. By accept-ing the coupling constraints that accompany a given ecotechnology, humans gain freedom of maneuver and reduce their capability constraints.

The major gain in capability and capacity from the adoption of a pastoral ecotechnology, however, is that it allows the exploitation of marginal environments unsuited for other forms of exploitation, notably agriculture. By accepting the coupling constraint between human and beast, the latter can be harnessed as energy converters in environments which would otherwise be less productive to a human population.

The environments thus exploited are not static; they fluctuate immensely in quality, over time and in space. It is through the particular mobility potentials associated with the herding of animals (as opposed to the cultivation of plants) that pastoral nomads are enabled to establish a stable basis for subsistence despite the vicissi-tudes of nature.

THE TIME-SPACE STRUCTURE OF THE NOMAD PASTORALIST HABITAT

Pastoral nomadism is a livelihood form that is ecologically adjust-ed at a particular technological level of the utilization of marginal resources. These resources occur in areas too dry, too elevated, or too steep for agriculture to be a viable mode of livelihood, and the nomadic pastoralist thus makes use of resources that other-wise would be neglected. Historically, pastoral nomadism is best described as a specialized off-shoot of agriculture that developed along the dry margins of rainfall cultivation. (Johnson 1969:2).

The special habitat of nomadic pastoralists is perhaps best defined in counterdistinction to that of agriculture. Roughly speaking, nomadic pastoralism is practicable in areas where the growing season is too short for domestic plants to mature due to lack of rainfall or where it is too short for other reasons such as low temperatures. Rough and sloping terrain may be another impediment to agriculture. As far as rainfall is concerned, it is not only the duration of the rainy season that matters, but also the amount, reliability and distribution of precipitation. As a habitat gets increasingly marginal, the shorter the growing season and the more patchy the vegetation becomes, patchy both in a spatial and temporal sense. The same is the case with water supply. This principal difference between agriculture and pastoral nomadism is illustrated in Figure 4:1. While agriculturalists use land by staying put in one local area, at least during the time of cultivation, the pastoralist's use of marginal regions assumes movement between local areas, often on a very short-term basis. By such movement, pastoralists can procure enough vegetation for their herds on a regional scale although there may not be sufficient pasturage locally for more than a short time. The path of movement, however, is not only a function of the availability of vegetation or water but even more so the coincident accessibility of these resources. It is thus common that plentiful vegetation is ungrazed for lack of water within range, or vice versa and worse, that local overgrazing and over-intensive use of land results from the accessibility to water.

As shown in Figure 4:1, the precondition for the exploitation of marginal niches and habitats is that of *mobility*: Those who tend the animals must lead a life with a more or less nomadic slant, and they must be prepared to allocate their paths to those times and places (regions and seasons) where this ecotechnology is applicable and viable, although this might entail forms of fission and fusion which are less desirable on purely social grounds. Johnson (1969) complained that in much of the ethnographic material on pastoral nomadism, the specific routes taken were often unaccounted for. Raising the level of ambition to cover both the route and the time-table of movement, materials are even harder to come by and the data probably have to be collected anew before such time-geographic desiderata can be met. However, let us use some of the scant empirical material available.

The time-space structure of movement and rest of one group of nomadic pastoralists, the Baluchi of Pakistan, have been described by Scholz (1974:285-89). He covers an eleven day period for a section of the Bangzulai tribe as they move from the mountains down to the plains via the Bolan pass, south-east of Quetta (Figure 4:2). The herd is composed mainly of goats and sheep while camels are being used for carrying luggage and riding. The nomads moved with their herds nine days out of eleven during the period of observation, resting for two days. All in all, during the period of observation, actual movement lasted for 55½ hours and the total distance covered was approximately 190 km, giving a mean speed of travel of about 3½

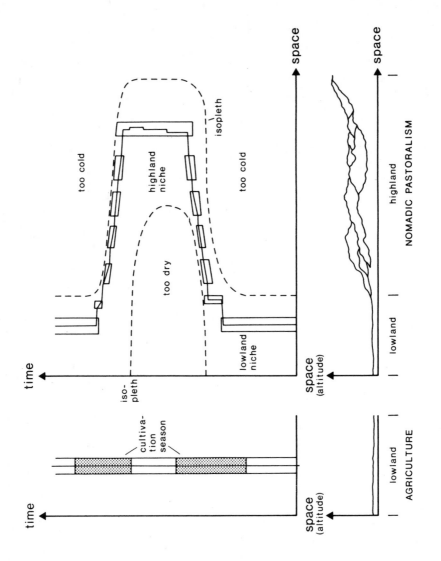

FIGURE 4:1   The difference in utilization of the habitat between tropical agriculturalists and nomadic pastoralists. For the former, the sequence of parcels used (parcel defined as a portion of space-time, cf p. 151) is essentially vertical, whereas for the latter it is vertical and diagonal in the annual cycle.

The right part of the figure shows a common example of the time-space structure of the niches/habitats available for pastoralism. These niches are delimited on the one side, by excessive cold preventing the growth of pasturage, and by excessive heat and dissication, on the other. The kind of nomadic pastoralism depicted is of the highland-lowland kind. (The fusion-fission processes associated with this annual cycle are not shown. Cf Figure 4.4 on the Basseri yearly cycle.)

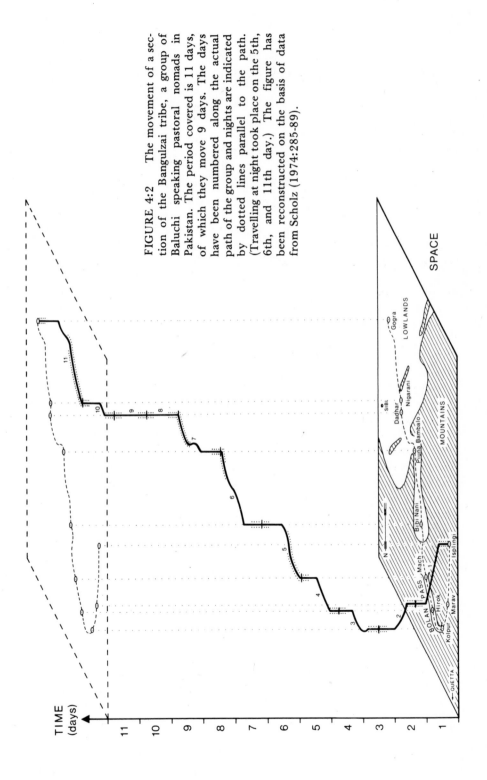

FIGURE 4:2   The movement of a section of the Bangulzai tribe, a group of Baluchi speaking pastoral nomads in Pakistan. The period covered is 11 days, of which they move 9 days. The days have been numbered along the actual path of the group and nights are indicated by dotted lines parallel to the path. (Travelling at night took place on the 5th, 6th, and 11th day.) The figure has been reconstructed on the basis of data from Scholz (1974:285-89).

TIME
(days)

SPACE

km/hour. Some of this travel took place at night, depending on the exigencies of the situation, such as the extent of crowding in the Bolan Pass. Including the two days of rest, an average of more than five hours per day were spent moving (cf Chapter 9), which is probably well above the average for most pastoral nomads during their most mobile seasons (cf Barth 1961:150-51), but in this particular instance it was induced by the shortage of suitable mountain passes and a desire to avoid mixing herds with other groups passing through the same bottleneck.

Sometimes the timing of nomadic movements can best be likened to the time-tables of a railroad system, an analogy used by Barth (1959) in his essay on the land use of migratory tribes in South Persia. In the province of Fars, the cultural heartland of ancient Persia, pastoral nomadism entails the planned sharing of time-space in such a way that collisions between different herding groups are kept at a minimum. Again, this is particularly the case at bottleneck points such as mountain passes or narrow valleys which must be passed through by several groups. To illustrate this, Barth employed the technique of plotting paths in time and space which is typical for railway scheduling (Figure 4:3). It is essentially the same idea as the time-geographic concept of paths or trajectories, and the figure shows when different portions of a given region are occupied by various ethnic groups.

On the basis of a new 1:100 000 scale map for the region combined with data in Figure 4:3 by Barth and supplementary data on time use for migration also given by Barth (1961), it was possible for our present purposes to reconstruct the entire annual cycle of the Basseri in a three-dimensional time-space graph (Figure 4:4). Although this graph may not be accurate down to the minute detail, it still gives a correct overview of the seasonal movements of the Basseri between winter lowland and summer highland pastures. This picture is typical for many other nomadic groups as well.

The Basseri live in tents throughout the annual cycle and move their sheep from the winter lowlands in March through a set of areas occupied by sedentary agriculturalists on their way to their customary mountain pastures in the north. The sheep of the Basseri are not as heat resistant as those of the lowland peasants, so the Basseri must move their sheep to escape dissication in the dry season (cf Figure 4:1). By utilizing pastures *successively* over time, yet avoiding congestion and competition with other tribes using the same route (a route known as *il-rah*), the Basseri can optimize the use of their environment and maintain a breed of sheep which is larger and more productive that those bred by the sedentary peasants. Upon reaching their summer quarters in June, the Basseri fission into small sections again (although not to the same extent as in winter), but

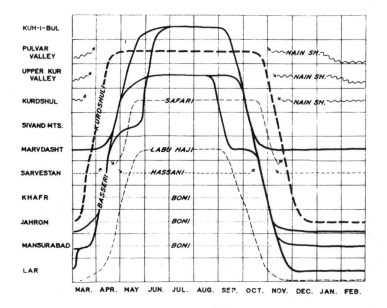

FIGURE 4:3   Barth (1959), from whom this graph is taken, explains it as the 'succession of tribal groups in the strip of land occupied by the Basseri during a yearly cycle... Solid line — schedule and route of the Basseri. Heavy dashed line — Kurdshuli tribe, thin dashed lines — Arab tribes, wavey lines — other shepherds. Small arrows — place and time where other tribes enter or leave the Basseri strip.' (Courtesy: F. Barth.)

during the summer these groups remain in one locality until the trek back to the lowlands begins in late August. On the way back, some groups stop temporarily and let their herds graze on the stubble in the harvested fields (simultaneously manuring these fields), while the majority of the Basseri move on to the lowlands, where they again fan out into smaller groups for the winter grazing.

ASPECTS OF SHORT-TERM AND LONG-TERM CARRYING CAPACITY

Carrying capacity can be associated with either the long-term population dynamics of livestock and people, or the short-term dynamics of mobility and its effects on the sharing of local prism habitats. Starting with the latter, the limited carrying capacity of local parcels of space-time* is one important factor behind the rate of movement of herds through different places. As Barth puts it:

*   The concept of parcel is defined on p. 151, Chapter 5.

FIGURE 4:4

The annual cycle of a pastoral nomad group, the Basseri of South Persia. Their system has been reconstructed in three-dimensional form with the aid of a 1:100,000 map of the region combined with the data furnished by Barth (1959, 1961), including that in Figure 4:3. Legend: Shaded areas indicate land 2100 m or more above sea level; dotted surfaces signify salt lakes and swamps.

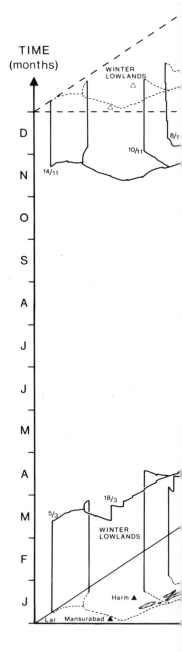

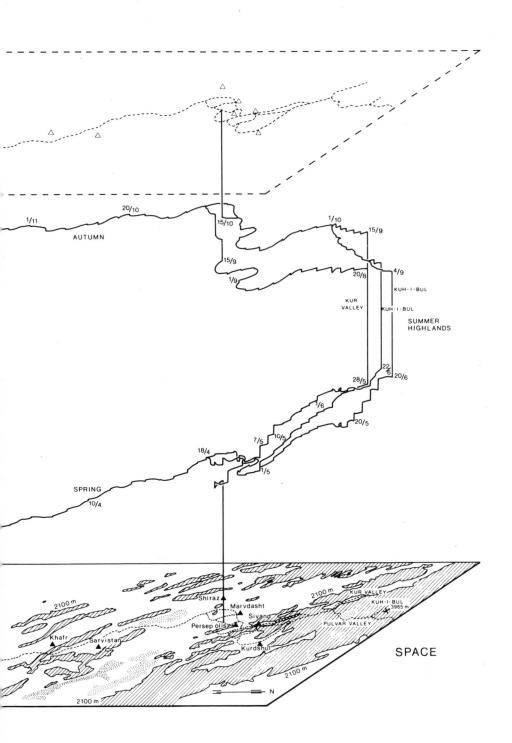

AUTUMN

1/11

20/10

15/10

15/9

1/9

1/10

15/9

20/8

4/9

KUH-I-BUL

KUR
VALLEY

KUH-I-BUL

SUMMER
HIGHLANDS

28/5

22
6

20/6

1/6

20/5

7/5

10/5

18/4

1/5

SPRING

10/4

2100 m

Shiraz

Marvdasht

2100 m

KUR VALLEY

KUH-I-BUL
3965 m

Sivand

Khafr

Sarvistan

Persep olis

Kurdshul

PULVAR VALLEY

2100 m

SPACE

2100 m

N

2100 m

...in addition to the variables of time and place implied in the rail-road-schedule analogy, there is also a seasonal variation in the carrying capacity of any one place... the pattern of succession /of herding groups/ in any one locality is adapted to the seasonal changes in the carrying capacity of that locality, so as to maintain a balance between the rate of utilization and the rate of production, between *load* and *capacity*. The 'load' increases when larger tribes rich in flocks pass through, or when they linger in a locality; it is reduced when smaller flocks utilize it, or when they travel more quickly. (Barth 1959:6) /Italics added/

The carrying capacity of a certain parcel, i.e. a piece of land for a given period of time, thus depends on the *period of exposure* to grazing by livestock, the size and composition of the herd, and so on. In other words, the load has both a herd size and a time dimension.

In the longer time perspective, the number of humans which can be fed and otherwise supported by a herd is a function of several basic factors:

The stability of a pastoral population depends on the maintenance of a balance between *pastures, animal population* and *human population*. The pastures available by their techniques of herding set a *maximal limit* to the total animal population that an area will support; while the patterns of nomadic production /utilization of products from herds/ and consumption define a *minimal limit* to the size of the herd that will support a human household. In this double set of balances is summarized the special difficulty in establishing a population balance in a pastoral economy: the human population must be sensitive to imbalances between flocks and pastures. (Barth 1961:124) /Italics added/

Of course, the volume of pasture that will be available depends to a great extent on the herding strategy itself as well as the mix of eco-technological elements (the choice of technology) in terms of the species of animals reared and their numbers, the combination of animals in different locations, and so on. As Johnson (1969) points out, it is for this reason that,

... all nomads follow regular patterns of movement closely adjusted to the seasonal availability of pasture and water in various zones throughout the year. As a result, at any time of the year, portions of the tribal territory may be empty while herds and herders, following recurrent patterns and rational principles, occupy those areas which offer the best seasonal sustenance for the flocks. (Johnson 1969:18)

This implies that while the intensity of land use at an average regional level may be very low, the local intensities found at certain times or seasons may be much higher, often extremely so.

Since the feasible solutions to environmental exploitation and the adaptation to a habitat with time-space fluctuations have to be found within a whole set of constraints (roughly analogous to a linear programming model), one single constraint may be critical in reducing the overall level of production and limiting the day prism range which can be occupied by grazing. The availability and accessibility to water is one. This was striking in the case of the Bushmen hunter-gatherers (Chapter 3), but applies even more to tropical nomadic pastoralists. To some degree, however, the latter are able to relax the constraint as far as domestic water supply goes due to their superior transport technology. Their pack animals can carry a supply of water for human consumption, so that their day-prisms can be extended. But this also assumes that the animals can get the water they need. This in turn is partly a function of the 'choice' of pastoral technology, i.e. the composition of herds and herding skills and strategies.

Water supply, prisms and the choice of animal technology

Few pastoralists rely on only a single animal species, although in the reindeer belt, for instance, the only other domesticated animal traditionally was the dog. But pastoralists further south generally have a broader range of choice in animals: camels, horses, donkeys, sheep and goats as well as cattle are ordinarily available in sub-tropical and tropical regions. These animals allow a variety of uses such as providing milk and carrying riders and freight, apart from being sources of meat, fibres and hides. This gives rise to a diversity of strategies in herding, in the composition of herds and in the sizes to which herds can be built up.

If households and other units are to be *viable* in surmounting the fluctuations in food supply, this requires choosing the right combination (bundle) of animals to herd at the right time and the right place (cf Stenning 1958 and Barth 1961). Although all domestic animals employed by pastoralists are energy converters which make marginal lands productive to humans, some species adapt better to given environments than others, and a pastoral group increases its viability by selecting animals that give satisfactory yields in their special habitat. In a more dynamic perspective, the environment harbours various forces and organisms which affect the mortality and hence productivity of herds. The use of a species with a short gestation period, such as sheep and goats, permits a fairly rapid build-up of herds that have suffered previous reduction, while in the case of camels which have other useful qualities, the incremental process is slow due to long female gestation periods. Using the camel for meat production would lead to a drastic reduction of herds that would be slowly compensated, while milking camels has no such detrimental effects in the long run. Sheep and goats propagate fast and are therefore more suitable for meat production, although goats are also well suited for supplying milk.

For pastoralists in marginal arid regions, the capability of a species to withstand shortage and irregularities in *water supply* is of major relevance. In this respect, the *camel* is outstanding because of its ability to endure prolonged periods without water, especially if the vegetation itself is moist. There is a difference, of course, between how long a camel can go without water before it perishes from thirst, and how long water needs can go unsatisfied before lactation is reduced or comes to a full stop. Lewis (1961) in his treatise on Somalia mentions that camels can go without water and still produce milk for over 20 days in dry grazing, while sheep and goats under the same circumstances require water every few days. For the Bedouins on the opposite side of the Red Sea, Sweet (1969:176) reports that 'with green, lush growth camels need not be watered for as long as 30 days, whereas camels, as well as other animals, subsist on the dried forage in summer, but require water once every three days or oftener, depending on the heat.' While sheep and goats thus may go unwatered for up to five to seven days with relative impunity, cattle need water more often, say every two to three days. There are, however, variations in these specific water requirements, a point to which we will return briefly.

The reason for emphasizing this capability constraint is, of course, that the extent to which marginal arid regions can be utilized is greatly a function of how water requirements can be met. The time-space distribution of water sources and vegetation in conjunction with the potential speed and endurance of the animals, largely determine the sizes and volumes of prisms within which grazing is feasible. It also bears heavily on the rate by which an animal population eats itself (and hence their human custodians) out of their local prism habitat, i.e. the time it takes to reach the level of carrying capacity or possibly transgress it by overgrazing. As Lewis (1961:59) puts it,

> The problem is therefore to find adequate pasturage and at the same time keep within striking /read prism/ distance of water. Often a choice has to be made /an allocation of paths/ between passable grass but fairly distant water, and poor and virtually non-existent grazing with water close at hand...
> In the dry season, the sheep and goats have to remain within a radius of a few miles from water.

This is virtually the same problem which faced the Bushmen (Chapter 3), although they had different constraints on mobility, storage and portability than pastoral nomads.

Prism size and interaction frequency

The varying size of prisms and grazing ranges of different species used by pastoralists can perhaps be most clearly discerned among those herders who are compelled to gather their herds around wells in the dry season. This is typical for the Kababish of Kordofan (in the Sudan), the Mutair (Saudi Arabia) and several pastoral groups in Cyrenaica (Libya), for longer or shorter periods during the year (Johnson 1969). Wells are physical investments which take time to build and maintain and they are also fixed in space and become *stations of return* for the livestock. Hence, there are prisms generated with wells as centres, but the sizes of these prisms are not so much a function of differential *speed* of movement as the *frequency* with which different animal species require water under various conditions. The horses formerly herded in such large numbers by the Kazaks of Central Asia (cf Forde 1934) certainly have greater capacity for fast movement than sheep, goats, cattle or camels, for instance, but what is more strategic to prism size and hence the carrying capacity of the local prism habitat is the frequency interval with which water is needed. On this point, authorities give a rather wide range of data for the same species, but we may assume that this reflects variations in external conditions (such as temperature, moisture in the vegetation, etc.) rather than variations among sub-species. When sheep, for instance, have to subsist on dry fodder, as do the sheep herded by the Baluchi tribes in Pakistan for a good part of the year, they have to be watered as much as three times a day in hot weather and every other day during the cool season (Johnson 1969:68 referring to Pehrson 1966). For the Kababish, Johnson relates that sheep have to be watered every fourth day during the dry season. Cattle might be watered every day during the hot season and equally frequently throughout the year if they are to milk well.

The impact of watering frequency on prism size and range is illustrated in Figure 4:5, where it is assumed that cattle and sheep need water every day, goats every other day and camels every eighth day. Although this may vary somewhat, the principal differences of prism reach among species stand out clearly.

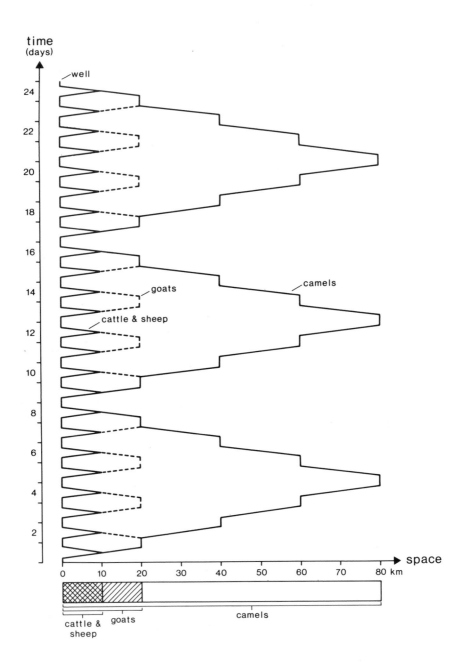

FIGURE 4:5   The grazing ranges of different species around a water source (e.g. a permanent well) as a function of the speed of movement and the frequency of watering. The figure shows only the approximate proportions of prism sizes and should not be interpreted as exact, since watering requirements vary with temperature and moisture in the vegetation, among other variables.

Camel pastoralism: The Bedouin case

The advantages of camel pastoralism are thus obvious: camels are able to graze within a much wider range than cattle or 'shoats' (sheep and goats), and marginal lands well beyond the reach of these other species can be exploited by camels. The Bedouins of the Arabian Peninsula are renowned for their camel herding. Their highly specialized ecotechnology is viable for manifold reasons. In the words of Sweet:

> ... within the arid regions in which their culture developed, the Bedouin camel-breeding tribes have maintained a distinctive pattern and a dominant position over other societies and settlements in their territory by virtue of their ability to *exploit the grazing ranges into which other local economies cannot spread,* and by virtue of their fighting strength, mobility and control of communication routes. (Sweet 1969:158-59) /Emphasis added./

The versatility of the camel places it in a special category. It can be used for many forms of transport, for riding, and for carrying domestic water and equipment, apart from its magnificent capability of withstanding temporary shortages of water. It can further be used in activities such as fighting and raiding or for trading purposes. Last, but certainly not least, it serves as a source of food:

> The size, strength, mobility, endurance and lactating capacity of the camels provide the subsistence base for Bedouin society; for many families, milk may be the only food for several months of the year. (Sweet 1969:173)

The capacity of the camel to carry baggage ('carrying capacity' in the literal sense) is in the order of 300 kg which is an impressive figure.

But camels too are subject to capability constraints, of course. Their major handicap is their long *gestation period* which has the effect that when herds for some reason are reduced in number, recuperation takes much longer than for sheep and goats which multiply fast (cf Dahl and Hjort 1966, for a thorough study on camel population dynamics). Long gestation periods also affect the balance that must be kept between lactating and non-lactating (fallow) animals in the herd. This need for an appropriate age-sex structure of herds in turn has implications for the *minimum herd size* necessary to support a given household size (cf below).

> ... A female camel is bred for the first time in her sixth year, and then only once in two years produces a single offspring ... In order to meet subsistence needs in milk, minimum herd size must therefore be maintained and the herd must be managed to ensure a continous supply; hence there is strong pressure to increase herd size. (Sweet 1969:158-59)

Relaxing the water constraint by the diffusion of wells and its effects on local and long-term carrying capacity

In arid regions, the pressure to build up herds and that to construct more wells are parts of the same strategy of pastoralism. In order to extend grazing ranges, pastoralists have for centuries been digging temporary wells and constructing permanent ones. This has also been the policy of modern governments, now that new technologies for securing a water supply have become available. With more wells, the increased sharing of the local habitat can be reduced and herds can be distributed more evenly. (Cf Figure 3:19 on density and the sharing of prisms). This policy seems rational enough in the short-term perspective.

Traditionally, the seasonal intensification of land use around wells typical for ethnic groups like the Kababish, has been based on the political control of wells as a means of controlling surrounding pastures. When water becomes the main scarcity factor, the emergence of regulatory constraints such as private ownership of wells is a common feature. This is a tendency analogous to that of land for agricultural purposes: increased shortage of land tends to induce greater regulation of access to land; usufruct rights are replaced by ownership rights vested in specific persons or social units. In arid regions, these constraints add to those imposed by Nature itself: a seasonal scarcity of places to which livestock can be moved with impunity. Both herds and herders become the victims of 'imprismment' at their water sources.

Government or private organizations who have assisted pastoralists by the construction of wells have sooner or later discoverd the long-term implications of their sometimes simple-minded policies. After a few years have lapsed, herds have been built up to fill the new capacity generated by more wells. The pastoralists then reach another ceiling which was previously only latent: grazing land itself becomes the major constraint. Due to local overgrazing and the reduction of perennials — not least by the browsing of goats — the vegetation becomes inadequate although there may be more than enough of water. Removal of the vegetation gives wind erosion the upper hand, and this dismal train of events terminates in the long-term reduction of carrying capacity. Herd population dynamics is not slow to respond; the animal mortality rates go up, often drastically, while the fertility rate swiftly shifts downward. *

*   It is beyond the scope of this study to give more than a hint of the problem, but the literature in this field is extensive. For a recent review and bibliography, cf Dahl and Hjort 1979.

Many of the contemporary management problems in pastoralism are centered around ways and means to control herd size so as to maintain ecological balance. Projects are needed to increase herd take-off and productivity; to stabilize the level of food production and to modernize the system of husbandry, for instance through ranching or cooperative schemes (Jahnke and Ruthenberg 1976. Cf Ingold 1978 on corresponding problems in reindeer husbandry.) In tradition-bound societies, a major obstacle in checking the expantion of herds is the syndrome of private herds and communal pastures (Jahnke and Ruthenberg 1978). Pastoralists acting as individual units only show responsibility towards their own groups and herds, not least because they are capable of controlling a localized herd of theirs but not the vast country crossed by them in the annual cycle. Hence it is more strategic for them to secure their individual shares in communally used lands than caring for the land in general. The less land or water there is, the more crucial it is to expand one's own share. The weak territoriality and land tenure situation thus tends to throw the pastoralist system of livelihood into whole-sale crisis. A political organization at a sufficiently high level must intervene to regulate land use (e.g. national or regional governments). In practice, this means a reordering of territoriality in time-space terms: Tenure systems must be timed and spaced to ensure the *viability* of pastoralism as an ecotechnology, not only at the household or minimal kinship unit level (cf Stenning 1958 on household viability), but also at the aggregate level of the tribe and region.

DIVISIBILITIES AND COUPLING CONSTRAINTS AFFECTING CAPACITY UTILIZATION

Divisibilities operate at different levels, from the micro-level of the individual (human or animal) who is indivisible, to the sub-population or total population which as an aggregate is *divisible*. Division at the level of the household, the task force or group organized around some specific project, or the local community, amounts to *fission,* while the corresponding aggregation or bundling constitutes *fusion.* Since the human being is indivisible (and cannot herd animals in line with Figure 8:22, for instance), divisibility as a source of flexibility in the use of a habitat must be implemented at a higher level than the individual.

Starting with the forces of *fusion*, there are several negative and positive (goal) constraints which work towards fusion in pastoral societies, both in keeping together herds and in uniting herds with people. A larger group of people are better able to protect their herds from outsiders, for instance, and the fusion of separate population sub-units facilitates a greater intensity of social intercourse. Keeping people and herds together is even more important due to the basic food bond. This is perhaps the most characteristic coupling constraint between people and herds, since the former thrive on the milk, meat and blood of the latter.

On top of this is the herd size factor. The herd must have a certain *minimal number* of lactating animals in order to adequately support a household of given size. (This is one variant of the familiar population base concept in economic geography.) Since not all animals are lactating to begin with, the herd must also include (fallow) cows or dams which are not lactating, as well as, in the case of cattle, calves, heifers, bulls and oxen. Moreover, the lactation volume in the dry season places the minimum limit on herd size, since this is the bottleneck season when the animals deliver the least milk for human consumption. What little milk they give must be supplied to their own offspring first and foremost, lest herd reproduction suffers. Consequently, herd size must be optimized according to lean season shortages which results in a much greater pressure on the vegetational capacity in the local prism habitat.

One basic mechanism for resolving this conflict between 'load' and capacity is *movement* in space, as already emphasized, but another is *fission* and separation of herds. Just as herds can be split up according to the capabilities of different animals to endure thirst, the carrying capacity of the marginal pastoral habitat is boosted by splitting herds according to the pasture requirements of different species.

Fission, fusion and movement in the Kababish annual cycle

The Kababish of Kordofan are an ethnic group whose members have to a noteworthy degree resorted to strategies of fission. Their ecotechnology which includes goats, sheep and camels enable them to use their habitat much more efficiently and intensively than if any of the three species had been excluded.

The annual cycle fo the Kababish is illustrated in Figure 4:6. In the seasons when the vegetation is lush and when there are temporary pools of water spread in the landscape, the Kababish can increase their range of grazing in time-space, while in the dry seasons, the Kababish and their herds concentrate around the permanent water sources. Starting with the hot and dry *saif* season (A in Figure 4:6), all categories of herds and people are focused on the *damar*, the well and its adjoining pastures (cf Johnson 1969:82-90). During this season, the population load on the local prism habitat is at its maximum, and everything feasible must be done to avoid overgrazing,

since the land can hardly take such intensive use, particularly towards
end of the season.

> ... the households and the herds /are/ in close proximity, the
> camels ranging up to four days journey from the wells, while the
> sheep, who have to be watered more frequently, go no farther
> than two days journey. This is the hardest time of the year both
> for the nomads and the animals and it is not until the ... rains that
> the pressure is relieved ... (Johnson 1969:89)

In this season land is used according to the structure shown in Figure
4:5 above. Although herds are split up and some members have to
leave the main household, they all unite at regular short intervals.

Towards the end of May, as the rains are approaching in the south
of Kordofan, one category of household members take the sheep and
the camels and move south to ease the grazing pressure around the
wells. The main household and the goats are left behind (B in Figure
4:6) and do not participate in this *shogara* movement (C). The
animals staying by the wells are typically the lactating ones, the goats
plus a few baggage animals. The less mobile sheep are not taken as far
south as the camels, however, and as the *kharif* rains spread northward, the sheep and camel herds turn northward in an oblique course
also (C). Meanwhile, the main households with the goat herds have
started to move north as well (B), and all the herds and household
members converge sometime in the middle of August, and a movement called *nashugh* ensues (D). An explicit part of the annual
strategy is to keep herds and households together for as long as
possible during this season (late *kharif*).

As the wet and cool season comes to an end, a point is inevitably
reached when the 'shoats' and the main household members must
commence their *mota* move (E) back to the dry season wells. If the
rains have been good, this movement will terminate around the end
of November. The camel herds, however, have gone further into the
deserts of Northern Kordofan and Darfur on the *gizzu* movement (F),
assuming the rains have been adequate for the growth of *gizzu* pasture. As long as these remain fresh, the camels can move on without
water (Johnson 1969:89). But after some time, the camel herds too
have to head south towards the dry season wells and the later they
arrive, the better, since this releases the pressure on pastures at the
wells.

In November when the 'shoats' have returned, the *damar* pastures
are still in good condition (G) and the vegetation does not dry up
until February. But as the dry and hot season progresses (G), the
pressure on the vegetation within the local habitat around wells
mounts up, and intelligence about incipient rains in the south is anxi-
-ously awaited. When positive news comes, a new annual cycle begins.

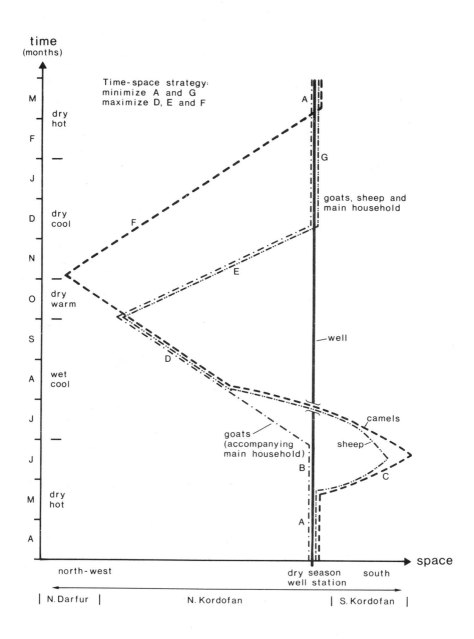

FIGURE 4:6   The structure of fusion, fission, herd separation and nomadic movement in the Kababish annual cycle. The path of the main household (elderly, women and children) coincides with that of the goats. The dry season wells and the surrounding pastures (segments A and G) form the hub (i.e. the main station of seasonal return) around which the system revolves.

It is thus only for some five months (segments A and D) that all herds and household members are together, and even in the hot and dry season (A), there is daily fission and herd separation according to watering frequency and prism range (as in Figure 4:5). These fusion-fission processes tend to work in a manner which is very different from how they work among the Basseri, for instance (cf Figure 4:1 and 4:4). The divisibility of herds and households is thus an important factor promoting better capacity utilization of the habitat when there are seasonal coupling constraints to wells.

A note on short-term path allocation and long-term herd dynamics

The time-space structure of fusion and fission at the household level and higher levels of aggregation is thus a question of striking a balance between the goals and projects of the people and the needs of the animals for pasture, water and protection.

In virtually all foraging and pastoralist societies, glut seasons — seasons when the environmental constraints are least severe — are the seasons facilitating fusion, while lean seasons are generally the times compelling fission. There may be exceptions to this glut-season-fusion rule, but it generally holds true. Closely associated with this syndrome, however, is Liebig's 'law of the minimum'. This is the principle of population dynamics related to the sizing of populations according to critical seasonal or perennial bottlenecks. The whole *raison d'être* behind the fission of herds and the separation of herds from people in seasons of shortage is the survival of both. But seasonal survival in itself would be an unsatisfactory level of ambition in pastoralist societies, and pastoral ideology and strategy must be aimed at the long-term build-up of herds. It is thus the *perennial* bottlenecks which most pastoralists aim to pass through. For this there must be a wide enough margin in the bad seasons of the good years, so that one can manage the bad seasons in the real bad years.

Formulating this in terms of population dynamics, we may say that the *fertility* and *mortality* factors affecting stock number and growth are closely interlinked with the short-term *migration or mobility* factor. This is because the time-space structure of mobility and access to water and pasture is closely associated with the lactation and fertility factor. When grazing or watering is bad, fecundity and fertility decline, and if conditions are really deteriorating, mortality increases. Both the productive and the reproductive capacity of the herd will diminish.

To mitigate some of the impending disasters to which pastoralists in marginal regions are especially liable, certain institutions can be used to regulate the access to herds, such as 'herding out', lending and borrowing, selling and buying, stealing or otherwise exchanging livestock. Consequently, there are many factors of a non-biotic nature which affect the composition of herds, apart from the factors of fusion and fission already mentioned. This often leads to a discrepancy between what may be called the 'management herd' and the 'ownership herd'. So far we have been discussing 'management herds', i.e. the animals actually herded by a certain group of people, regardless of whether they own or owe animals elsewhere. These institutional means of altering herd composition are very strategic in the recuperation of individual herds after periods of disaster.

Given that both mobility and divisibility (fusion-fission) of herds are vital components in the time-space implementation of herding projects, it must further be pointed out that the division of herds is also a function of *the divisibility of households* or other herding units. The extent to which different categories of animals can be sorted out and herded in an efficient way depends on whether there are household members available to do the job. The normative allocation of activities by age and sex obviously impinges on pastoralist time input and 'labour utilization', a problem discussed further in Chapter 9. We will now instead turn to some other factors concerning mobility, fusion and fission, namely those associated with the coupling of pastoral activities to agricultural.

AGRO-PASTORALISM AS A MIXED ECOTECHNOLOGY

However weak, the famous umbilical cord between nomadic pastoralism and agriculture expresses itself in various kinds of coupling constraints which structure pastoral activities in time and space. Johnson (1969:164) points out that:

> Only for those mountain tribes like the Beni Mguild or Marri Baluch for whom agricultural activities are important — so important that they undertake farm labour themselves — is the pattern of movement significantly affected by the need to be in certain places for the sowing and harvesting of crops.

In this case the link between agriculture and pastoralism is more than one of intermittent trading or the temporary leasing of agricultural land for grazing and manuring purposes. The Marri Baluch consist of groups which range from predominantly pastoral nomads, although actively engaged in agriculture at certain seasons, to cultivators who use pastoralism as a supplementary livelihood. While this may lead to conflicting pulls and time-space locations of different activities, these can be resolved partly or largely by adaptation and timing of pastoral movements and by mechanisms of fission and fusion already described. Part of the population or sub-population units such as households may be more stationary and tend to their agricultural land, while the remainder herd livestock.

The mixed agro-pastoral systems have probably been the most common since the inception of agriculture and pastoralism, and just as the pastoral element can be more or less developed in predominantly agricultural systems, cultivation activities can be more or less represented in predominantly pastoral systems. It is dangerous to the understanding of any kind of ecotechnology to think too much in pure types rather than transitional and mixed types. Diversification is frequently a better strategy of survival in human ecology than specialization. Both from an evolutionary and a structuralist viewpoint, the mixed cases are essential in explaining how various extreme and highly specialized types of systems have emerged.

Storage versus mobility, case I: The Karimojong of northern Uganda

The agro-pastoral Karimojong described by Dyson-Hudson (1966) exhibit yet another combination of projects and constraints. Their particular compromise between pastoralism and agriculture contains many elements that are reminiscent of the contradiction between storage and mobility among hunter-gatherers (Chapter 3). At the *storage* end of the spectrum are the agricultural activities conducted in what Dyson-Hudson terms 'permanent settlements'. These are places of storage for agricultural inputs and implements as well as for outputs, mainly sorghum stored in granaries. These stockaded settlements are located at perennial wells in the river valley bottoms which is the only place where cultivation is feasible in this semi-arid region.

At the *mobility* end of the continuum are the pastoral activities. Throughout the irregular wet season, the local prism habitat around the permanent settlement is adequate both with respect to pasture and water. In the subsequent dry period, however, the main herds must be split from the main households, and some members must assume a nomadic life and move with the herds to 'seasonal camps' in the highland areas (Figure 4:7). Unlike the Kababish who were compelled to gather around permanent *water* in the *dry* season, the Karimojong assemble in permanent settlements in the *wet* season, when there is sufficient *pasturage* within daily range of these settlements. Since the entire domestic unit consumes the produce from the animals in the form of milk, blood and meat, its basic pastoral projects are best sustained when there is as short a distance as possible between the food supplying and the food demanding units, i.e. between the herds and the people. This is the case in the wet season and the Karimojong conform to the typical glut-season-fusion rule.

The Karimojong start eating themselves out of their local prism habitat in the dry season, or rather, their herds do so on their behalf. At this juncture, one option would have been to move the entire community, to abandon the stockaded permanent settlements and enter into nomadic pastoralism on a full-scale, albeit seasonal basis. The Karimojong, however, perfer a solution of *seasonal fission* in which the main households consisting of the old men, the women and the children stay behind with the milch-herd or 'settlement herd' (cf Figure 4:7), while the majority of adult men, male youths and some young women move out with the main herd or 'camp herd' to relieve the grazing pressure around the permanent settlements. These settlements are thus largely the sphere or domain of the women, and even in the wet season when households have fused, the men spend most of their time (except when eating or visiting their wives) outside the permanent settlement itself.

As for the fusion and fission of herds, the Karimojong strategy is again similar to that of the Kababish, although the latter had camels instead of cattle. The goats and the milch-cows are kept with the main households as long as possible during the year and the sheep are also kept behind. The goats browse by preference while the sheep graze, so the 'shoat' combination works out well, just as their faster rate of breeding permits more frequent slaughter than with cattle. The milk and meat thus forms a valuable supplement to the beer and porridge diet of the women and children in the dry season.

The main herd of cattle (the 'camp herd') by contrast consists of the non-lactating (fallow) cows, the bulls, oxen and heiffers and calves, plus lactating cows, since the camp group is almost entirely dependent on animal foods (milk, blood, fat and meat).

The conflict between storage and mobility, between agriculture and pastoralism, is thus resolved through seasonal fission:

> The determining factor in the movement of both camp-herds and milch-herds ... is the ... need to compromise ... The requirements of the herd as livestock pull it outwards towards better grass and sufficient water in the drier months. But the requirements of the family, which subsists on the herd and which is tied for reasons of security to the central settlements, exert a strong pull in the opposite direction ... To keep his family in animal produce for as long as possible, yet move his herd before the beasts lose condition, is the basic problem... (Dyson-Hudson 1966:56)

But the length of the nomadic displacement is rather short; the Karimojong camp groups do not move far and every opportunity of delaying the move in the dry season is seized upon:

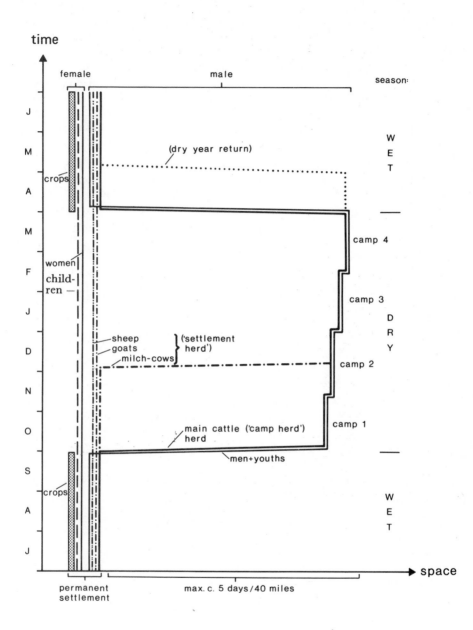

FIGURE 4:7   The structure of fusion, fission, herd separation and nomadic movement in the Karimojong annual cycle. The agropastoral Karimojong practise seasonal fission. Herds and people are split into two 'camps', one remaining in the permanent settlement and the other moving seasonally away. The young men (mainly) take the major herd (the 'camp herd') to the highlands in the dry season, while the women, children and older persons stay behind.

> Other things being equal, a herd-owner favours water sources closest to his particular settlement area, and then it is possible to make trips from camps to settlement with sour milk, fat and blood ... /Otherwise, when distance is greater/... the family must subsist as best it may on beer and porridge, or join the herd at camp for shorter or longer periods. (Dyson-Hudson 1966:59)

The fact that most Karimojong camps are not more than four or five day's trek from their permanent settlements (some 60 km or 40 miles), implies that it is possible to keep communication links open. As the milch-cows that initially stay behind go dry, they are sent to the main (camp) herd and are replaced by lactating cows, but as the dry season progresses, all milch-cows and their calves must be sent off to join the camp herds where grazing is better (cf Figure 4:7). In dry years, of course, the return of the camp herd may have to be delayed a couple of months or so until pastures near the permanent settlements have improved.

The agro-pastoral Karimojong thus combine the structures illustrated in Figure 4:1 and can do so more readily because of the shorter seasonal movements in space.

Storage versus mobility, case II: The Scandinavian seter system of transhumance to satellite herding settlements

From an agro-pastoral system like that of the Karimojong with one fixed and one seasonally mobile segment of people and herds and a short distance variant of nomadism, the step is rather short to a system designated as *transhumance*. Johnson draws the line sharply between nomadic pastoralism and agro-pastoral transhumance:

> /Nomadic pastoralism/... is not to be confused with transhumance. Transhumance is a term used to describe a *spatially limited pattern of movement* in mountainous areas which was first recognized in the Alpine regions of Europe. The literature indicates that a *village of permanent buildings* occupied by all or part of the population all of the year, rather than a *mobile tent camp*, forms the nucleus of a transhumant society. Although pastoral activities are one of the concerns of a transhumant community, *agriculture nearly always remains the dominant interest*. In other words, *pastoral movements are limited in scale*, usually take place in one valley system, and are undertaken by only *a small proportion* of the total population. None of these features are shared by pastoral nomads. (Johnson 1969:18)/Italics added/

In practice, however, the scale is a sliding one between pastoral nomadism and agropastoralism, as in the case of Marri Baluch and

Karimojong. Furthermore, in the Nordic or Scandinavian systems of transhumance, the agricultural component should not be exaggerated. Pastoralism formed the main pillar of the mixed ecotechnologies that once predominated in parts of Fennoscandia, and the fodder consumed by the animals in the winter was not cultivated but *gathered*.

The central feature of the Nordic agro-pastoral system was the transhumance to outlying pastures around satellite settlements. Such a place was called a *seter* and the milch-animals, cows and goats, were taken there for summer grazing (Frödin 1930). Traditionally, the little land worth cultivating was usually found in valleys or on the shores of lakes, while the surrounding forests were taken into possession for grazing. As long as the resident human population and their flocks kept within certain limits, a system of *daily return* of livestock was feasible (as in Figure 4:8a). But once herds grew beyond a certain size, the animals would deplete the local prism habitat. The size of this prism habitat was small, since the animals moved slowly and if driven far in the morning and evening, the beasts would give less milk, so overcoming distance was a drain on both time and milk. In summer, it was also hygienic to keep the animals and the insects attracted by them at a comfortable distance away from the main settlement. Satellite summer pastures, often located uphill in areas less good for agriculture, thus had more than one advantage but also the drawback that the milk had to be fetched from the *seter* morning and evening and brought back to the village (Figure 4:8b).

However, as the livestock populations of the permanent settlements increased even more, so did the shortage of *seter* pastures at close range from the villages. The satellite pastures had to be moved further outward and with the increased commuting distance to milk the herds and bring back the produce, the daily routines were changed to save time. Rather than transporting the milk home twice per day, the women left for the seter in the late afternoon, milked the animals and stayed the night, returning home after the morning's milking. In this way, only two moves per day were necessary (as in Figure 4:8c). But this solution assumed that some dwelling was built at the *seter*, together with the barns and milking shelters already erected there. We get what Frödin (1930) calls the *semi-seter* system, characterized by being used from evening till morning only. The milk was processed in the main village.

In northern Jämtland the semi-seter developed in a direction where the milk was also processed at the *seter*. In other parts of northern Sweden, the women stayed behind one or more days every week to boil cheese and churn butter, the number of such days depending on the milk supply.

The evolution towards a *full seter* system took place when the

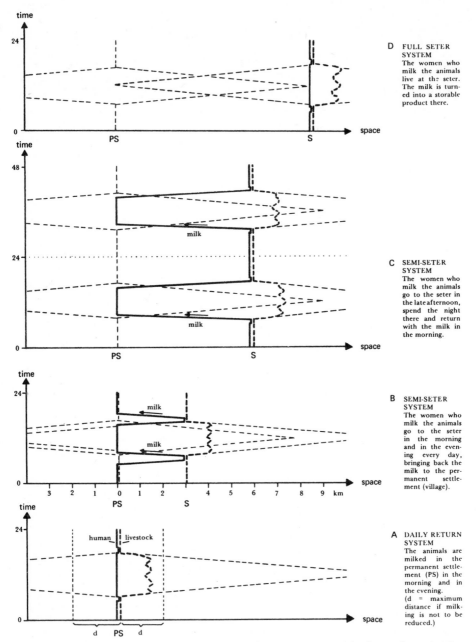

FIGURE 4:8   Variants of the Scandinavian *seter* system. As far as the animals
are concerned, all cases shown are of the 'daily return' kind, but with increasing
distance from the permanent settlement (PS), more and more activities are
transferred to the satellite settlement or *seter* (S). Case B does not show a
'mature' version of a semi-seter and is perhaps better characterized as a summer
pasture with a cowshed, but here all satellite settlements with adjacent pastures
have been referred to as a *seter*.

distance between the main village and the satellite pastures was so long that travelling was perceived as a great waste of time and energy that could be better spent on other tasks. In this system the herd-girls only used to go back home during those parts of the summer when their labour-time was particularly needed, as in the harvesting season. Sometimes the pasture requirements were so high that two sets of *seters* emerged, one nearby used in spring and autumn, and another farther away for use in the summer. Typical for the *full seter* system was the erection of storehouses for the various milk foods produced at these satellite settlements. The actual processing had thus also been shifted outside the permanent settlements which in turn emphasized storable milk foods even more (Figure 4:8d).

Storage versus mobility, case III: The storage of winter fodder

Generally among agro-pastoral peoples, a major problem is that agriculture is fixed in space, at least during certain seasons, while extensive pastoralism relying on natural fodder requires large areas and hence spatial mobility. In this sense, our previous examples indicate that agriculture is associated with storage and pastoralism with mobility. In many other respects, however, this proposition is an oversimplification. Let us give one example.

Pastoralism itself contains the seeds of a contradiction between mobility and storage, particularly the storage of lean season fodder. For centuries in Scandinavia, and in the northern parts till the early present century, agriculture was but a supplement to pastoralism, an ecotechnology which was also mixed with fishing, hunting and gathering. Adaptation to northern conditions with a short growing season and an enduring snow cover introduces its own constraints on pastoral and other projects. It was in the short but intensive summer season, especially up north, that animal production had to be maximized, and the milk had to be converted into storable products, mainly cheese, butter and soft whey cheese. Storability also facilitated transferability and trade in these products which were one of the cornerstones in the 'storehouse economy' typically associated with northern areas where people had to tide over the winter. In most places, to milk the animals in winter was virtually unthinkable. The problem was rather to ensure that a sufficient number of milch-animals, cows and goats, survived on the meagre fodder rations that they were often given. Scandinavian folk-lore contains many a vivid story of how the beasts had to be more or less carried to the first spring pastures when the tyranny of winter at long last gave way. Late autumn slaughtering of surplus (especially male) animals was one means of squeezing through the winter bottleneck; the gathering of fodder in the summer was another.

Storing fodder for the winter was an effective way of opening up a niche for pastoralism on otherwise hostile northern latitudes. But it assumed that people would spend enough time accumulating fodder, an ecotechnological variant which gave rise to certain day-prism effects in the allocation of space and human time. The more livestock per capita and the longer the winter, the more time had to be put into the gathering of hay and leaved branches in the short and hectic summer season. *

In this situation of peak season time shortages, economizing on travel time was strategic. Human time is more efficiently used in pastoral projects if the *animals locomote themselves* to peripheral pastures rather than if *people have to carry fodder* from the periphery to the permanent settlements (Figure 4:9). In northern Scandinavia settlements were located in valleys or near water courses by pre-ference, since these locations were not only favoured in terms of agriculture or transport but they also yielded the best fodder. By collecting the necessary winter fodder closer to the barns, much less time had to be spent transporting the fodder or travelling to the meadows in the summer peak season. And if there was not enough fodder nearby, satellite fodder houses could be erected close to natural fodder sites, and the harvest could be transported back by sleigh in the slack winter season. (Even when fodder came to be cultivated, fodder barns were still dispersed in the fields in many parts of Sweden, especially the north, to save time in summer.)

There were thus other reasons than herd size and the rate at which a herd would 'eat itself out' which affected the utilization of the local prism habitat. An important logic behind the *seter* system was to keep the animals *outside* the close-to-village cultivation and fodder zone (cf Chapter 6 on spatial zoning, infield-outfield systems, etc.) By adopting this land use structure (Figure 4:9), both the gathering of fodder for winter and the summer application of manure from the stalls or barns was conducted with much less waste in travel and transportation time. Transport was better done at leisure in the winter season when snow and frozen waters generated a much im-proved transportation landscape. In winter, even the manure from distant *seters* could be taken back to the permanent settlements, assuming, of course, that it had not already been applied to improve the pastures around the *seter* houses.

---

*   Human time supply was thus a critical constraint on the number of animals that could be supported. In many other forms of pastoralism, human time con-straints on herd size are much less pronounced, since it takes as much time to shepherd 10 animals as 30 animals. (Cf Chapter 9 below and the point made on divisibility on p. 126 above.)

While later developments towards the cultivation of fodder represented *intensification* of land use, the earlier agro-pastoral systems
relied mainly on spatial *expansion*. In the *seter* variant of this system,
the expansion culminated in northern Sweden in the 1870's. In
southern areas, agricultural intensification had started a century
or two earlier. It is interesting to note that even in the extensive
agro-pastoral systems, manure was used to improve the fertility of
fields and pastures, since manure accumulated anyway at the places
of summer milking or winter stall-feeding. (Cf chapter 6 on manuring and short fallow cultivation.)

Given that this form of pastoralism based on stall-feeding assumed stable settlement anyway, it was quite compatible with a comparatively stationary activity such as agriculture, except up in the
north where the winters were longer and the soil less good than
in southern Sweden. Then the *seter* system allowed spatial expansion and the contradiction between fixed agriculture and mobile
pastoralism was resolved through *seasonal fission*. But in the Scandinavian case, the sexual and spatio-temporal division of activities was
reversed; it was the *young women* who moved with the livestock to
the *seters*, and the men and other categories who stayed behind.
This is a sex role solution which would have been abhorred by many
an African and Asian pastoralist.

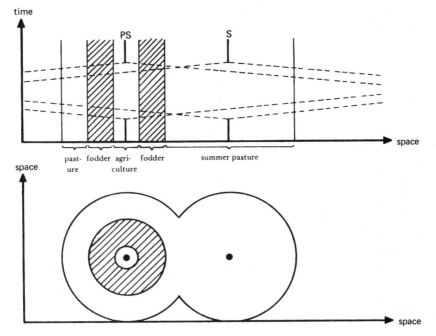

FIGURE 4:9 The structure of land use around an agropastoralist village, land use
intensity decreasing with distance from the village (PS). (Cf p. 211-20 on zoning.)

WATER SCARCITY VERSUS SNOW ABUNDANCE: THE POLAR CONDI-
TIONS OF PASTORALIST ECOTECHNOLOGY

The reindeer-breeding *Lapps* of Scandinavia and Finland did not
store fodder and stall-feed their animals in winter; their store-house
was the natural habitat itself. Since these fodder stores were obviously
dispersed, mobility was their chief pastoral strategy. We are then
back again in the pastoral nomadic *genre-de-vie*, but we have come
a far way from the tropical conditions of water scarcity to the sub-
arctic predicament of snow abundance. However, Figure 4:1 is still
as applicable to the Lapps as to the Basseri, except that in Lappland
it is the summer swarms of insects more than dried out pastures
which drives the herds into the fells.

Traditionally the Lapps were divided into three main sections,
the nomadic Mountain Lapps, the Forest Lapps and the Skolt Lapps,
the latter two being only seasonally nomadic. Given a number of
transitional types, the Lapps in fact represented most positions on
the sedentary to nomadic continuum.

Concentrating our interest on the Mountain Lapps, they used to
practice a nearly year-around nomadic life-style up till the first
few decades of this century, and their migratory paths have been
meticulously recorded by Manker (1953). The migration of each
siita, a group of households herding deer together, at one stage used
to reach from coast to coast nearly, i.e. more than 300 km in a NW-
SE direction, from near the Gulf of Bothnia across the alpine
Scandes to the Norwegian coast. The system was centered on the
autumn and spring settlement or *viste*. This station was located just
below the tree-line in the sub-alpine fells, and this was the 'trans-
shipment point' where the winter boat-sled caravans pulled by deer
oxen were converted into summer pack caravans, and where gear and
food was stored close to the turf-cot dwellings found there.

Even in the wild, the reindeer is known for its strong migratory
instincts and prior to the rise of pastoralism, Lapp hunters pursued
the deer on their migration routes, using tame deer as draught and
pack animals and also as decoys to attract wild reindeer. As long as
domestic herds were small, the pressure on adjacent pastures was
slight and induced little migration. With specialized herding and
large herds, movement in search of fresh pastures became essential.
The Mountain Lapps thus made optimal use of the different vegetation zones, shifting between the herbacious alpine summer pastures
and the winter lichen pastures found in the broad coniferous belt
between the Scandinavian mountains and the Gulf of Bothnia.

The Lapp annual cycle was structured into the famous *eight
seasons*. In late winter ('spring-winter') when snow conditions made
the digging for lichen hard work for the deer, the move was started

to the sub-alpine region where bare patches of accessible vegetation could now be found on the mountain slopes. Having reached the spring station, people lived in turf cots by the tree-line while herds were grazed in the higher ranges. *Spring* (May) was the calving season when the cows had to stay put until the calves had grown strong. This used to be a season of intensive herding in the old days when herds had to be protected from marauding bears, wolves and wolverines. As the *spring-summer* season was heralded by swarms of mosquitos and dangerous gadflies, time was ripe to move higher up into the treeless alpine region where herbs were maturing quickly in the waxing mid-night sun. The boat-sleds had been stored away at the spring-autumn station and instead pack caravans were used into the high ranges. There the herds could be let loose for the season and required little tending. This was the glut season when families in different *siitas* had leisure to meet and socialize. It was also the milking season, and in the old days when herding was *intensive* (cf below), the animals were milked throughout the *summer* and well into the autumn. The milk was turned into storables such as cheese. As summer reached its peak, herds had to be moved up to the very mountain tops to escape from the flies, the deer resting on the snow in the day and foraging in the short nights. By *autumn-summer* (end of August), nightfrost had reduced the insect pest and the downward movement with herds and pack caravans could begin. The deer were now fat, the calves big and the cows gave plenty of milk. By *autumn* the Lapps had returned to their spring-autumn station in the birch zone, and the separation of herds began, just before the rutting season. With the shorter days and the imminence of snow, the boat-sleds were prepared. The first snow cover then paved the way eastward to the lowland pine region and its awaiting lichen pastures. This was the *autumn-winter* season, the slaughtering and the market season. Then came *winter* when water courses froze to facilitate the further trek eastwards. The people lived in their conical tents moved by boat-sled. Winter was a season of intensive herding, since large herds finished pastures fast and had to be carefully protected from wolves, who could do much damage and disperse herds which then had to be painstakingly rounded up again. In the beginning of winter when the snow was soft, digging for lichen was less arduous for the deer, but with a growing snow-cover, conditions became more burdensome. It was in *spring-winter*, when the snow got heavy and crusted with ice, that the deer caused most work to the herders, since the deer could move quickly on top of the snow. As the days got longer, it was again time to head for the mountain pastures, and with this move, the annual cycle began anew (Skum 1938, Manker 1963).

It can be seen that migrations took place mainly in the transitional seasons — spring-winter, spring-summer, autumn-summer and autumn-winter — while during winter, spring, summer and autumn, a 'daily return' kind of system predominated (cf p. 144-45 below). But there were also great variations among the Lapps, for example those inhabiting the Jokkmokk region studied by Hultblad (1968; cf Figure 4:10). The distinction made between sections according to ecotype into Mountain and Forest Lapps, although largely valid, must be used with caution. The Forest nomads (E2 being the only case in Figure 4:10) seldom went beyond the forest zone, and were thus different from the Mountain nomads who did. But there were also variations among Mountain Lapps in zonal patterns. So regardless of whether nomadism is classified by ecotype or length of movement, it is wise to map migration paths before sorting groups out into more or less watertight compartments.

From intensive to extensive ('predatory') pastoralism in Lappland

In the early 1600's, the Skolt Lapps in the east still practised reindeer *hunting* while the new and intensive technique of reindeer *breeding* had been practiced in western Fennoscandia for some time. Reindeer herding was a specialized activity which reduced the role of hunting and fishing, and this original form of reindeer pastoralism was intensive. Cows were milked twice a day from the beginning of June till the end of September and these cows grew very tame from this intensive interaction with people. Since there were limits to the number of animals that could be milked by a given household, herds were smaller in this system. But milking had certain adverse effects on the health and reproduction of herds. The concentration of cows and oxen at milking time defecated local pastures, and human needs frequently left calves with too little milk (Hultblad 1968:136-40).

FIGURE 4:10 (Opposite page)
The nomadic pattern of different *siitas* of Lapps in the region of Jokkmokk, according to Hultblad (1968). Only one group depicted (E2) conforms to the Forest Lapp pattern, while all the others (except C5) are Mountain Lapps. They differ, however, in the extent to which they traverse different ecological (altitude) zones, the alpine region being located to the right of the dashed line on the map (A). All groups spend the winter in the forest zone, but vary in the extent to which they occupy the sub-alpine and alpine zones. The *visten* indicated in Figure A are as follows: 1: spring residences, 2: summer residences, 3: autumn residences, and 4: winter residences.

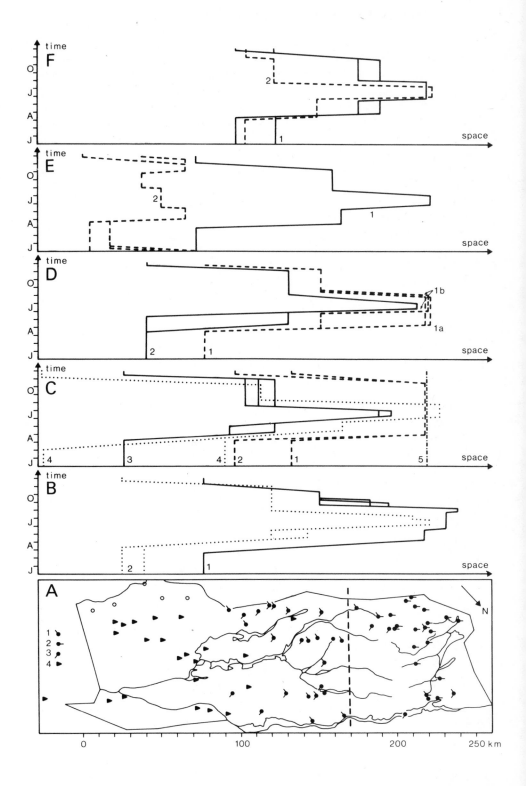

By the end of the 19th century, however, a growing sector of trade had reduced the need for self-sufficiency and created a market for meat. The gradual elimination of predators like wolves and wolverines by means of modern firearms had also made the concentration of animals for protection less crucial, while milking was less important now that other foods could be bought instead. More milk to calves in turn meant better rates of survival and herd growth. But one effect of this gradual extensification (or deintensification, cf Chapter 6) was that the deer were less easily approached and much slaughtering had to take the form of shooting-off the animals. This makes the borderline between herding and hunting very thin, and the system is transformed into what Ingold (1978) calls *predatory pastoralism*.

The extensification of herding also made it less necessary for the women to come along into the summer fells to do the milking. So they and the children and aged stayed behind at the spring-autumn station which had generally been hub around which the pastoral activity system had revolved:

> Present conditions are largely as follows: In the north... the summer, autumn and spring residences have been combined into one central residence... This is situated in the summer area near some large watercourse to which the people can comfortably come with a boat or near some transportation route... At such places, the conical tent has practically disappeared and the turf cot, after a period of domination, has largely been eliminated by more house-like turf huts and, eventually, by European-type houses...
> In the south, ... the traditional summer and winter residences have largely disappeared. Instead of centering in the summer area, however, the population is concentrated in the autumn-spring residences. The Lapps settle more and more in small homesteads where the family remains throughout the year, possibly with some short trip out to a summer residence, while the active reindeer herders follow the herd and lodge with the sedentary population during the winter and in cots, cottages or conveyable tourist tents during the summer. (Manker 1953:40)

Reindeer husbandry has thus changed character as a result of the growing interaction with the outside world, and nowadays the former Mountain Lapps practice a kind of transhumance covering only part of the population. Their ecotechnology has also been modified and mixed up with industrial components such as field radios, modern fences and various new means of transport such as motorboats and snow scooters. Pelto (1973) gives an interesting report of the 'snowmobile revolution' affecting Lappland, but that is another story.

INTENSIFICATION, EXPANSION AND SEDENTARIZATION

By and large, pastoralism in general and nomadic pastoralism in particular is a *low intensity* form of land use (space-time occupation).
Although the carrying capacity of a pastoral ecotechnology such as
European sheep rearing may be higher than that of hunting-and-
gathering, this does not imply that a pastoral area is densely packed
with people and animals. In the case of nomadic pastoralists, large
portions of space-time are left unoccupied (or fallow, cf Chapter 5)
for long periods of time.

However, the land use intensity, not least of nomadic pastoralists,
varies a great deal with the season. In the tropical dry season when
herds are concentrated around water sources, land use intensity is
manifold higher than in the wet season when herds are dispersed.
Dry season intensities may even reach a level of overgrazing resulting
in serious land degradation. It is also evident that land use intensity
varies with distance from a prism centre (cf Figure 4:5), being higher
close to stations of return such as wells or fixed residences.

Among pure pastoralists with comparatively large herds, *spatial
expansion* tends to be the main solution to increased space(-time)
requirements rather than temporal intensification (cf Chapter 5).
One means of expanding the size of pastures in arid areas was to
relax the water constraint (more wells), thereby reducing local intensities. Another perhaps more common solution throughout history
has been territorial expansion through conquest. Pastoralists are
fortified in this pursuit by their capacity for movement, and the
Mongol hordes that once swept across Asia come readily to mind.
These Asian pastoralists had an advantage in their equestrian ecotechnology, similar to that of their camel-breeding Bedouin counterparts (or to the Plains Indians hunting from horses, cf Chapter 3).

Harris presents an interesting argument on expansion versus
intensification:

> Most nomadic or seminomadic pre-state pastoral societies are
> expansionist and extremely militaristic, but strongly patrilineal or
> patrilocal rather than matrilineal and matrilocal. The reason is
> that animals on hoof rather than crops in the field are the pastor-
> alists' main source of subsistence and wealth. When pre-state pas-
> toralists *intensify production* and, as a result of population
> pressure, invade the territories of their neighbours, the male com-
> batants do not have to worry about what's going on back home.
> Pastoralists usually go to war in order to lead their stock to better
> pasture, so "home" follows right along behind them. Hence the
> expansionist warfare of pre-state pastoral peoples is characterized
> not by long-distance raiding from a home base, as is the case

among many agricultural matrilineal societies /cf Figure 3:21/, but by the migration of whole communities — men, women, children and livestock. (Harris 1977:62)/Italics added/

The point is that it is often technically infeasible to relax the constraints which keep pastoral nomadism in its extensive mould, especially in marginal habitats or when agriculture and sedentarization is an undesirable alternative. Then, as herds build up, new pastures must be sought and in many cases this can only be achieved through encroachment.

Alternatively, when herds dwindle for reasons such as disease, droughts, theft, raiding or similar calamities, the need for expansion into new pastures is reversed, and in many cases *sedentarization* and the adoption of agriculture may be the only recourse open. Pastoralists who have lost their herds need not migrate, and sedentarization — at best with the remnants of former herds — simply becomes a by-product of the situation. This does not preclude other reasons why nomads become sedentary and have to practise a more intensive form of land use (notably agriculture), reasons such as the application of superior force by governments wanting better control over 'belligerent' or 'uncooperative' pastoralists.

Conflicts between high intensity agriculture and low intensity pastoralism

Although many forms of pastoralism and cultivation are compatible and even mutually supporting, e.g. agro-pastoralism or mixed farming * (cf Chapter 6), there are also strong latent conflicts between the two ecotechnologies. A prime instance is when agricultural settlement expands into intermittently used pastures. Then the best watered and most fertile lands are usually occupied first, and if these are the pastures in bottleneck seasons, the blow to pastoral activities is much greater than is indicated by the sheer size of lost area. One typical case would be the introduction of irrigation schemes into river valley pastures. Another common instance is that of Turkey:

The history of peasant nomad relations in Turkey has been a story of a steady expansion of the peasant /agricultural/ economy at the expense of the nomadic, so that at the present time, large scale, highly organized nomadism — at least in western Turkey — is a phenomenon of the past. That pastoralism was formerly more

---

*   Another case of compatibility is when the livestock herded by pastoralists graze in the fallow fields of cultivators. This invasion of fallows conforms to Boserup's definition of intensification, since land is then more continuously occupied (cf p. 156-7 and 190-91 below).

highly developed in northern Turkey that it is today seems un-
deniable... (Johnson 1969:21)

... by a growing demand for citrus products /and increased malaria
control/ ... The result was a steady expansion of the agricultural
population into the traditional winter pastures of the nomads and
a consequent irrevocable alienation of grazing land for the commercial agricultural uses... (Johnson 1969:25)

The encroachment of agriculture into pastoral lands is a main theme
in the agrarian history of a wide variety of places:

... as population increases, there is a definite tendency to increase
the amount of land devoted to agriculture and at the same time,
encroach on the common pasture traditionally reserved for all
Marri /Baluch in Pakistan/... (Johnson 1969:68) /Cf the role of
population growth in intensification theory in Chapters 6, 7
and 9./

In the stages preceding the industrial revolution in England, there
was similarly a land use conflict between agriculture and pastoralism
(cf Chapter 10) which was part of the general land use and energy
crisis eventually giving rise to the use of coal (Wilkinson 1973).
In northern Sweden, prior to government regulation, clashes between
settlers and pastoral Lapps were legion in the agricultural colonization process at the end of the 19th century (Bylund 1956).
Later, in the industrial era, Lapp pastoralism was less at odds with
settlers practising agriculture than with modern forestry methods
and with government dam-building and hydroelectric power projects.
Elsewhere, in the Sudan, for example, the expansion of mechanized
farming has pushed the pastoral nomads into marginal environments were overgrazing induces desertification (Rapp 1979). It is the
subtleness and intermittence of pastoral land use, i.e. the low intensity, which makes the land seem 'empty' to uninitiated observers or
to those having political gains to be made from defining pastoral
lands as 'unoccupied'. In no region in our century has this process
had greater historical significance than in the Palestine part of the
Fertile Crescent, where Arab pastoralists and other categories have
been displaced by immigrants from elsewhere bringing along more
advanced and intensive technology.

NOMADISM, MULTIPLE RESIDENCE AND DAILY COMMUTING: SOME
CONCLUDING REMARKS ON PASTORALISM AND MOBILITY

Let us end this chapter with some brief comments on the time-space
structure of pastoral ecotechnology. To classify pastoralism by its
movement pattern into full nomadism, semi-nomadism, semi-sedentary nomadism and sedentary pastoralism may be useful as an
introduction to the pastoral theme, and little else. Not only is a
further understanding of the spatial mobility pattern necessary, as
proposed by Johnson (1969), but also of a good number of other
factors and relationships. In this chapter an attempt has been made
to outline a dynamic-structural model of pastoralism and its articulation and praxis in time and space in order to interrelate different
features of pastoral society and ecotechnology, mobility being only
one such feature, albeit a very important one.

A strategic point of departure in explaining both mobility and the
overall structure of pastoralism is the *local day-prism habitat* and its
limited carrying capacity. The time it takes to reach this limit depends on several factors such as the mix of livestock used, the size of
herds, and the qualities of the natural habitat within day-prism range.
Generally, a habitat in a temperate zone with lush vegetation is able
to sustain a given herd for much longer than a corresponding herd in
an arctic or arid region, but the less rich the vegetation, the faster a
herd will eat itself out of its local prism habitat.

The explanation as to why a *daily return system* (or daily commuting system) as opposed to a *seasonal return system* (i.e. a nomadic seasonal migration system) is implemented, rests on a mechanism
of choice and path allocation similar to that facing hunter-gatherers
(cf Chapter 3) or shifting cultivators (cf Chapter 5). Remaining in
the old station of residence as nearby food or fodder resources get
depleted would involve a greater expenditure of time and energy
to secure daily sustenance than shifting the base station itself (i.e.
the residence serving as prism base). And when group size or herd
size increases, so does the rate of eating one's way out of the local
prism habitat (but not at a proportional rate, as explained on p. 90-
92, since the spatial area increases exponentially with distance).
Hence, some pastoralists are able to practice a daily return system
from a permanent residence and still find fodder enough within
day-prism reach (Figure 4:11A). Other groups need supplementary

FIGURE 4:11 (Opposite page): Daily and seasonal return systems as basic
variants of a common pastoralist theme. The basic generating mechanism (cf
Chapter 2 on structuralism) is the daily prism habitat and constraints affecting
this fundamental module. 'Sedentary' pastoralists have only daily return or
symmetric modules (A) in their annual cycle, but the more nomadic a pastoral
people is, the greater is the frequency of non-return modules (asymmetric
prisms) in their annual cycle. (For further explanations, see the text.)

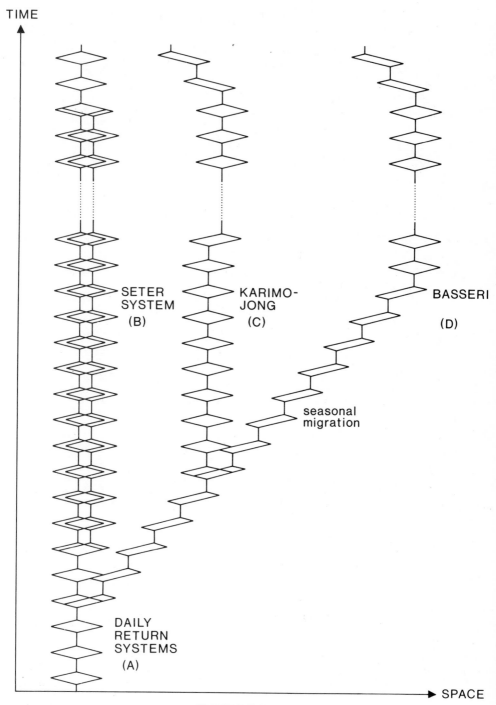

FIGURE 4:11

satellite pastures, as in the *seter* system (Figure 4:11B), and have to have double or multiple residences on a seasonal basis. But in both of these systems, pastoral mobility is based almost exclusively on *symmetric prisms* (cf definition in Figure 3:6), except for the less than half a dozen days when the livestock is shifted from one settlement to another. Turning then to the Karimojong case, the daily return system still prevails during the wet vegetation season. In the dry season, however, as the 'camp herds' are taken to highland pastures, a short series of *asymmetric prisms* prevail, while at the camps once established, the daily return function (e.g. around water sources) again predominates (Figure 4:11C). So here we find multiple stations and shifts of day-prism bases. The Karimojong are pastoral and to some extent nomadic, but not to the same extent as the Basseri, for instance. The Basseri live in mobile tents the year around and only conform to the daily return pattern for limited periods, notably in the summer highland habitats. The Basseri thus have a much greater proportion of asymmetric day prisms in their annual chain of prisms (Figure 4:11D). And rather than having an essentially dual or triple residential station structure, they have a *multi-residential station system.*

The main point conveyed in Figure 4:11 is that these different systems (type A-D) represent permutations of the same basic day-prism module. There is consequently a continuum between the so called 'sedentary' *daily return systems,* on the one hand, and the more or less 'nomadic' *seasonal return systems* with daily return components, on the other. From an evolutionary perspective, it is thus understandable how 'nomadic' pastoralist systems may gradually have emerged through the extension and/or extensification of daily return pastoral systems, or how nomadic pastoralists who lose their main herds in periods of disaster may become daily return pastoralists with the remnants of their original herds. This is roughly what Johnson alludes to in his statement that,

> ... at the sendentary end of the continuum some groups are always dropping out of nomadism and become sedentary (the dominant process today) while other groups are increasing the size of their flocks and becoming more nomadic. (Johnson 1969:17)

In the next chapter we shall turn to an ecotechnology which also operates with a mildly 'nomadic' slant; not by virtue of animals eating fodder, but by people 'eating forest'. This is the case of shifting or swidden cultivation, where forest becomes the fodder for cultivation activities. Here we encounter a new set of projects, constraints and permutations of basic time-space mechanisms, giving us new food for thought.

5 SHIFTING CULTIVATION

THE SHIFTING CULTIVATION ECOTECHNOLOGY

Shifting and swidden cultivation is a food production system, which is very demanding on space-time in the settlement system and therefore allows relatively low population densities compared to other cultivation systems. Yet it is a system with a greater food production and carrying capacity of land than either hunting-gathering or pastoral nomadism. While food collectors essentially depend on the natural rate of animal and plant reproduction, agriculturalists have learned to control the rate of plant reproduction, and this permits an intensification of food production (Harris 1977). As with other cultivation systems, a major capacity constraint lies in the maintenance of soil fertility and land-cum-labour productivity, but many other conditions and constraints join in to give this population, activity and settlement system its overall structure.

The term *shifting cultivation* itself gives a hint of spatio-temporal structure and variability. As with food collectors, shifting cultivators must be mobile in space and time, but with different frequencies, durations and relative locations. Not only are individuals mobile but the settlement system itself undergoes spatial changes and shifts which reflect on society and habitat. Some of the previously outlined mechanisms are found also with shifting cultivators but in new combinations and with different magnitudes and significance. When stationary plants are cultivated, land (space-time) requirements also become more obvious perhaps and the cultural landscape more visible. In this context, carrying capacity and packing problems in a settlement space-time budget are more easily represented, although they are crucial in all systems and ecotechnologies.

Conklin (1961) gives the following minimum definition for shifting or swidden cultivation: 'any continuing agricultural system in which impermanent clearings are cropped for shorter periods in years than they are fallowed'. Pelzer (1945) has given a more inclusive definition by listing the following characteristics:

1) rotation of fields rather than crops,
2) clearing by means of fire,
3) absence of draught animals and of manuring,
4) use of human labour only,
5) employment of dibble-stick or hoe, and
6) shorter periods of soil occupancy altering with long fallow periods.

Shifting agriculture has several names such as swidden, slash-and-burn, field-forest rotation, and cut-and-burn agriculture. The term swidden was introduced by Izikowitz (1957:7) using an obsolete English word and it has since become widely used. It illustrates the most marked feature of this cultivation system: the use of fire as a tool for clearing and as a method of fertilization. The term 'shifting' emphasizes the space-time feature of relative impermanence of land occupation. It may sound objectionable to some to call this type of cultivation system 'agriculture' instead of horticulture or gardening, since 'agri'-culture is generally associated with permanent fields under plough tillage. All the same, at the general level of analysis employed here, this distinction will be dropped.

From the angle of time demands on the population, shifting cultivation is rather efficient and productivity per person-hour and day is generally high. This latter factor — not the productivity per hectare-day or hectare-year — is often decisive for retaining shifting cultivation in societies which also practise sedentary and more land-use intensive forms of agriculture at the same time, a fairly common situation in Africa, Asia, Latin America and Oceania (Cf Netting 1968, Scudder 1962, Leach 1954, 1961, Lewis 1951, and Pospisil 1963.) In terms of space-time requirements, swidden agriculture is wasteful if the land lying fallow necessary to keep the system viable is included in the calculation. By that token productivity per hectare-year is low, while on the very plots used in a given year it may be high.

Before the 1960's, shifting cultivation systems were quite neglected compared to Asian wet rice cultivation, plantation agriculture or the cultivation systems of Europe and North America. The only materials available were a few pioneering studies by Linton (1933), Richards (1939), Allan (1949), Peters (1950), Izikowitz (1951), Conklin (1954, 1957), Freeman (1955, 1970), DeSchlippe (1956), and a few others. From having been a relatively disregarded field, an upsurge of interest in the 1960's and 70's produced numerous penetrating and theoretically relevant materials. Apart from comparative studies such as that by Spencer (1966), Boserup's theory (1965) on the relation between agricultural change and population growth, placed shifting cultivation and transitional systems at the forefront. Her basic formulation of intensification theory had a strong catalytic effect on other work on tropical agrarian systems. Anthropological and geographic case studies also became available for comparative analysis, such as those by Scudder (1962), Bohannan (1968), and Stauder (1971). There were also new approaches in human ecology, systems theory and quantitative geography by scholars like Brookfield and Brown (1964), Brookfield and associates (Brookfield ed. 1973), Rappaport (1968), Christiansen (1975) and many others. Various economic and development dimensions of shifting cultivation systems were brought forward excellently in the Tanzanian materials edited by Ruthenberg (1968). New Guinea is perhaps the region where shifting cultivation system dynamics has been subject to the deepest probing enquiry. These mentioned studies constitute but a small sample of all the materials available.

RETHINKING RURAL SETTLEMENT AND LAND USE

Shifting cultivation systems have an intrinsic form of land use dynamics which serves as a useful starting point for rethinking settlement and land use. Unfortunately, little has been done in geography to produce a concise statement on the topic since Chisholm published his volume in 1962, although so many more ideas and materials have emerged in recent years. What is needed is a more synthetic approach which can encompass all the various elements of land use, settlement, land tenure, time economy, domestic organization, and so on. An objective of this chapter is to make a modest contribution to a conceptual reorientation in this field. While the preceeding pages took up prisms and mobility aspects, the dimension of terrestrial space as a source of room to accomodate populations, activities and resources will now be added. In so doing, we try to move away from static notions of land use towards a more *holochronic* perspective encompassing the major aspects of time as a location matrix, a resource, and a dimension of structural change. This is the only way to counterbalance the view of the world as something temporally flat.

*   One scholar who has placed shifting cultivation systems in a grand systemic and structuralist framework of analysis is Jonathan Friedman (1979). Friedman looks into the dynamics of social change, explaining the relations between social structural change in general and technological changes, such as agricultural intensification, i.e. the transition from shifting cultivation to more intensive systems, his examples taken from the Burma region. Since his study appeared very recently, it is only possible to deal with a few aspects of it here (cf Chapter 2 on structuralist approaches).

Packing activities in a regional space-time budget

Various processes of turnover, growth and decline are influenced by the fact that the component actors and resources occupy space for various periods of time. Single stations (e.g. buildings or fields) as well as the entire time-space region in which they are placed constitute a resource which provides room or accomodation for human populations and their activities, artifacts, domestic organisms, and so on. *Space-time* in the settlement system is a universal input in all human activities, and just as there is a *limit* to the packing of activities in a population time-budget, there are corresponding limits to packing in a regional *space-time budget*. To say that an *activity* consumes space-time is merely a short hand for expressing that the component individuals, materials, artifacts, and organisms occupy space-time. (Various space/time concepts and appropriate terminology is discussed on p. 153-4 below.)

What is in essence propounded here is that the space or room available for occupation is not simply a function of areal size or floor-space area. It also depends on the period that this spatial area is or can be occupied as a fraction of the total time span considered. Occupation is not a timeless concept but is something with duration. Thus the area occupied for a few hours in the day or a few months of the year is roomwise smaller than an area occupied all day or all year − in spite of the land or building area being the same. So when assessing the capacity of a region to 'carry' or accomodate different activities, it is not simply a matter of land area but *space-time volume*. The maximum volume is the land area multiplied by the observation period.

This time geographic view of land, buildings and occupied space gives rise to a time allocation problem with respect to space. Just as there is a limited amount of time for an individual to allocate, any specific portion or parcel of space can be said to have limited time available for occupation or use (e.g. for plant cultivation), not human time, of course, but settlement space-time. The temporal allocation of space is crucial to an understanding of human ecology. But the reader familiar with general ecology is also well aware of how populations of organisms exhibit temporal differentiation of niches in the habitat, such as nighttime and daytime birds of prey. This time allocation problem can be regarded as a packing problem in a limited space-time budget. A region over time may be more or less densely packed with quanta of space-time demanding activities, and the structure of space-time occupation is profoundly related to overall socio-cultural capacity and performance. If enough space-time cannot be supplied various human activities cannot be performed.

Taking the example of an agricultural population, people have a set of cultivation projects which they must carry out in order to produce food. This incurs a space-time demand for crops, and, if aggregated for a whole region, the volume demanded in terms of hectare-days may be distributed in the regional space-time budget as in Figure 5:1. Different crops have diverse cultivation periods, growing seasons, and maturation periods and are temporally constrained by factors such as the availability of water. Each crop in a field can be looked upon as enclosed not only by spatial boundaries but also by temporal. In this way, blocks, pockets, or *parcels* of space-time required for activity and living form the elements of the regional or local space-time budget.* This temporal allocation of space is parallelled by the temporal allocation of human time.

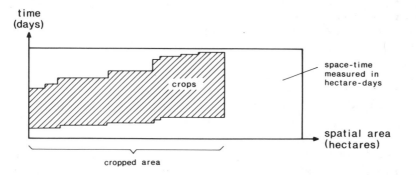

FIGURE 5:1   The space-time budget for a small agricultural region. The demands on space-time for crops is indicated by the hatched surface.

A time-space region may be more or less densely packed with parcels that occupy parts of it. The degree to which parcels are packed along the time dimension will be referred to as *temporal intensity*. The extent to which parcels are packed along the spatial dimensions is termed *spatial density*. Both terms are fairly well in accord with customary usage when dealing with issues of population density and land use intensification (Figure 5:2). The carrying capacity for parcels in a regional space-time budget is a function of both density and intensity, which is an important feature in the time-geographic perspective on land use.

*   The carrying capacity of a regional space-time budget is thus related both to the carrying capacity *for* parcels and *of* parcels. *Parcel* means both something which contains something else and something which is an allotted portion of a greater whole. The verb *parcel* means to divide, apportion, allot and even allocate. A parcel can thus be a unit of allocation. But the term parcel as used here refers to a space-cum-time entity, not merely a spatial unit.

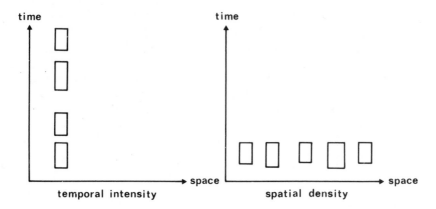

FIGURE 5:2   Two forms of packing in settlement space-time, one being along the temporal dimension (left), the other along the spatial dimension (left). Packing can increase along both dimensions in the same region, of course.

A given volume of space-time can be distributed in several ways. This applies both to the number of parcels and the volume of each parcel. Parcel volume can thus be measured in square metre minutes for a room in a building or in hectare-days for an agricultural region. The parcel occupied by a crop is generally a function of both the length of the cultivation season and the land area. Measured in space-time terms, the same 12 hectare-month volume can be distributed in several ways such as 4 hectares for 3 months, 2 hectares for 6 months or 1 hectare for 12 months. Since different crops have varying maturation periods, cultivators in all parts of the world manipulate the packing of cropped parcels both by choosing crops and by regulating supplementary inputs such as irrigation water so as to speed up or extend cultivation periods.

In between parcels covering the same area but at different times, there is often unoccupied or vacant space-time. This we may generally call *fallow space-time* or fallow room, as in Figure 5:3a. The volume of a parcel is partly a function of the physical size of people, organisms or objects and the duration of occupation, partly dependent on the room for micro-local movements and uninterrupted manoever necessary to perform given activities. This is referred to here as *elbow room*. Shortage of elbow room generally implies crowding, which in turn tends to increase human time consumption.

But the amount of elbow room which individuals demand in various situations is not just determined by biotic or physical conditions but also by technical and social norms as to expected or desired performance. This makes it difficult to sort out in a clear-cut way how

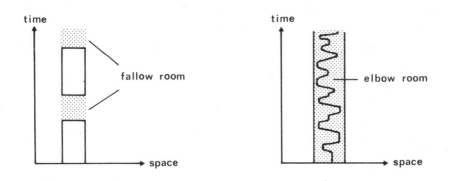

FIGURE 5:3   Fallow room (left) as opposed to elbow room (right).

much elbow room is incurred by physical crowding and how much is due to normative desiderata. The consequences for surrounding parcels tend to be the same regardless of these causes. However, these norms must be understood as to how they have evolved as micro-adaptations to macro-environmental situations over a longer time period.

A brief note on terminology and space-time concepts

Unfortunately, there has been a good measure of confusion of terminology with respect to time-cum-space in sociology, geography and allied disciplines. The term *time-space* in the present book — congruent with *time-geography* — denotes time and space as an existential-locational matrix. All objects and events are located relatively to one another in time-space, regardless of whether some form of time and space can also be looked upon as resources.

On the other hand, when discussing *resources* or inputs such as human time, vehicle-time, machine-time or building-time (room-time), resources which can be temporally allocated, it is obviously more consistent to denote land or room to accomodate humans, organisms or objects as *space-time* and the total space-time in a region of some kind thus becomes its *space-time budget*, as the term is used here. It would even have been logically consistent if one prefers to use the concrete term land, to speak of 'land-time'.

Since time-geography and other time-cum-space approaches in the social sciences (cf Carlstein, Parkes and Thrift 1978) have not been inspired by Relativity Theory in physics, the fact that physicists use the concept of (4-dimensional) space-time is of little concern to social science structured along time and space, since the Einsteinian relativity effects are not felt at the slow speeds with which humans, organisms and artifacts move on Earth. It is perhaps necessary to point this out, lest some social scientists might be tempted to designate the time-geographic approach as a variant of 'social physics' which it is certainly not.

Our terminological considerations do not end here, however, since there is also the conventional concept of a *time-budget* (cf also Chapter 9). This has been used by sociologists and others for years (e.g. Sorokin and Berger 1939) to denote a systematic record of a person's activities and time use. In sociology, the time-budget was not very often defined in an *ex ante* perspective, as a resource to be spent in the future, or as the supply side of human time (cf Chapter 8), or as a potential resource to be *allocated* under various constraints, but as an *ex post facto* record of actual time use. The latter ought really to be referred to as a time account or time use record, but the concept of time-budget is now so widespread that we probably have to live with it. It is still important to notice, however, whether it, when used, refers to potential time supply or to a time use record.

Returning then to the concept of space-time budget, a few geographers who seized upon the sociologically defined time-budget, wanted, rightly, to add the spatial locational coordinates and the place dimension. They thus called such a specified time-budget a 'space-time budget'. But the latter term is a real misnomer. To the extent that it should be used as a kind of description of the individual-path and its activity content, it would be much better referred to as a time-place account. It does not, for instance, treat space as a resource, nor does it consider the space-consuming properties of various entities and activities.

The whole field is not free from terminological inconsistency, and it would be most valuable if researchers in the field eventually could reach some consensus as to appropriate terminologies. So far, the time-geographic approach seems to be the most general time/space approach in social science, since many of the sociological or economic time allocation or time-budget approaches have failed to incorporate space, location and habitat to any greater extent, nor do they deal with the temporal and spatial allocation of other resources than human time. There is, however, a spatial and temporal allocative dimension to the use of all resources.

LAND CROPPING CYCLES AND THE INTENSITY OF SPACE-TIME OCCU-
PATION

A basic cycle in practically all agrarian systems is the year and its
seasons, i.e. *the annual cycle of activities.* The variations in solar
radiation, temperature and precipitation largely regulate the growth
of plants and heavily influence cultivation activities both in location
and volume. In shifting cultivation societies, there are generally two
other cycles of longer duration than the year: the land cropping cycle
and the residence cycle. These cycles in the settlement system are
harmonized in various ways depending on a number of factors, such
as local population size and dispersion, transport technology and
prism range, the choice of main and subsidiary crops, and the size of
the regional territory.

The *land cropping cycle* is primarily responsible for the fact that
swidden cultivators shift residence at perennial intervals. The land
cropping cycle can be defined as the period that pieces of land are
cropped and fallowed before they are cropped again.

The first year a plot is brought under cultivation, it has to be
cleared and burned. The amount of time allocated to clearing depends on the previous state of the land and its vegetation. If forest is
cleared, this involves considerable labour-time. The larger trees are
either felled or killed off by ringbarking, while smaller trees and
bushes are taken down altogether. In some forms of shifting cultivation trees are pollarded in a wider area and the branches assembled in
the field, as in the Central African *citemene* system. After clearing,
burning ensues which has two main functions: it supplies the soil
with ashes thereby raising fertility, and it kills off potential weeds
that would compete with the crop. In most cases the fire is controlled and does not escape into the surroundings, while this is not uncommon in regimes with a marked dry season.

The land once cleared can be used for more than one year in most
cases, as in systems with permanent cultivation of fields, but since
fertility is commonly not artificially maintained through manuring
or application of fertilizer, it drops quickly after the first year. The
second year gives a poorer crop unless the soil is exceptionally fertile, and by the third or fourth year or so the land is taken out of
cultivation and left fallow. The primary reasons for the reduction in
yields are, partly the loss in fertility caused by leaching, partly that
the weed germs or roots have not been rendered harmless by intensive burning and therefore compete more successfully with the crops.
This makes the field or garden more time-consuming to cultivate.

When the cultivation of an old plot is not longer kept up, a new
piece of land is cleared to replace the old plots left fallow. Each
household in most swidden cultivation societies has some land which

belongs to each age category, e.g. some which was cleared two years ago or more, some from last year, and some cleared this year. Different crops are often planted depending on how long ago the land was cleared. In many cases, the new plots are farther away from the village or dwelling than the plots of the preceding year. The relative location of dwelling and field in the day-prism thereby alters over the years.

The land cropping cycle of shifting cultivation systems can be regarded as a special case of cropping cycles for any system of cultivation, if the various systems are ordered according to the *temporal intensity of space-time occupation.* This is what Boserup (1965) called the *frequency of cropping* in her dynamic model of land use. For convenience of exposition she classified the temporal intensity of cropping, i.e. the duration of the land cropping cycle, according to the period of fallow during which space-time was not occupied by any crop. This can be depicted as in Figure 5:4, where it is assumed that land is cropped for two years in succession before it is left in fallow.

The least intensive system is that of (1) *forest fallow* cultivation, in which space-time is occupied by crops for one or two seasons and then is left fallow long enough for forest to grow up, a process of at least some 20 to 25 years. (2) *Bush fallow* cultivation implies a somewhat higher temporal intensity of cropping, where land is only left fallow long enough for bush vegetation to reappear. The fallow period may vary considerably but is inadequate for the regeneration of large trees. (3) *Short fallow* cultivation hardly allows anything but wild grasses to get a foothold before the land is cropped again. This latter system does not really belong to the shifting cultivation category, as the total land cropping cycle has been reduced to about four years or so. When the fallow period is this short, residence tends to become permanently located over the years and there is only rotation among fields. (4) *Annual cropping* is when the same land is cultivated every year, on the whole, with merely a seasonal fallow period, although the crops cultivated in a given field may be rotated. The land cropping cycle as previously defined lasts only a year and this system is characteristic of the sedentary agriculture found in large parts of Asia (as well as temperate European agriculture). The system with the highest temporal intensity of cultivation is the (5) *multi-cropping* system in which even the seasonal fallow period is reduced, and the land gives two or more harvests per year. This system often employs some form of irrigation when most intense and allows very dense populations.

The tremendous difference between the systems in the occupation of space-time by crops can readily be detected in the above model. In a multi-cropping system, a plot of land is occupied about

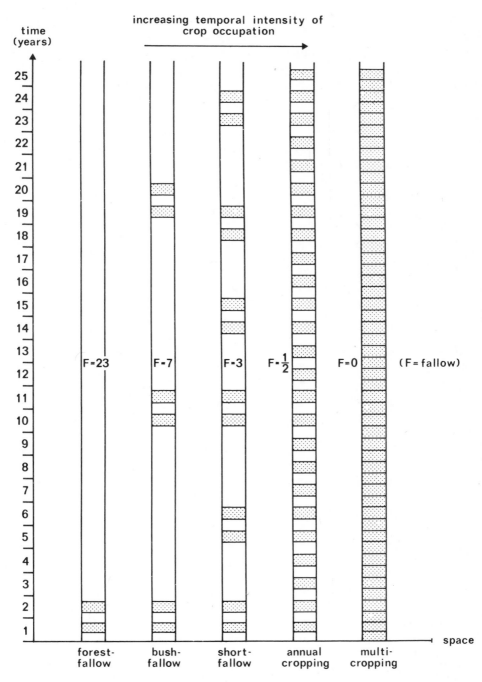

FIGURE 5:4   The intensity of occupation of settlement space-time by crops.
Each pillar symbolizes a given plot of land over a 25 year period. The classification according to intensity follows that of Boserup (1965).

fifty times as intensively as in the forest fallow system. Figure 5:4 thus shows very emphatically that 'land requirements' for a population of cultivators must be looked upon in space-time terms rather than as a simple function of spatial area.

The residence cycle, space-time requirements and local territorial size

The land cropping cycle, the residence cycle and the size of the local territory are clearly interrelated. Although the shifting cultivation ecotechnology generally gives a higher carrying capacity of land than food collecting, a process by which shifting cultivators gradually deplete the land resources within local prism range can still be discerned, as was the case with hunter-gatherers and pastoral nomads. The main difference is in the rate and frequency by which the shift of residence occurs. For hunter-gatherers or pastoralists it may be a question of weeks while for swidden cultivators it is a matter of years.

Since land usually has to be cleared each year in swidden agriculture, this new land mostly has to be located farther away from the dwellings than previous plots. This goes on for some years until the travel distance from dwelling to field becomes too costly in human time. At that stage the dwellings are shifted closer to fresh land. The period that the cluster of dwellings or village remains in the same place can be defined as the *residence period*. The duration of the *residence cycle*, on the other hand, is the time it takes before residence is again taken up at an old site in the local territory.

Before looking into the mechanisms by which a local population exhausts a local prism habitat and adapts to local space-time shortages, let us regard space-time requirements as given and examine how the frequency of residential shifts and the duration of the residence period relates to the size of the local territory. The latter is the space-time domain to which the local group has customary claim, and is not the same thing as the local prism habitat, of course, which is the land within reach given the various determinants of prism size. For shifting cultivation societies in ecological balance, the local territory is generally larger than the effective local prism habitat. This means in practice that in spite of local shortages of space-time for cultivation activities, sufficient space-time is available to the local population in their broader territory. Most shifting cultivators operate in some kind of local territory.

The activity structure of a shifting cultivation population can thus be looked upon as a space-time packing process, where crop-cum-fallow parcels are packed into a territorial space-time budget. This can be depicted in a simple graphical model (Figure 5:5). A

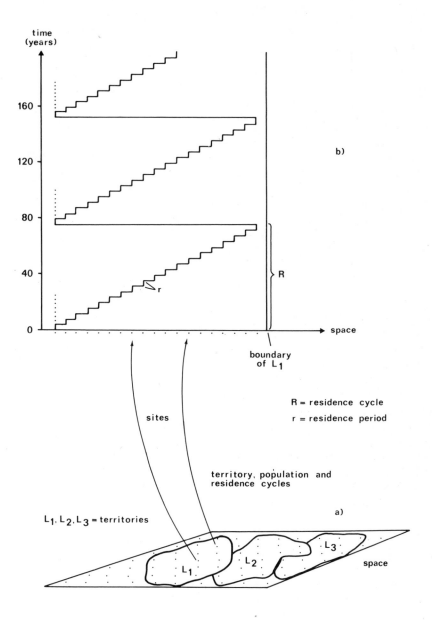

time
(years)

b)

160

120

80

R

40

r

0

space

boundary
of L₁

R = residence cycle
r = residence period

sites

territory, population and
residence cycles

a)

L₁, L₂, L₃ = territories

L₁

L₂

L₃

space

FIGURE 5:5   Territory, population and residence cycles. The path in the top figure illustrates that of a village population moving in a territory over a 200 year period.

given area contains three local territories with a set of potential sites suitable for location of villages (each site represented by a dot). Focusing on one territory, its sites are occupied in some time sequence and for the sake of exposition they are ordered in this sequence along the spatial axis of Figure 5:5b.

A village of cultivators can now be illustrated as a path, but a divisible path unlike that of single individuals. The path is vertical when the village remains in one place throughout a residence period. After all sites (or most of them) have been used, the village returns to a previous site again and a residence cycle is completed.

The structure of the village path and the packing of crop-cum-fallow parcels in territorial space-time is shown in Figure 5:6. It can readily be seen how the parcels are distributed in space and time. The white area is also fallow land, of course, but fallows which are not absolutely necessary to maintain the system. The minimum fallow space-time to reproduce the system is indicated by the dotted area.

Given the same population size, what would happen if the size of the territory was smaller for some reason? In Figure 5:7 the territorial space-time budget is less than half the previous size, and the parcels are consequently much more densely packed. This implies a much shorter residence cycle and the village returns to an old site much earlier than in the preceding case. Yet, this does not necessarily lead to pressure beyond carrying capacity, since fallow requirements are still met.

Territories may be reduced in size for several reasons. It may be a consequence of overall population growth and concomitant fission of villages so that two local populations now occupy an area formerly reserved for one such group. Or unfriendly neighbours may simply have invaded part of the territory. But regardless of the causes, the adaptive response has the same effect on packing: it leads to an increase in packing intensity and a shortening of the residence cycle.

It is also apparent from this simple model that if the territory is further reduced in size for some reason, some other kind of response will emerge, and this is generally a reduction in fallow periods and/or extension of the cultivation period or some similar change in eco-technology, as indicated by numerous cases throughout the world discussed by Boserup (1965) in her theory of agricultural intensification. It is obvious that a reduction in the fallow period would give the greatest gains in space-time for cropping, and this would be a suitable recourse to take, were it not for the associated problems of maintaining soil fertility. (Cf Chapter 6 on intensification aspects.)

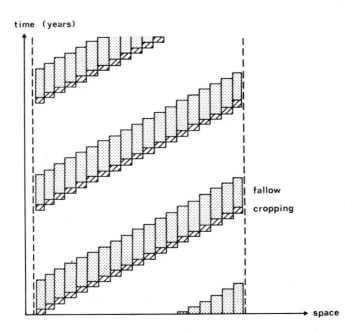

FIGURE 5:6   The packing of crop-and-fallow parcels in a territorial space-time budget. Hatched portions indicate cultivation; dotted portions the minimum fallow period necessary to maintain the system at its given level of land use intensity.

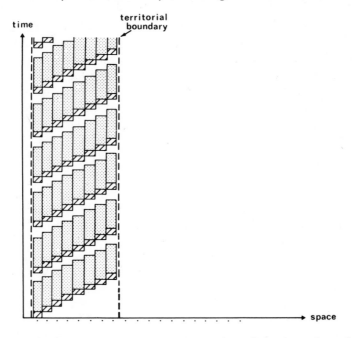

FIGURE 5:7   The same system as in the previous figure but where the territory is less than half the size. Had it been even smaller, the minimum fallow period might have to be encroached upon, resulting in land use intensification.

THE CARRYING CAPACITY OF THE LOCAL PRISM HABITAT

The carrying capacity of the local prism habitat is conditioned by a multitude of different factors, negative and positive constraints which intersect to give a set of feasible activity alternatives within which the system can operate. The determinants of prism volume and range immediately come to the fore, such as the speed and mode of travel, which are not only a function of individual capability but also the topography of the landscape. Prism size and structure is partly regulated by: the amount of time which can be spent on travel when allowance has been made for both activities inside the village centre and in the peripheral fields; the human time resources that can be mobilized depending on population composition (age, sex, etc.); and the division of activities by population category. Finally, thought must be given to the space-time requirements per person and consequently to the size of the local population sharing the same prism centre and competing for land (space-time) within prism range. We shall pay further attention to the population size factor below, but regard it as a constant for the time being.

The gradually increasing distance to fields with the progression of the perennial cropping and residence cycle is a common theme in many studies of shifting cultivation systems. An early but typical account is that by Linton (1933:39-40) on the Tanala of Madagascar:

> Since the newly cleared land produced the best crops, the usual... method was to utilize all the original jungle, which could be profitably exploited with the village as a center, then move the village to a new locality and begin the process again... it is necessary to allow the land to lie fallow from five to ten years between plantings. The growth which springs up during this time consists mostly of bamboo, fern and low bushes and second planting never yields as good crops as the first. Because of this, the /Tanala/ try to clear new land each season, going farther and farther away from their villages.

The shift of village location can be aptly illustrated by a series of day prisms representing a succession of years in a residence period (Figure 5:8). In the first year, the main fields are located near the village. After a year or so, new fields have to be cleared farther out and in the third year still farther out. After a few years of residence, it is less costly in human time to move the entire village to a new location than to walk to the fields almost daily (or even more frequently in the peak seasons). By the eleventh year, for instance, the village has moved to a new location where the fields are again close to the dwellings and a new residence period commences.

Had the abandoned dwellings in the old site been large and built

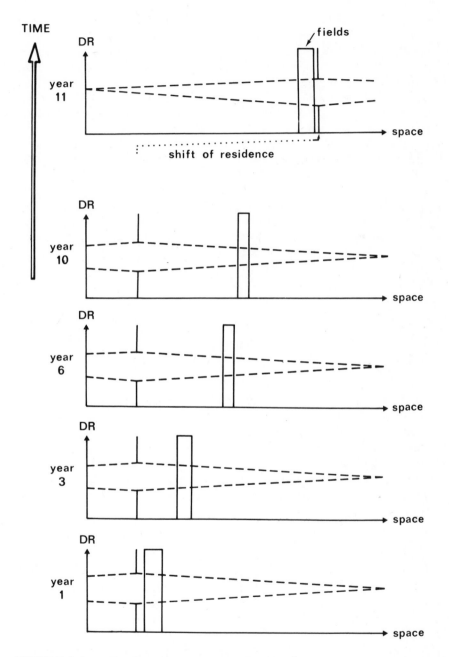

FIGURE 5:8   The day-prism and the shifting of fields and residence. A field is envisaged to shift farther away from the prism centre with the progression of the perennial cultivation cycle (this effect is mildly exaggerated in the figure). This process goes on for some 10 years before the village (residential location) is shifted to a new site. (Cf Figure 3:11 above.)

of permanent materials, such as the more prestigeous kimberly brick buildings adopted by better-to-do households in Zambia (cf Kay 1964 or Long 1968), these dwellings would have represented a sizeable investment in labour-time. The time costs of deserting good buildings combined with those of reerecting similar structures elsewhere would be considerable and perhaps prohibitive. In temperate climate zones where buildings are less perishable this may be so, but in the tropics pole and dagga or wooden dwellings with thatch generally do not last for much more than five or six years anyway. By the end of a residence period, houses are worn out and have lost their attraction.

But the analogy of eating ones way out of a local prism habitat is as appropriate for shifting cultivators as for hunter-gatherers and pastoral nomads (cf Figures 3:12-13 and 4:1), except for the difference in rate. Because of the large quantities of forest consumed by many shifting cultivators, such as the Iban of Sarawak, they live up to the title of forest eaters or *mangeurs de bois* (Freeman 1970:286). This is especially the case if they operate under a marked dry and wet season regime, where fires may escape during the dry season burning and devastate much larger areas than those cultivated. Linton, for instance, claims this is not uncommon in Madagascar (Linton 1933).

Adapting to local space-time shortages by double residence

There are several ways of adapting to local space-time shortages which can be taken advantage of either singly or in combination. The shift of residence has already been discussed, but another common method is to extend prism range, whereby accessible space-time tends to increase exponentially with distance (cf Figure 3:14). This can be done by means of faster transport (which is less common), by increasing travel time and hence working hours, or by adopting double residence. The latter implies temporarily reducing village size and dispersing the local population over a wider area, which automatically reduces the space-time shared by all.

The Iban of Sarawak described by Freeman (1955, 1970) are cultivators of hill rice in swiddens in an area of steep and rugged terrain. Precipitation in the region is abundant which makes the timing of the burning somewhat difficult but the cultivation season is very long, so delayed burning has few consequences. For a shifting cultivation system, the level of labour input per hectare and person is comparatively high, and much time must be spent guarding the crops. The settlements are located in the river valleys to enhance capacity for travel and communication. Travel on foot is slow and arduous and time can be saved by using canoes in the riverine system pervading the country. Some rice is exported via these channels and all rice is brought from the fields to the settlements on the completion of harvest. It is less time and energy demanding to carry these loads downwards from the fields along the hill slopes than it would have been to carry them up to hilltop villages, as is typical of some shifting cultivators (for instance, the Lamet of Laos or the Siane of New Guinea).

The Iban have more than double residence; they have a triple-tier hierarchy of settlement. The largest settlement unit is the village or *long-house*, so called because of the way families combine to erect one large building. Such a unit may have from 140 to 300 members, depending on the location. Each long-house consists of a number of *bilek*-families ranging from 2 to 14 members, with a mean size of 6, and from 25 to 50 families (Freeman 1970:1-11). It is the family which is the agricultural production unit and the size and composition of the *bilek*-family largely defines its level of production.

> When a new /long-house/ settlement is founded, the forest in its immediate vicininty is felled first, and as long as farms lie within a radius of about *half a mile to a mile* (or slightly farther, if they can be reached by canoe), activities are conducted from the main long house. After a few years, however, as more distant land is brought into cultivation, accessibility becomes a problem, and when this stage is reached the community breaks up into a number of distinct groups — each group /of a few families/ farming a different part of the long-house territory. (Freeman 1970:161) /My italics./

This results in the formation of satellite settlements called *dampa*, which are occupied for about five to six years, and as soon as the nearby slopes have been used, the Iban again move. The process of depleting the local prism habitat of the *dampa* settlement is largely the same as for the central long-house village, although the residence period is longer for the smaller satellite settlement, as might be expected. Only after some 15 years from the establishment of a long-house has the secondary jungle nearby regenerated to an extent that the various *bilek*-families can move back for another round of cultivation. Thus a grand settlement cycle has terminated and a new one begins.

In this manner the Iban practice double residence, but even around the *dampa* sites there is a seasonal dispersion of the population. These temporary shelters and watch huts are located by the fields themselves. This is crucial as the rice approaches maturity and has to be protected from monkeys, pigs and other vermin. So, in effect, the Iban have triple rather than double residence, in a spatial and temporal hierarchy.

Double residence in the annual cycle is extremely common in shifting cultivation societies, both as a means of extending day prism reach and as a way of securing the agricultural produce for human benefit. Linton (1933:318) reports, for instance, that the Tanala return to their villages after they have been abandoned for six months, but the literature is full of similar cases (cf Figure 5:9).

Double residence offers one possible way of keeping a main village fairly permanently located while still allowing the practice of shifting cultivation with forest fallow. So instead of shifting prism location at several year intervals, the shift is made twice per annum or so. The Iban have chosen to do both.

Time resources and the sensitivity to distance

Geographers and economists in the field of location theory often refer to how different activities are sensitive to distance or how inter-action between places decreases as a function of distance. So does Chisholm in his excellent work on rural settlement and land use, and he furthermore relates this to costs and the currency in which they are paid /italics added/:

> At the small scale of phenomenon - the farm and the village - it was noted that the cost of overcoming distance arises largely, if not exclusively, from the *amount of human time* expended, both in accompanying all inputs and produce and in transferring from one site of operation /station/ to another. (Chisholm 1962:76)

Behaviour in space is intimately associated with the expenditure of human time, a resource which is scarcer at some times and places than others depending on the demand for it. However, it can become particularly scarce if spatial behaviour is highly irrational and inefficient. Although peoples of various cultures and settings may perceive the duration of diverse activities very differently, time for travel (as for other activities) is also parallelled by the expenditure of energy, and travel is often more energy demanding than a good many other activities, particularly if people have goods or perhaps children to carry. This partly explains why the level of awareness of travel time expenditure is generally high.

Freeman furnishes some striking examples of sensitivity to distance in terms of time use (1970:197):

> If a farm be distant more than about twenty minutes /sic/ from the *dampa*, it generally happens that the family builds a substan-

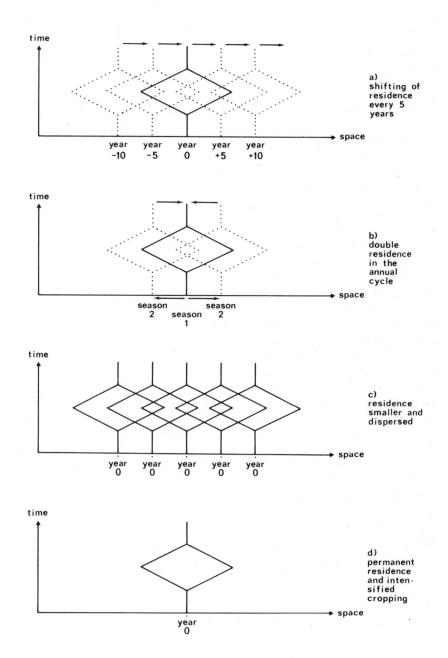

FIGURE 5:9   Four ways of adapting to local shortages in settlement space-time.

tial /farm house or temporary shelter/, and uses it as a semi-permanent base during most of the agricultural cycle.

The long-house was only used as long as the fields lay within a radius of up to a mile or 1.5 km, which covers an area of 7 km^2.

> During the course of the agricultural year a man visits his farm from 150 to 200 or more times. And the women, on whom the heavy burden /and time expenditure/ of weeding and reaping falls, may be even more frequent visitors. By having a strategically placed base, a family can, in the course of a year, save themselves many hundreds of hours of arduous walking over difficult and hilly country. (Freeman 1970:161)

Time allocation in the daily round and its variations throughout the annual and perennial cycles is thus a key to the structure of the settlement system, although the daily round and its time-space structure has been a rather neglected topic.

Leaving aside the general analysis of how human time resources are mobilized in shifting cultivation societies (cf Chapter 9), an important parameter in this context is the division of activities by sex and age. If the heavy burden of cultivation is placed on the female half of the population, their time resources may wield a greater influence on the settlement system than those of the men, or otherwise, if it is structured around the activities of men, the women may simply have to pay the extra time costs that this entails.

The Siane of New Guinea described so interestingly by Salisbury (1962) may serve as an example. * A typical daily round for a Siane man and woman can be depicted as in Figure 5:10. It was the lot of the women to feed the children in the morning before setting out for the gardens, a walk which might take up to an hour. Siane villages are located on hilltops in this very broken country. This siting is the opposite of that of the Iban who live in the valleys and use streams as transport routes, walking up to the fields along the hill sides. The only other way of travelling with less effort is along the hilltops and down to the gardens, which was the Siane solution. After five hours of work in the gardens, each woman brings sweet potatoes and other crops, firewood and water home to the village in the early afternoon. They then start to prepare food for the men, who — on the days when they do garden work — leave later and arrive home later, but who are free from work and domestic tasks on their return and can

---

*   Salisbury's materials on the Siane have been reinterpreted along time-geographic lines elsewhere (Carlstein 1978c), with special regard to the innovations affecting this ethnic group, for instance, the steel axe.

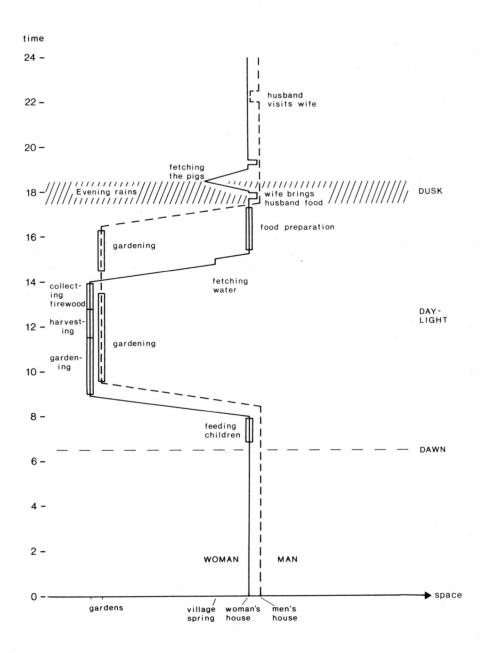

FIGURE 5:10 The daily round of activities on an average gardening day for a married man and woman. They spend most of their day with members of the same sex and do not share the same house. For lack of data, no Siane children or aged are included.

devote their time to social activities. Women then feed their men-folk and children, themselves and go to call home the pigs. Later they fetch the dishes from outside the 'men's house', where all men in a ward group sleep, and retreat to their own individual huts.

This is the daily round which affects the average active woman for something like 275 days per annum, when she is not sick or visiting a neighbouring village. Due to the time-saving effects of steel axes, the men nowadays only go to the gardens every third day, which means about 120 days per year. The Siane village with an aver-age population of some 200 inhabitants is thus primarily a social and ritual centre, while the major subsistence activities of producing sweet potatoes, pigs and other food is located in the periphery along the hill slopes.

With this daily activity programme, we can readily see how the volume of the local prism habitat is reduced quite drastically (Figure 5:11). The maximum day-light prism delimited by the speed con-straint and dawn to dusk working hours outside the village is unreal-istically large. Women would then lack time for tasks in the village itself. In practice, they can only leave after their morning chores and must be back around 15:00 hours to cook. Since they must also spend up to five hours in the gardens — planting, weeding, harvesting, and so on — they cannot really use land so far out that time for gardening is consumed by wasteful travel. This factor also reduces local prism reach considerably. Adding the final constraint that some of the land immediate to the village is reserved for communal gardens producing food for occasional ceremonies (including fallows), the portion that can be used of a maximal 12-hour prism habitat is indeed minimal.

It is interesting to note why these gardens for ceremonial pur-poses are closer to the village. On the grand festival occasions the women have to cook for many more people, both guests and hosts; thus, they must spend longer working hours in the village and have less time to travel. On these days, their prisms are even smaller than ordinarily and the relative location of parcels in the settlement system is adapted to this situation.

A major reason why the Siane get away with maintaining village populations of 200 individuals and still keep permanent residence is that their ecotechnology is based on root crops (in combination with pigs for proteins) rather than on cereals such as rice or millet. Siane agriculture is thereby more space-time intensive, since tubers give higher yields per hectare and deplete the soil much less. Root crops also tend to be less labour-time intensive, whereby the Siane can afford to work shorter hours and accept longer travel time to the fields·that the Iban can, for instance (cf p. 199-205).

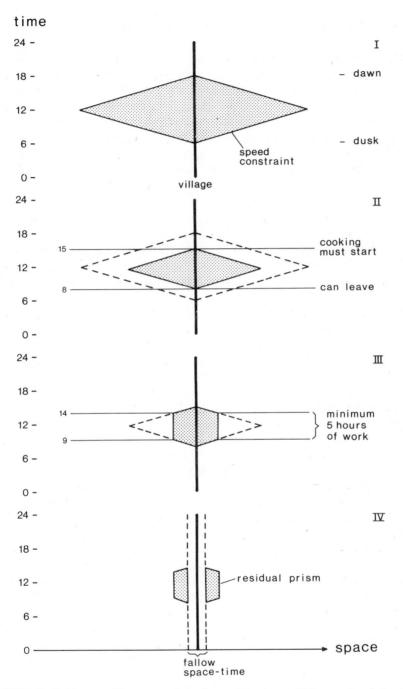

FIGURE 5:11   The space-time for cultivation within reach of the Siane women as a function of the aggregation of various constraints. Only the residual prism in IV remains of the local prism habitat for the women, in practice.

Habitat size and the mechanism of split day prisms

Another mechanism reducing the size of the local prism habitat and hence its relative carrying capacity is that of split day prisms. The splitting of a potential 12-hour day prism has considerable negative multiplier effects on the space-time accessible to a village population. Like the previous case of cumulative constraints acting on the Siane (Figure 5:11), split day prisms largely explain the somewhat paradoxical fact why shifting cultivators have to change residence although the land within range increases exponentially with time distance from the prism center.

Unlike the Siane living in a mountainous area, people in many other tropical societies are less satisfied with working in the middle of the day when temperatures become intolerable. Moreover, they may need to be home for some hours at mid-day to perform other activities. This return in the middle of the day splits a larger potential prism into two (or even more perhaps) smaller ones, the effects of which can be simply illustrated as in Figure 5:12. We assume a speed of movement as low as 3 km/hour, which is what Izikowitz (1951: 40) reports for the Lamet, who admittedly live in a hilly region also, but the figure probably applies anywhere for a person carrying a child or some burden. We also make the realistic assumption that four hours of work have to be spent in the fields regardless of travel time. With an un-split day prism, an area within a radius of 12 km or 452 km^2 would thus be available for cultivation. But only if as much as eight hours were spent on travel.

This is clearly beyond tolerance and reason, since very little time would remain in the day for other activities in and around the village. Time use studies such as that by Richards (1939) on the Bemba of Zambia, show that women, who do most cultivation, have numerous other time consuming chores, such as fetching water and firewood, cooking, pounding grain, taking care of children and cleaning. They cannot waste time on travel to such an extent, regardless of all the ideas circulating about surplus time in 'primitive' or 'underdeveloped' societies (cf Chapter 9). Even with four hours of gardening and four of travel (making an area of 113 km^2), the daily time allocation to other important activities would be upset.

With garden work in the morning and afternoon within two sub-prisms, as in Figure 5:12, we get down to a *3 km radius* and a 28 km^2 area for shifting cultivation, providing this habitat does not overlap with those of other local groups, which would certainly be the case for the Siane or the Lamet. But even this is under the rather bold assumption that people are willing to invest four hours of travel for the same period of cultivation.

What is more likely to happen under the above circumstances is

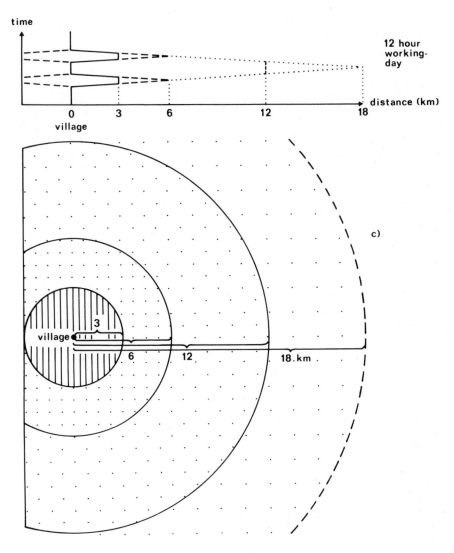

time

12 hour
working-
day

distance (km)

0   3   6   12   18
village

c)

3
village
6   12   18.km

spatial area for cultivation

FIGURE 5:12   Division of a day-prism into sub-prisms and its effect on the
size of the spatial area for cultivation. (Cf Figure 3:14)

that instead of working two and travelling two hours in the morning and repeating this in the afternoon, people would do all four hours of work in the morning and save themselves two hours of travel. Especially if we consider peak season time inputs, four hours' travel is more than enough to induce double residence. Travelling to distant fields — which many shifting cultivators definitely do — can only be afforded at a much lower frequency, such as once every *other* day rather than twice daily. (The important adjustment of travel frequency to distance is dealt with on p. 219-20 below.)

For a population of swidden agriculturalists to enjoy the advantages of a bimodal working day, fields have to be very close to the dwellings. Even in areas where people can comfortably walk at 5 km/ hour, the corresponding area would be cut down to a 2 km radius or so from the village or about 7.5 km^2. As mentioned before, the Iban would walk about a mile which similarly amounts to a 7 km^2 area.

We can now judge how these travel distances tally with the annual land or space-time requirements and the perennial cropping and residence cycles. The Iban, for instance, cleared around 0.33 ha per head and year, and a long-house village of 140 persons would need an area of 50 ha anually or 0.5 km^2-years. Were all land arable, they could reside at a long-house for 14 years, and deducting some 30 per cent of the area as unsuitable for cultivation, the residence period is cut down to 10 years. For a village double the size, the residence period is 5 years before this local population of Ibans would disperse to their *dampa* settlements.

This estimate is not too far off the target, but extending the idea of Liebig's 'Law of the Minimum' — to the effect that the seasonal bottlenecks tend to limit the capacity of a system — it is clear that the Iban are not inclined to allocate even as much as one hour daily in each direction when walking to fields. Neither are the families uniformly exposed. As Freeman points out (1970;218-75), the *bilek* families are the autonomous units for cultivation, and due to demographic chance their age-sex composition varies cross-sectionwise and with the stage in the developmental cycle of the domestic group. Some families experience time shortages more severely during the cultivation cycle and are more sensitive to travel time spent. It may also be that the whole *dampa*-group accomodates its residential shifts to the requirements of the most exposed families, assuming the general tolerance level in this respect is not so low as to trigger off early residential shifts anyway. This discussion shows how the structure of the daily round of time allocation is of utmost importance in relating the activity system to the settlement system.

It must also be remembered that societies combine different ecotechnologies, and hunting-gathering has not been displaced by shifting cultivation but rather *superimposed* on it. Freeman notes that:

> Again, a house near to relatively unfrequented jungle represents far better opportunities than does the main long-house for that foraging of wild fruits, fungi, and other edible plants, at which Iban women excel. (Freeman 1970:161)

For the Iban the combination of cultivation and food collecting, women collecting and men hunting as seems to be the general rule, is a force pushing in the direction of dispersed settlement and more frequent residential shifts. This is, of course, counteracted by the extra time costs of maintaining contacts within the population that dispersal entails.

It may seem contradictory at a superficial glance that hunter-gatherers such as the Bushmen have to shift residence at a manifold higher frequency than swidden agriculturalists, in spite of the fact that these food foragers operate within a much broader prism range than do cultivators. The basic reason, of course, is one of difference in carrying capacity with ecotechnology, but hunter-gatherers also tend to be less sensitive to distance insofar as their level of time pressure of work is generally lower, it seems (cf Chapter 9). They can, in a way, afford to spend more time on travel and foraging is much more of a mobile than a 'stationary' activity anyway.

Although these time allocation arguments are undoubtedly basic to explanations of residential shifts, and so on, this is not to claim time resource utilization gives the whole story. There may be numerous other reasons why a group shifts residence or splits up, or conversely why they fuse or stay put in one locality, but even so, this too has time allocation, land use and settlement consequences which sooner or later give rise to various feedback loops to all sectors of society.

VILLAGE SIZE, FUSION AND FISSION

The size of the local population cluster (or village for short) has been given only cursory treatment so far, but the number of people sharing the same local prism habitat is a strategic variable in activity, population and settlement dynamics. This brings us back to the basic contradiction between concentration to reduce the time costs of interaction within the population, and the increased sharing of a prism habitat that this entails (the problem defined on p. 92-95.)

One set of factors work towards increased village size and local population concentration, such as fusion, while another set militates against this, such as fission or dispersal. Different categories of the population, men and women, for instance, may be pressurized in opposite directions, or the same category may be exposed to contradictory tendencies during different seasons. It is not uncommon that many social, political or ritual activities are broader in recruitment and have the village centre as their optimal location, while agricultural or other food quest activities that take place in the periphery are more space(-time) consuming and involve smaller groups. Whatever the case, it is essential to look into the forces behind village size more carefully.

The influence of warfare on settlement size

Historically, warfare and defence have had a substantial impact on the structure of settlement. Fortified villages, for instance, are more permanent in their location than ordinary ones. They also differ in their siting. But hilltop locations that are excellent for defence purposes, are often less favourable to agricultural activities. Settlement relates to warfare in numerous ways, and the arrows of causality cannot be determined in a simplistic fashion. A given system of settlement can contribute to disputes and feuds in some situations and be the adaptive outcome thereof in others. We shall only explore a few of these issues here.

In his expanded theory of population growth and intensification, Harris (1977) points to some of the ecologically adaptive functions of warfare, mainly as a mechanism of reducing population densities and relieving pressure on local and regional resources.

> The reasons why some anthropologists deny the reality of high levels of combat among band and village peoples is that the populations involved are so small and spread out as to make even intergroup killings seem utterly irrational and wasteful... Hence we must look to the contributions of warfare to the conservation of favourable ecological and demographic relationships in order to understand why it is practised... The first contribution is the dispersal of population over wide territories... Raids, routs, and the destruction of settlements tend to increase the average distance between settlements and thereby lower the overall regional density of population. (Harris 1977:35-8)

Without subscribing to a view that ecological reasons are behind all forms of warfare, the relationship is still interesting although it is controversial. And, as Vayda (1961) points out when dealing with expansion and warfare among swidden agriculturalists, it is in healthy contrast to psychological explanations of 'bloodthirsty' primitive patterns of warfare.

Starting at the other end, however, the common response of local groups who are attacked by expansionist neighbours or invaders from remote regions is to *aggregate in larger numbers*. With an increased village size, a larger defence force can then be mobilized in a shorter time and perhaps even act as a deterrent. The ethnographic literature abounds with examples to the following effect:

> When the external pressures increased as a result of Ngoni and Bemba raiding in the area, the form of settlements underwent a radical change. Many of the small, impermanent settlements were abandoned and large stockaded villages... were built instead. (Long 1968:80)

An alternative response to increasing village size is to reduce it and scatter the population even more. This was the solution chosen by another group of shifting cultivators, the Majangir of Ethiopia:

> Nor do the Majangir see an advantage in nucleated settlement as a possible means of defence against enemies. Given their lack of substantial property to defend /no cattle, for instance/, the dispersed settlement system of the Majangir in fact facilitates their traditional defence against attack by superior force: to hide by fleeing into the covering forests. (Stauder 1971:107-8)

For shifting cultivators, such dispersion renders the additional advantage of facilitating more permanent residence while still having land within short walking distances. The Majangir are well aware of these benefits, in fact, and 'see no advantages in other systems over their own, which indeed gives maximum protection to their fields with minimum of bother' /i.e. time expenditure/ (Stauder 1971:107).

On the other hand, one consequence of larger village size for defence or other social reasons is that nearby lands are more rapidly exhausted. A village with a stockade also represents a greater time investment and becomes more costly to abandon. The greater sharing of the same local prism habitat tends to increase pressure on human time resources, not least because of the increased travel distances to fields. Although an extended prism reach is amply rewarded in terms of additional land or space-time, the day still only has 24 hours and the additional work load can certainly be felt. This explains the reaction to the introduction of peaceful conditions so often shown.

Socially, life in a larger settlement where more people are easily accessible is often regarded as more interesting and attractive, but it is also more complex and generates more rigorous forms of social control, coordination and even exploitation. Lineages tend to fission less readily since potential 'big men' have greater difficulties in acquiring adequate resources and safe places to move to, but similar concomitants of village size are beyond the scope of this study. But another effect on settlement may be a push towards larger and more communal gardens, since the insecurity caused by hostile groups makes it useful to have many people working in the same place in case of surprise attack.

It is in this light that *Pax Britannica* or a similar peace imposed by some superior military power can be seen to have a profound effect both on settlement and activities. The impact of *pax* on the use of time and space resources has been discussed by numerous anthropologists and others dealing with Africa, Asia and Oceania; for instance: Bohannan and Bohannan (1968), Epstein (1968), Kay (1962, 1964), de Schlippe (1956), and Scudder (1962). In Zambia, the process of fission has been well documented, and Long, for example, reports the following for the Lala:

> With the arrival of the British and the subjugation of the Ngoni and Bemba peoples in the 1890's, the stockaded villages began to break up quickly... In July 1906 attempts were made by the British South Africa Company to put into effect a policy of amalgamation... into units of about 100 huts each /some 250 individuals/ ... /The Company/ found it impossible to prevent fission of these villages... an important precipitating cause /being/..., of course, sheer ecological pressure now that there were no need to stay together for protection. In relatively short time the soils and woodlands within convenient walking distance of the village were in short supply and it was more practical for some groups to hive off and build their own settlements where natural resources were still plentiful. (Long 1968:80-83)

This fission progressed steadily and the average village size decreased from 166 in 1924 to 55 in the 1940's, when Peters made his pioneering survey of shifting cultivation systems (cf Peters 1950), and further to 48 persons in the mid-1960's (Long 1968:83-90). In terms of village paths, the history of fission in a small area would look as in Figure 5:13).

Pacification is reported by Kay (1964b) to have had the same effects in a nearby area of Zambia, not only as an outcome of ecological adaptation but also due to witchcraft and sorcery. Many people who felt insecure in the larger villages of the Ushi chiefs chose to move out. Depending on the people and the place, there are numerous social forces that may contribute to fission, of course. But again, the time allocation and settlement consequences of dispersal are often similar regardless of the particular causes.

The ban on wars as an innovation may also be directly time-saving in a way which is less related to settlement. With *pax* the menfolk need no longer devote much time to defence mobilized in regiments. This time can be diverted to other activities such as cash cropping.

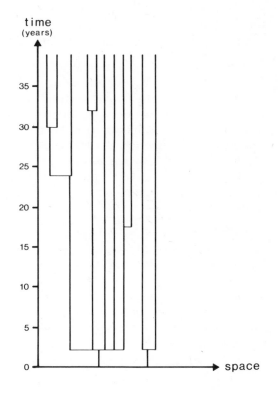

FIGURE 5:13 The fission
of two villages over a forty
year period. (Adapted from
Long 1968:85.)

The facilitation of permanent settlement through dispersal

The fission and reduction of local population concentration makes
it possible for a people like the Majangir to have plenty of land
within reach with little travel (cf Stauder 1971). Even extensive
forms of swidden agriculture with cereals as staple crops can be compatible with permanent residence, provided the settlements are small
and dispersed (the solution depicted in Figure 5:9c). But when
villages are large and the cultivation system is extensive in land use, it
is very difficult to keep villages permanently located, unless dual residence is practised. Alternatively, the system must be intensified by a
change towards cultivation of tubers, for instance, as was the case for
the Siane of New Guinea. (Cf intensification aspects in Chapter 6.)
    Although dispersed residence has the ecological advantage of
minimizing travel time to gardens, social life often suffers, since it
contains too few people with whom the extra leisure time can be

shared. Among the Majangir whose homesteads are dispersed within limited areas, for example, the sharing of one anothers' company at mealtime is sharply curtailed by time distance:

> Only domestic groups which are spatially close will be able to share food, accomodation or labour regularly. Males, for instance, may share the same sleeping hut and the same meals on a regular basis only if the /sleeping hut/ is near enough to the cooking sites of both their female partners to send the men food at night. This range is restricted to ten minutes' walk ...
>
> /Certain forms of kinship behaviour/... tend to be more definitely manifested the nearer the residential propinquity, whereas they tend not to obtain at all when persons live further than fifteen or twenty minutes' walk from each other... (Stauder 1971:111, 110)

There are thus limits to fission and dispersal inherent in the detrimental effects on social life and sharing time with others that this has for swidden agriculturalists with long fallow systems. In more intensive cultivation systems, where dispersed settlement is a way of saving time that needs to be put to productive uses in farming, the density of population is often high enough to make people closer in range in spite of the absence of villages. The Kofyar of Nigeria would be a case in point (Netting 1968), as would, by and large, the Chimbu of New Guinea (Brookfield and Brown 1963, Brookfield 1973).

Site specific and internal forces of nucleation and fusion

There are also, of course, numerous cohesive forces which work towards fusion and the maintenance of village size. Village life has many attractions which dispersed settlement would not provide. The very company of others, who are generally relatives or friends, ritual leaders, work partners or collaborators is much desired for its social meaning and personal value, and the possibilities of encountering traders or interesting outsiders increase with settlement size. Richards (1939:28) working in what is now Zambia relates that,

> my first impression of a Bemba village was one of constant vivacity, chatter, and quick laughter, and it remains to this day my most permanent memory...

Though it sounds somewhat idyllic and pictures a situation which is culture specific, examples are legion. Wilson describes how the Nyakyusa of Tanzania form age-villages to enjoy the 'good company' of age-mates (1951). The Majangir, who in fact have a rather dispersed settlement system, appreciate the company around their homesteads to such an extent that they are willing to relinquish other favourite (but spatially scattered) pursuits, such as apiculture:

> ... the honey a man collects in a year depends largely on the time and energy he devotes to apiculture... /But/... In addition to agriculture, there are also the social demands and attractions of settlement life, especially drinking parties... which limit the time a man can devote to apiculture /i.e. dispersed honey production/. (Stauder 1971:23)

Political dominance by some groups in a village over others may also serve to maintain its nucleated form and prevent fission. Even internal competition and the need to keep an eye on potential competitors for women, livestock or prestige may act as a force of fusion. Likewise, the scale of ceremonial activities and competitive display can be increased when settlements are larger, thus allowing more of conspicuous consumption and the like, which can be converted into prestige, power and enhanced control of resources.

The compromise between fission and fusion can be achieved through pulsating residence or avoidance to disperse the entire population. In spite of peak season labour time demands at a distance from the village centre, aged, children and incapacitated members may remain behind while the more active categories disperse in a way similar to how the Prarie Plains Indians left parts of their population behind during their seasonal hunts (cf p. 88 and 100). Freeman notes for the Iban that:

> It is during the critical few weeks just before and after reaping begins that these precautionary measures reach a climax /i.e. of guarding fields from intruders/. At nights the *dampa* in the valleys are inhabited only by the aged and the decrepit, by the sick, by nursing mothers and young children, for all the able-bodied members of the family are away. (Freeman 1970:201)

In New Guinea it is very common that the concentration of population varies with the stage of the ceremonial pig cycle. As the work load increases when pig herds build up, there is a tendency towards more dispersed settlement (not least for the women). In other cases dispersed homesteads or hamlets may be the rule, while villages are used temporarily for ceremonial activities (cf i.a. Brookfield 1973, Rappaport 1968 and Waddell 1972). This is yet another example of a cyclical residence-cum-settlement system.

Apart from these factors, there are also *site-specific* forces of attraction, such as a place of good dry season water supply or one where fishing may be particularly rewarding and thus allows nucleation without the discomforts and time drains of excessive travel in the daily round. Long notes a similar precondition for the Lala of

Zambia (again this is but one instance among many):

> As greater reliance is placed on hoed gardens, the advantages of remaining together on the spot where there are good soils for secondary gardens and plentiful water supplies, may outweigh the advantages of hiving off to establish smaller settlements within close proximity of good woodland but where living conditions are less hospitable. The spread of cassava is especially important, as it allows for a more settled community. (Long 1968:96)

The actual siting of an agglomeration of residences is very important in terms of optimizing the carrying capacity of the local prism habitat and hence the size of the population that can be supported within it.

External forces of fusion and nucleation

Chapman (1970) describes an interesting case of the contending forces of fusion and fission on Guadalcanal, an island in the British Solomon group to the east of New Guinea.

The Melanesians living in the two villages of Duidui and Pichahila on the south coast of the island practise a typical form of shifting cultivation with long fallow of about 20 years. The available land is quite sufficient since 10 to 15 year fallow periods would suffice to replenish the soil. In the Weather Coast region, 75 per cent of the villages have less than 100 individuals and only three villages have more than 200 inhabitants. The principal crops are yam and taro, followed by cassava, bananas and other tree crops, and sugar cane, with taro providing the most consistent food supply (Chapman 1970: 29, 37). In the 1960's, the garden plots were located as far as one and a half hours' walk from the villages, which is a good indicator of prevalent settlement concentration.

The inland village of Pichahila had a *de jure* population of 110 resident in 18 households, which by local standards is a large village for a bush or non-coastal location. In the old days, these islanders practised double residence with the main villages situated either on ridgetops or on the river banks. The small impermanent hamlets were located in the periphery. This hamlet-village system was called 'two places' by the people themselves. As with other tropical cultivators in forested areas, the life span of dwellings was short, ranging between a typical span of five to eight years (Chapman 1970:62). The system is highly reminiscent of the Iban long-house—*dampa* system.

> This persistent pattern of dual-residence, with its ebb and flow between large settlement and small, was intimately related to a

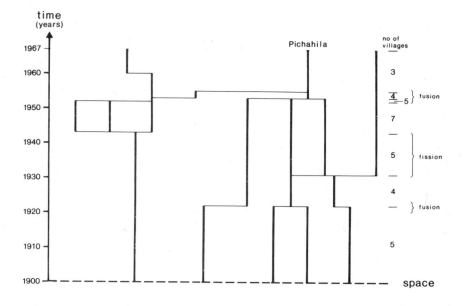

FIGURE 5:14 The fission and fusion of villages associated with the main village of Pichahila during a seventy year period. Periods of fusion and fission succeed one another, the period beginning with five villages and ending with three. At most there were seven villages. (Unlike individual-paths, village paths are divisible, of course, symbolizing as they do, many individuals. The graph is adapted from Chapman 1970.)

> type of shifting cultivation for which the period of bush fallow
> was lengthy ... and the consequent difficulty of securing garden
> land within easy walking distance to the /village/.
> (Chapman 1970:62)

This is the backcloth against which the imposition of a settlement policy of enforced fusion must be gauged. The case history of the main village of Pichahila and nearby villages is depicted in Figure 5:14, although the dispersed hamlets are not indicated for lack of data. The role of these hamlets can be expected to increase during periods of consolidation and fusion of the main villages, however.

The first period of consolidation of main villages enforced by the District Officer is 1916 to 1924. The reason behind these measures was to make it easier administratively to control the people, to reduce the problems of proselytising a dispersed bush population, and administrative ambitions to achieve 'ordered communities'. The selection of sites for villages was also guided by a desire to 'ease the access for administration patrols' by locations closer to water and intermittent shipping, which was the only travel alternative to walking.

This initial consolidation movement was largely nullified by sub-sequent fission, both for ecological reasons and to alleviate epidemic disease. In the early 1950's another era of more 'successful' nuclea-tion began, and the seven villages of the 1940's were merged into three.

The story points to the struggle to reduce time expenditure for travel which took place between the administrative patrols, who wanted to boost their own capacity to cover as large a population in as short a time as possible, on the one hand, and the ordinary villagers on the other, who did not want to waste time and effort commuting on a daily basis. A time cost of immense proportions was dumped onto the localites in order to save the time of outsiders, administrators, policemen and missionaries, or at best to facilitate introduction of schools or medical services. (Cf the case of enforced fusion in Tanzania in Chapter 6.)

Christiansen (1975) gives a similar account for another island in the British Solomon Protectorate, namely Bellona or Mungiki island (located north west of Rennell island and south of Guadalcanal). In this case it was mainly the introduction of Christianity and churches which inspired the fusion of settlements. Some 90 homesteads and hamlets in 1938 had merged into about 33 settlements by 1943, the largest having around 40 inhabitants while most of them were smaller, each having a church. By 1962 these had become consoli-dated into seven villages plus a few scattered homesteads, all concen-trated along the central road which formed the artery of the island. The new villages contained about 50 individuals with two larger ones of about 100 and 250 individuals respectively.

The time allocation changes manifested themselves as follows:

> Traditionally distances to gardenland were short, often people moved temporarily when distant gardenlands were cultivated, primarily to reduce the time spent in walking. This is surprising, especially since walking distances were normally small, seldom exceeding 500 m... The intense church activities /after 1950/ im-posed heavy demands on normal daily work patterns...
> ... in 1965-66 /worshipping activities/ interfered little with sub-sistence work: usually everyone joined in the morning and eve-ning prayers in church... Here it is only worth noticing that estab-lishing of a large village ... really meant an extra expenditure of work ... In the peak periods of work ... the extra walking has re-duced effective daily work hours considerably... (Christiansen 1975:151-3)

In some parts of the island, otherwise fertile land is also being aban-doned as a result of being too remote, according to some evidence.

These cases clearly indicate the contradiction between different activities in terms of time allocation and settlement. A structure of settlement which is favourable to some activities or to some categories of individuals may be less advantageous to other activities and population categories. Enforced settlement and nucleation is one process which can have a strong impact on the activity system of a local population, leading to intensification and other structural changes to which we shall return in the next chapter.

SHIFTING CULTIVATION SYSTEMS: SOME TENTATIVE CONCLUSIONS

The population, activity and settlement sub-systems of a society- and habitat are interlocked and structured in a number of ways, which up till now have received all too little theoretical attention in either social anthropology, human geography or allied disciplines. In the present chapter, we have tried to lay a foundation for coming to grips with some of the important structural problems involved, but also for dealing with more land use intensive cultivation systems, such as the short fallow (medium intensive) systems of next chapter.

In his excellent description of the Majangir, Stauder (1971) wrote that his topic was the 'social aspects of land use' or 'the spatial aspect of their society', and yet he feels strongly the inadequacy of similar designations:

> /The monograph on the Majangir/ deals, first, with Majang *ecology*, or 'adaptation to the *environmental context*'. It goes on to deal with aspects of Majang economy, and of Majang *demography*. But none of these words, and no simple phrase I know, adequately summarizes the *unity of content* of this study. Yet *shifting cultivation, domestic organization, settlement patterns* and *local grouping* are very much interrelated phenomena — and they can be treated naturally as a whole. (Stauder 1971:8)
> /My italics./

It is proposed here that the time-geographic models of shifting cultivation society and habitat do bring out the *unity of content* between the sub-systems mentioned by Stauder. On the basis of a time-space structural approach, it is not only possible to build up a family of sub-models dealing with demography, person-to-person interaction, daily and seasonal activity systems, settlement, and resource utilization and allocation, but also to integrate these different sub-models into a larger, composite formation. It is also possible to deal with different forms of shifting cultivation ecotechnology as variants on the general theme of swidden agriculture, which in turn can be looked upon as a variant of general agricultural ecotechnology. In the space of this chapter, we admittedly cannot do justice to the many kinds of shifting cultivation societies described in the literature; the chapter should rather be regarded as a mere beginning and much fruitful work remains to be done.

In the next chapter, we shall look into how shifting cultivation systems can be intensified and how they undergo structural change with intensification. It would have been more satisfactory to look into this from an innovation process viewpoint, i.e. taking concrete cases of how various innovations and changes have resulted in more intensive cultivation systems. However, these innovation theoretic aspects will have to be left aside in this study. Instead, a number of agrarian systems at different levels of land use intensity will be examined in time-space structural terms.

6 SHORT FALLOW CULTIVATION
Reinterpreting the structure of
local and regional intensification

ON THE ECOTECHNOLOGY OF SHORT FALLOW (MEDIUM INTENSIVE) CULTIVATION SYSTEMS

Medium intensity cultivation systems are found in many forms in tropical and temperate areas of the world, but here we will only call attention to a few of the structural features of these systems in this limited chapter with a bias to the tropics.

With the reduction of fallows and the extension of the cultivation period, which was Boserup's main definition of intensification, new problems of fertility maintenance have appeared. With any transition from forest fallow to bush or grass fallow, and hence a shortened period of regeneration, there would be less vegetation to burn or use for mulching, so this source of soil enrichment is diminished. But if the fallow period is so short as to allow only the regrowth of grasses, the land is not very readily prepared by swiddening either, since grass roots survive or may even be stimulated by such treatment. To reduce competition from grasses and other weeds, the land has to be dug up by some technique, such as hoeing and mounding. Mounding may be used in pure long-fallow systems as well, but becomes particularly necessary when fallows have to be kept short.

It is also important to consider that different crops have specific rates of soil depletion and vary a great deal in their nutritional demands. Cereals, for example, tend to be more demanding than root crops or tree crops. But just as nomadic pastoralists exercise choice of technology by selecting certain animals in given situations, so most cultivators have knowledge of how to grow a wide variety of plants and can choose certain combinations that are more viable in defined contexts than others. The carrying capacity of land is to a considerable extent a function of similar choices. The kitchen gardens so common throughout the world are often the locus of different experiments with cultigens as well as a crop-bank containing a greater variety of plants than the other kinds of fields or gardens used. With a change of external circumstances, subsidiary crops in kitchen gardens sometimes attain the status of staples.

Choice of technique in the form of combinations of animal and plant species is no less important for fertility maintenance or even fertility improvement. In some systems, such as the agro-pastoral ecotechnologies in East Africa, agriculture is practised on a basis parallel to pastoralism, but largely without the mutually supporting relationship typical of 'mixed farming', at least as far as soil fertility goes. There is only complementarity with respect to human consumption of products from both sectors.

But agriculture and livestock tending can also be combined with advantage in the actual process of cultivation. One direct form of linkage is the provision of manure from animals such as cattle, goats or pigs. Sometimes this is achieved by pastoral nomads such as the Fulani of Nigeria, who pass through the agricultural areas of other peoples at some seasons and are invited to graze their animals on residues in the fields in order to leave manure there. In other instances, livestock is owned within the same community of cultivators. Especially when animals are stall fed for a season or more continuously, manure accumulates in restricted areas and can be carted or carried to adjacent fields. Prior to the diffusion of artificial fertilizer, manure was in many systems essential to maintain a given level of intensity of agricultural production. In New Guinea, where a combination of tuber cultivation and pig husbandry is widely practised, the pigs definitely contribute to the fertilization of current and fallow gardens by breaking up the soil when browsing as well as when dropping manure, although this system is hardly mixed farming in the conventional sense.

As humans increasingly depend on animal manure for crop fertilization, so animals depend more on cultivated and delivered fodder: the animals have to be fed rather than tended while feeding themselves. For instance, horses, oxen and bullocks used to pull ploughs must be fed on grain and not just natural grasses. In New Guinea, the pig population often competes with the human population for cultivated tubers, a feature that tends to increase with the intensification or 'involution' of cultivation systems there. This increasing reliance of pigs on tubers partly explains the shift away from yams and taro to sweet potatoes, since the latter is more palatable to both people and pigs.

Finally the role of animals for traction and transport in mixed farming systems cannot be underestimated. Animals harnessed to the plough serve to increase the level of production by augmenting labour power (and sometimes saving labour time) in tillage, and draught animals can also be used to transport manure to fields on a scale that would be prohibitive if porterage was used. The services rendered by animals thus compensate for the extra food needed for their maintenance.

With the reduction of perennial fallows and the extension of cultivation periods, *new operations or activities* become necessary for preparing the land, maintaining yields and catering for domestic animals. Medium intensity cultivation systems often require more careful weeding than long-fallow systems do, and weeding is time consuming as is the application of manure and the feeding of animals the year around. The fact that the conspicuous activity of clearing trees is less prominent in short-fallow cultivation is amply counterbalanced by other time-consuming activities.

The extra time demand associated with more intensive cultivation systems is the main factor behind Boserup's general thesis that productivity of labour time has a strong tendency to decrease with the shortening of fallows (her definition of intensification). Although there is a great deal of evidence in favour of such a proposition, contradictory evidence is also available, and her thesis is somewhat oversimplified (cf i.a. Grigg 1979 for a recent critique). Many issues of theoretical and methodological nature remain to be sorted out, not only with respect to human time inputs in agriculture but in regard to time allocation and agricultural intensification in general. Time resources in agriculture will be particularly dealt with in Chapter 9.

As in many other chapters in this book, the entire field to cover is vast, and it is impossible to do justice to the many forms and combinations of cultivation systems, their level of intensity and ecotechnological features. However, we shall try to illustrate a number of general mechanisms related to intensification, rural settlement and land use, etc. again with the objective of rethinking some current ideas and placing them in a sharper focus.

Specifying the structural scale of analysis

Processes of intensification are operative at different structural levels, *scales* or *levels of aggregation*, such as the region, the local community or the household-cum-farmstead. While some generalizations about intensification in the occupation of land (space-time) are valid for several levels, others only apply to the local community or village level. For example, Boserup's (1965) theory of intensification dealt with population and agricultural systems at an average regional level and many structural features at the local or household level were therefore smoothed over (cf Chapter 1). Wilkinson (1973) addressed himself to problems at the same scale, while Harris (1977) was more versatile, covering all levels. Sahlins (1972) chiefly considered problems of the mobilization of human time and land resources at the domestic unit level, although he related this level to that of the local community in his discussion of local groups, carrying capacity, village size, fusion-fission, and so on. But rather scant attention was bestowed on the farmstead as a local unit. None of the above mentioned scholars were very aware of the spatial dimensions of settlement, and they did not relate intensification to any multi-level rural settlement theory, such as that of Chisholm (1962). On the other hand, many of the long-term dynamic aspects of land use characteristic of Boserup's approach were absent in that of Chisholm. Brookfield (1968, 1973, with Hart 1971) did combine intensification theoretic aspects with a multi-level spatial theory, but was less explicit on the topic of human time resources and their allocation than we aim to be in this study.

In this chapter we will begin at the regional level and work our way down to the local community and farmstead levels. However, throughout this entire discourse, a conscious attempt is made to interrelate the levels from the region down to the single individual in a consistent manner. This is an essential prerequisite to building up a more general multi-level theory of intensification in the use of time and space resources.

INTENSIFICATION AT THE REGIONAL LEVEL

By intensive cultivation economists conventionally mean the input of comparatively large amounts of capital and labour to relatively small amounts of agricultural land. A distinction is also made between intensity and efficiency, as pointed out by Brookfield and Hart:

> Rigorously defined, intensity is a measure of the addition of inputs up to — and beyond — the margin where application of further inputs will not increase total productivity. In agriculture, intensity is measured by the use of inputs against constant land area: addition of inputs of land without corresponding increase in other inputs is a move away from intensive towards extensive operation. Yield /output/ is not significant in the measurement of intensity, but is involved in the measurement of efficiency: the most efficient system is that in which the average productivity per unit of input is maximized. (1971:90) *

---

*   We return to the problem of intensity versus productivity in Chapter 9.

Boserup was displeased with similar classical definitions of intensity with respect to land (use), since they cover 'only the use of additional labour /and other inputs/ per hectare of cropped area, while the change to more frequent cropping of a given area is not regarded as a kind of intensification' (Boserup 1965:43).   She defined intensification in agriculture in a new and different way,

> namely as the gradual change towards patterns of land use which make it possible to crop a given area more frequently than before. To redefine intensification in this way is almost tantamount to pointing out that the scope for additional food production in response to population growth is larger than usually assumed. (Boserup 1965:43)

The latter definition adds another dimension to intensification which is based on a dynamic conceptualization of land use, and in essence it is a space-time notion (reinterpreted time-geographically as in Figure 5:4).   Increased frequency of cropping or temporal intensity is also a form of expansion in space-time terms but along the temporal dimension. In the tropical areas to which Boserup directs her main attention, the immediate response to population growth (and increased population density) was increased demand for land, but rather than this taking the form of spatial expansion of cultivated area, it was in the reduction of fallows that new space was gained.

As mentioned in Chapter 1, a similar view of land use intensity and population density at the regional level has also been put forward by some geographers. Grove, writing on Nigeria, observed that:

> nearly everywhere there is a broad correlation between population density and intensity of agriculture. Certain systems of shifting cultivation such as those followed by the Bemba of Northern Rhodesia /Zambia/ rely on lengthy resting /fallow/ periods between a few years of cropping and are incapable of supporting more than a few people, say 5 to 10 to the square mile ... At the other extreme, there are large areas in West Africa and smaller ones in the East where numbers exceed 300 to the square mile and these densities are generally associated with much more efficient use of the land, giving higher production per unit area. The greater number of people both require and allow farming to be more intensive. (Grove 1961:115)

For a given population in a given region, the demand for agricultural land may rise for a number of reasons. Boserup (1965) took population growth as the major reason and it is certainly an important one occurring frequently historically and geographically. But it is far from the only one. The Tiv of Nigeria is but one interesting compounded case both in terms of requirements and responses:

> Tiv say that population increase is natural, and cite their genealogies ... to prove it. Although Tiv are in fact increasing, this is only one of several factors leading to the increased demand for land... One of the most important additional factors is that Tiv have steadily increased production of beniseed for the 'international market'... When beniseed became a cash crop, its acreage increased manifold. (Bohannan and Bohannan 1968:98)

Another *effective* rise in demand for land occurs if part of a region is occupied by invading neighbours. The remaining land base is thus reduced for the original population, while their land requirements are the same as before. The introduction of plantations producing for export in a region has the same effect, as does the imposition of a more extensive form of land use (such as cattle ranching) by colonizers. When the earnings from cash crops are not used to import food replacements, cash cropping is generally a land (space-time) demanding innovation (unless it can be fitted comfortably into the fallow period without causing a decline in fertility in the subsequent cultivation period). Rather than uniformly attributing intensification in land use to the effects of population pressure, it is a more correct generalization to regard intensification as one response to increased *effective demand pressure* on land stemming from a multitude of sources.

But responses can vary and need not always amount to intensification. One response is to 'export' people or labour-time (emigration, seasonal migration or commuting). Another is to import food or other land-saving commodities. Other responses may be land acquisition by means of conquest, purchase or land reclamation, or increasing the productivity (yields) of the existing land base by application of artificial fertilizer or more scientific management of the land.

Reclamation of lands through better drainage of swamps and river basins or through terracing so that previously uncropped hilly land can be used for cultivation are some of the true forms of expansion of cropped land, while colonization and territorial conquest are forms of expansion by one population (or sub-population) at the cost of another. The Tiv solved their land hunger problems by a combination of spatial expansion and intensification. Their kinship system is a model case of a social sub-system conducive to what anthropologists have called 'predatory expansion'. But since the inception of British rule and *Pax Britannica* in Nigeria, boundaries between many ethnic groups have become more rigid, thereby accentuating the Tiv land shortage. But the Tiv, of course, are only one among numerous cases of similar processes in operation.

Intensification through the reduction of perennial fallows

Simple spatial expansion has the advantage of not requiring any real change in ecotechnology, as for instance when virgin forest is cleared. If spatial expansion cannot relieve demand for land to crop, intensification through the abandonment of long fallow systems is the common alternative. Even the expansionist Tiv were compelled by circumstances to adopt this solution:

> Tiv in MbaGabor said that they had extensively reduced their fallow periods — too much to allow for proper revitalization of the soil. Iyon people said that they had nearly done away with fallow. (Bohannan and Bohannan 1968:63)

When long fallow cultivation systems are in ecological balance, the fertility of the soil is generally restored during the fallow period. If, however, the cultivation period can be extended and/or the fallow period reduced, then additional space-time for cropping is obviously gained. * But the major constraint on such temporal intensification is that of declining fertility. So while encroachment on fallows becomes a near automatic form of adaptation to increased demand pressure on land in lieu of other outlets, short fallow systems are riddled with problems of fertility and yield maintenance and require a considerable modification of long fallow ecotechnology.

Shifting cultivators generally prefer secondary forest for cutting (i.e. forest on land which has been used for burning and cultivation before), rather than climax forest with large trees which are time and energy consuming to take down. When virgin forests are used up, and additional demand pressure has built up, cultivators will sooner or later have to return to forest which has not reached the mature secondary state. Some groups like the Lala of Zambia (Long 1968) have reacted by cutting larger acreages than before in order to get the same amount of vegetative materials to burn. But this strategy rather accelerates the overall shortening of fallow periods.

Yields can be maintained, however, by changes in ecotechnology such as better routines of weeding. More careful weeding may not improve the fertility of the soil as such but it does eliminate competition from non-cultigens, and similar efforts are often accompanied

---

* We are at present discussing the reduction of *perennial* fallows, not *seasonal* fallows. The primary advantage of irrigation systems (discussed in Chapter 7) is that they reduce seasonal fallows and may also serve to expand cultivation into areas previously uncultivated for lack of adequate moisture. But Nature itself may also serve the cause of reducing seasonal fallows in areas of bimodal or more continuous rainfall throughout the year, for instance in equatorial areas such as New Guinea. In reducing seasonal fallows, the problem is not declining fertility but adequate moisture or tolerable temperature ranges.

by the elaboration of previous kinds of weeding tools, for example. More complete tillage and mounding are other kinds of changes that maintain or improve yields. More timely planting, better guarding and new crop mixes or rotation schemes are often part of a larger package of ecotechnological changes that are introduced.

It is not only by shortening of fallows but also by prolongation of cultivation periods for each garden or field that the balance between fallowed and cultivated land is altered in favour of the latter. A similar gain in space-time is often associated with new schemes of crop rotation. Cultivators are usually well aware of the fact that yields may be higher if, for instance, a cereal is sown before a root crop is planted, and some kind of crop rotation is the rule rather than the exception in tropical agricultural systems. Thus, cultivation periods may be extended from two to five years by suitable sequences and mixes of crops, so that the soil is less depleted and weeds are more effectively choked. When the average cultivation periods exceed the fallow periods and when forest can no longer regenerate and only small bushes and grasses grow, then by definition we no longer have a shifting cultivation system but one of medium intensity (cf Figures 5:4 and 6:1).

Grass fallows cannot be cleared by means of swiddening procedures, since this would leave the root systems intact. These either have to be dug up, dried and burned or else removed or dug down to

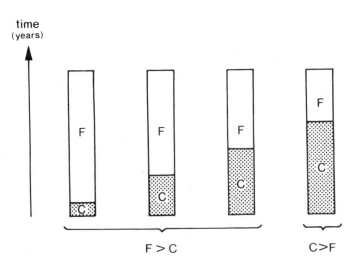

FIGURE 6:1   The relation between cultivation and fallow periods in the land use cycle of a particular parcel. When the cultivation period exceeds the fallow period, the system is no longer of the shifting cultivation type but one of medium intensity.

a depth where they are rendered harmless. The soil must thus be prepared by more intensive hoeing, tilling or ploughing which introduces an additional operation which is hardly necessary in long fallow systems.

It is only when land is cropped on a highly frequent or near continuous basis that tools such as the *plough* become strategic in land preparation. In the more extensive systems there are good technical reasons for not using ploughs, as this would, for one thing, make stumping imperative. Removal of stumps is hard and time consuming work, an investment that is less desirable to undertake if land is plentiful enough for long-fallow cultivation. But to combat grasses, the plough may be a time and energy saving innovation in spite of the extra time inputs that the keeping of draught animals entails. Boserup makes several interesting points on the plough as a means of agricultural intensification:

> The need for a plough is for short-fallow cultivation so compelling that cultivators usually avoid the stage of short fallow if they are unable to use ploughs, owing to the lack of animals or for other reasons. Such cultivators prolong the periods of cultivation under bush fallow up to eight years instead of shortening the period of fallow. By recultivating land year after year they avoid the excessive spreading of wild grasses, and by keeping a relatively long fallow period when the cultivation periods are over, they give the bush a chance to cover the land and thus prevent its becoming too grassy. The result is the type of intensive bush fallow which can be observed for instance in many parts of Africa...
> (Boserup 1965:25)

Parallel cultivation systems and deintensification

Throughout the temperate and tropical areas of the world, parallel systems of cultivation are commonly found to be practised by the same local population. In New Guinea, for instance, the Kapauku carry on a long fallow form of shifting cultivation on mountain slopes, intensive grass-fallow swidden cultivation and 'intensive complex cultivation' in valley bottoms, the latter including ditching, mounding and green manuring (Pospisil 1963). The Raiapu Enga in the New Guinea Central Highlands have two distinct cultivation systems, one involving intensive and continously used gardens with mounding and green manuring (the 'open fields'), the other slash-

and-burn cultivation in what Waddell (1972) calls 'mixed gardens'. Leach (1961) describes the situation in dry zone Ceylon (Sri Lanka) where multi-cropped and bund irrigated *paddy* is cultivated by the same villagers who also work *cheena* gardens by swidden procedures. In Africa, the Gwembe Tonga had intensively used inundated fields in the river valleys of the Zambesi, but in recent years they had also adopted shifting cultivation in nearby woodlands (Scudder 1962). Netting (1968) reports similar developments among the Kofyar of Nigeria. This list could be very long indeed, and we only mention the few cases here which are dealt with elsewhere throughout this study.

From the perspective of Boserup's model, it seems logical that if increased population (or rather demand) pressure on space-time should lead to intensification, population decrease (e.g. through out-migration, epidemics or general decline) or, alternatively, the availability of new land, should swing the pendulum in the direction of 'extensification' or 'deintensification'. Boserup (1965:62-3) was well aware of this possibility and historically, it is anything but an uncommon occurrence. (Cf Brookfield with Hart 1971:122-23 for Melanesian examples.) Boserup's theory is thus far from deterministic with respect to the direction of agricultural evolution.

The increased availability of land by means of colonization, reclamation or the like may facilitate deintensification, while *impaction*, when a given population finds itself in a smaller region or when a large population is shifted into a smaller region through warfare, for instance, works in the direction of intensification. This is what happened historically to the Kofyar of the Nigerian Jos Plateau, when they moved to the plateau in order to be less vulnerable to enemies than in the plains. The greater local land demand which arose fostered a very intensive system of cultivation with manuring as a major means of maintaining soil fertility (Netting 1968). Decades later with the introduction of *Pax Britannica,* some land which had previously been unused for fear of enemy raids could be opened up for cultivation. The Kofyar then quickly adopted shifting cultivation parallel to their old intensive system: (Netting 1968:113, 210)

> there is some evidence that the demographic pressures have substantially lessened in the last fifty years. Not only was the hill Kofyar population apparently much higher in the past, but bush fallows were strictly limited because of the possibilities of raids and ambushes... Hill villages have joined in the migration to bush lands in direct proportion to their proximity to the plain...
> Most Kofyar have chosen to maintain both /types of/ farms... the economic trees and domestic animals of the home territory cannot readily be transplanted to the migrant locations...

In the Zambesi valley, prior to the building of the Kariba Dam, the Gwembe Tonga similarly moved away from intensive valley cultivation towards an increased proportion of extensive swidden cultivation (Scudder 1962). Traditionally, several factors had maintained valley impaction and discouraged swidden agriculture. Large game such as elephants and monkeys had made guarding in woodland areas a difficult task and a vicious circle of frequent famines and raiding (partly for food) had made denser settlement in valley floor areas advantageous. Here local intensity could be higher and inundation of gardens gave two harvests per annum, at the same time as there was no shortage of domestic water in the dry season. This compact form of settlement also allowed a better social life with constant visiting back and forth between kin, a life style which was greatly appreciated.

With pacification, the defence factor diminished in importance at the same time as modern guns reduced the game populations in the region to a considerable extent. Population growth also stimulated migration to the less populous Plateau in response to declining yields and overcultivation. It was 'in the Upper River region where land pressure was greatest, /that/ the *temwa* /swidden agriculture/ developed first...' (Scudder 1962:60). This innovation process was rapid:

> During the past ten years, the importance of *temwa* in the Middle River region has increased at a spectacular rate. In 1956-7, *temwa* harvests exceeded those of other garden types. (Scudder 1962:43)

There were factors other than sheer demand pressure on land which lay behind the speed of shifting cultivation as an innovation process. Swiddens opened up new possibilities in an otherwise tight land tenure and social mobility situation. The younger ambitious men could now gain access to land which had previously been a limited resource turned over slowly at major life-cycle events such as birth, marriage, divorce or death. Few occasions for land transfer were generated in the traditional structure.

Not only were productive resources made available but also a kind of cultivation that probably required less time input per output unit. A 'reversion' towards a more extensive cultivation system thus serves both as a time saving and capacity releasing innovation, albeit this newly gained capacity may be fed back into essentially traditional projects.

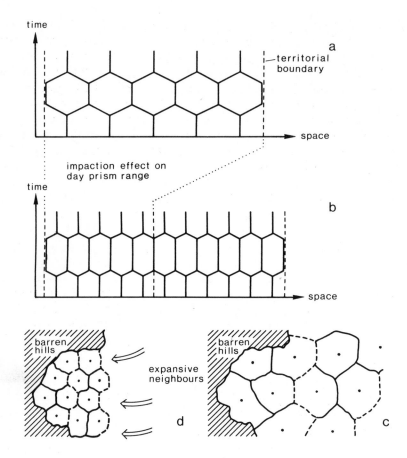

FIGURE 6:2   A hypothetical case of an impaction process and its effects on territorial size and day-prism range. Cases *a* and *c* show a population and its sections and sub-units prior to impaction. As the result of expansive neighbours on the one side and barren or inhospitable hill country on the other, both segments of the affected population are pushed into a much smaller region, with consequent effects on settlement and level of land use intensity. Cases *b* and *d* show the situation subsequent to impaction.

(A similar pattern of development may also occur when population growth leads to the gradual segmentation of villages and territories are subdivided into smaller units, although this is a different kind of impaction. Cf also Figure 5:7.)

INTENSIFICATION AT THE LOCAL LEVEL

In practice, increased demand for space-time at the regional level is an *aggregate outcome* of increased demand at the *local* level, i.e. in the various settlements scattered throughout the region. By the same token, there is no growth of population at the regional level which is not a reflection of corresponding processes at the local scale. However, local intensification may emerge without any population growth at the regional level at all, since local settlements may also grow for reasons of fusion of sub-populations and redistribution of the population in space. Moreover, there are many causes of a rising local demand for space-time other than an increase in population numbers. It is thus crucial to make structural relations at the local level explicit, if we are to grasp the causes, mechanisms and consequences of intensification in the use of space-time and human time resources (Figure 6:3).

Pseudo-intensification by means of land saving crops

Increased demand for space-time in the local prism habitats of agriculturalists commonly leads to intensified occupation of land (space-time) as well as intensified use of human time. But increasing space-time requirements can be offset by other forms of adaptation, for example a greater reliance on other cultigens, notably crops which *yield more* per space-time unit and thereby effectively function as land saving devices. An important group of comparatively high yielding crops are the tubers, such as yams, sweet potatoes, manioc (cassava) and ordinary potatoes. Just as the potato was an innovation having a substantial impact on food production in temperate Europe, manioc is a root crop which has diffused to an impressive extent in the tropics, especially perhaps in Africa (cf Jones 1957, 1959, Mabogunje 1972; for case studies, Long 1968 and White 1959). In Central Highland New Guinea the spread of sweet potatoes was of such fundamental importance that Watson (1965a, 1965b, 1967, 1977; cf Brookfield with Hart 1971:83-5) argued that it was an Ipomoean Revolution (*Ipomoea batatas* being the Latin name for sweet potato).

Cropping systems in which tubers form the staples tend to raise the carrying capacity of the local prism habitat and thereby permit greater local population densities. In spite of this, some scholars, particularly those with experience from New Guinea (e.g. Waddell 1972) are reluctant to regard the introduction of tubers, or their amplified role in a cultivation system as intensification in the strict sense (cf Brookfield's definition above). For one thing, intensity

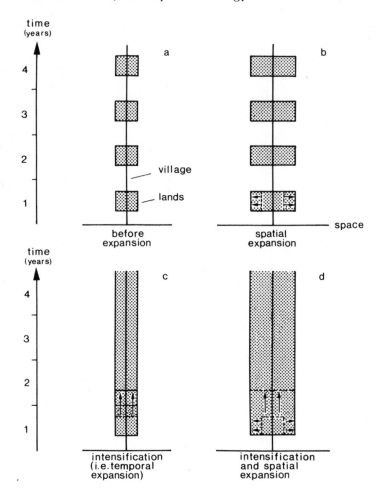

FIGURE 6:3   Different forms of expansion in terms of the occupation of space-time. In temperate zones, expansion by means of cultivating more land is the most common form, since seasonal fallows are difficult to eliminate (case b). In the tropics, expansion may take the form of 'colonization' of the dry season, generally by means of irrigation (case c). Sometimes both methods are combined. These forms of expansion are responses to increased pressures of demand.

defines a relation between inputs rather than between input and output, the latter relation being a matter of productivity or efficiency. For another, tubers are not only land (space-time) saving compared to crops such as cereals, but their cultivation is commonly also time saving. This does not correspond very well to Boserup's tenet that the intensification process in the 'typical case'... 'reduces output per man-hour, or, ... that agricultural employment increases at a higher rate than agricultural output...' (1965:43), i.e. that although the productivity of land may rise, that of labour-time declines. So there are sound reasons for agreeing with Waddell's conclusion that relations are more complex than this, given that many medium intensive cultivation systems where yield per unit of human time is greater than in some of the more extensive systems can be found in Melanesia (quite apart from innate variations in the biotic-physical environment).

On the other hand, a movement away from cereals towards root cropping is freqently accompanied by intensification in Boserup's sense of higher cultivation and lower fallowing frequency. Since tubers generally deplete the soil less rapidly, cultivation periods before resting can be greatly extended, and in many of the intensive African bush-fallow systems acknowledged by Boserup (cf quote on p. 195 above), root crops are often of considerable or decisive importance. Moreover, tubers are in many cases strategic in permiting *local* intensification around large size villages. Here they allow a denser packing of parcels within the acceptable day-prism range, although this type of crop may not be called for in terms of space-time supply at the average regional level. Tuber cultivation may thus pave the way for larger and more stable settlements by raising the carrying capacity of the local prism habitat (cf the case of Lala on p. 182 above). Since a changeover from cereals to tubers otherwise assumes a minimum of technical innovativeness in many cases, it is a common way of adapting to local space-time shortages. * For these mentioned reasons, tubers do contribute to denser settlement and more intensive use of land, and their introduction may thus be referred to as *pseudo-intensification* taking *ad notam* some of the special features of this variant of agricultural intensification.

White describes a case typical for Africa of how tuber cultivation can turn a long-fallow system into a medium intensive one:

In the last 25 years there have been important changes, due to the great decrease in cultivation of bulrush millet... Cassava is planted without a preliminary crop of bulrush millet... replanting is done as soon as a plant has been cleared out completely, and goes on

---

*   Some crop innovations may already be known and grown as subsidiary crops and have thus already diffused to the given locality in a qualitative sense. But still their changed status to a staple crop is an innovation process. In my time-geographic innovation studies (cf Carlstein 1978a and forthcoming), I have referred to this as *innovation by amplification* as opposed to *innovation by diffusion*, where in the latter case I see the innovation (e.g. a new crop) as a new and previously inexperienced culture trait.

continuously throughout the year ... year after year ... In place of various degrees of shifting cultivation there has thus emerged a semi-stable type of agriculture based on cassava and intermittent resting, groundnuts sometimes being alternated in the resting periods. (White 1959:24)

Cassava growing leads to greater *stability of villages* for several reasons... Cassava is much *higher yielding* in terms of bags to the acre than grain crops grown under customary methods, so that *less land* is cultivated, and shifts in villages due to land are not a feature of Luvale life, in contrast to the Lala, southern chitemene /i.e. swidden/ cultivators, who quote lack of trees for cutting as the commonest reason for the mobility of villages... Cassava is clearly not a crop which exhausts the soil... A cassava garden is commonly cultivated for *ten years without resting*... (1959:19) /My italics/

Taking another medium intensive African system with a similar history, that of the Tiv in Nigeria, the typical effects are again that fallows can be shortened (cultivation periods extended) and that the average size of fields-gardens (per household) can be reduced (Bohannan and Bohannan 1968:52-62). In this case, cassava expanded at the cost of millet and guinea corn in land hungry areas, where the cultivation systems became medium intensive and space-time allotments per capita were adjusted downwards, mainly in response to population growth and increased cash cropping. (A few of the above features are epitomized in Figures 6:4 and 6:5).

Boserup was not well enough aware of the role tubers played in her general intensification theme, although she does observe that:

... new crops were eagerly accepted both when the Portugese brought new crops to Africa and Asia and much more recently. Crops like maize, *cassava* and *potatoes* have spread very rapidly in primitive communities in recent decades also where they were unaided by government propaganda. Commercial crops have also been accepted readily when production was profitable. (Boserup 1965:68)/My italics/

She further observes that 'high yielding root crops' become more important in areas of high population densities without quite realizing how this observation ties in with her intensification model.

... food consumption may change to *cheaper* types of food... There is evidence from *densely* populated countries in Asia and elsewhere that agricultural labourers and members of other low income groups change their consumption from the more *expen-*

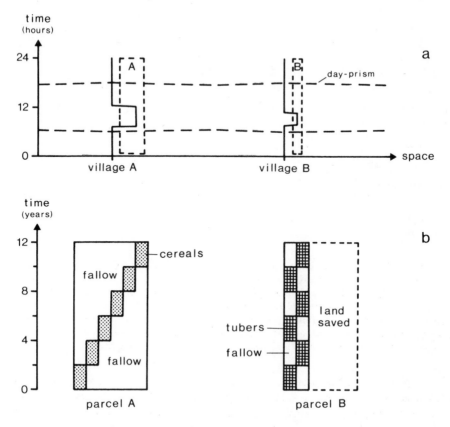

FIGURE 6:4   The graph shows the effects of a shift from cereals to tubers as staple crop. Since cereals generally require more land per unit of output as well as longer periods of fallow (parcel A), a shift to tubers both lowers land (space -time) requirements and permits a denser packing of parcels (parcel B). Land is thus saved in a process of 'pseudo-intensification'. Low land requirements also make distances to fields shorter (village B) due to the denser packing of parcels in space-time. Tubers thus tend to yield more, require less land and take relatively less time to cultivate. By permitting shorter village-field travel distances, tubers are thus doubly time-saving in relation to cereals. Lower time and land requirements may in many cases facilitate larger village size and a higher level of sharing of the local prism habitat. (Only one sample individual living in the village and working in the fields is shown in the top graph.)

sive cereals like wheat and rice and the types of cereals which can grow on *poor land* and to *high-yielding root crops* like tapioca /cassava/ and potatoes when *population becomes more dense* and food prices increase or wages decline. (1965:109) /My italics/

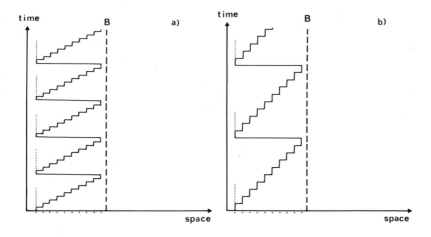

FIGURE 6:5   Following Figure 5:5, the effect of tuber cultivation (b) as opposed to cereal cultivation (a) is indicated by the prolonged residence period that is made possible by an increased reliance on tubers. Extended cultivation periods at each site reduces both the total need for sites and the necessity to encroach on minimum fallow periods (indicated by dotted lines for one sample site). The general effect at the territorial level is also a saving of space-time, so that with tubers territories can be smaller. (Cf also Figure 5:6 versus 5:7.)

But there are structural reasons *why* root crops have often spread at the cost of cereals, or why a high-yielding cereal such as maize has partly replaced less yielding cereals, although both maize and tubers are inferior for many purposes of consumption and are relatively deficient in proteins. They cannot be used for brewing beer as millet can, for instance. It is because they place relatively low demands on the local space-time budget that these crops spread, apart from the fact that tubers also allow more continuous cultivation and harvesting in the annual cycle, thereby bridging the seasonal hunger gap common in some African systems, for instance.

These dimensions of space-time packing within local prism habitats are no less relevant to changes in settlement size and other settlement reforms. Many countries have experienced programmes of resettlement and/or villagization both under colonial and independent governments. The settlement changes in some regions of Zaire (former Belgian Congo) constitute an interesting example. This process has been documented by Gullstrand.* In the precolonial period,

_____

* The regions studied by Gullstrand were mainly Bandundu and Equateur. Unfortunately, the untimely death of Gullstrand prevented him from finishing his doctoral thesis in Human Geography at Uppsala. Hopefully it will be published posthumoulsly, since it was near completion at the time of his tragic decease.

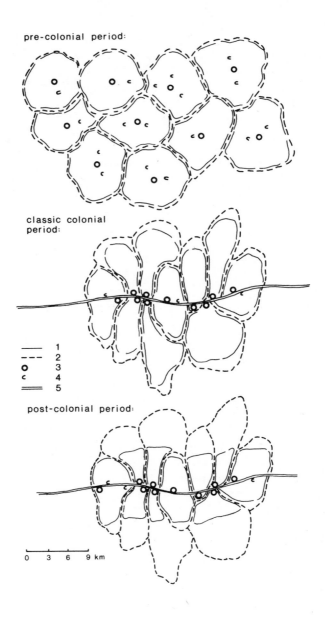

pre-colonial period:

classic colonial period:

1
2
3
4
5

post-colonial period:

0   3   6   9 km

FIGURE 6:6   Settlement change in parts of Zaire at different stages of external interference in local conditions. The distortion of territories and *terroirs* when villages were located along the roads during the classic colonial period paved the way for a later pseudo-intensification process and corresponding adaptations of action fields, according to a graph made by Gullstrand. (Legend: 1: limits of *terroirs*, 2: limits of territories, 3: villages, 4: site of abandoned villages, and 5: motorable roads.)

settlements were fairly evenly dispersed and shifting cultivation prevailed. In the classic colonial period, resettlement along motorable roads was imposed for reasons of political control, taxation, crop export and trade, regional integration and other similar interests. A distortion of territories and *terroirs* (action fields around villages roughly corresponding to day-prism ranges) resulted, with villages becoming increasingly permanent in their location. By the post-colonial stage, the *terroirs* had become much more compact and land use was more intensive, largely thanks to the wide adoption of root crops (cassava in particular). The use of tubers as staples at the cost of cereals can thus be one form of adaptation to the increased *local* space-time demands brought about by the concentration or fusion of settlements (Figure 6:6).

The Tanzanian case of enforced villagization

The previously mentioned cases of enforced resettlement and fusion of dispersed settlements seem insignificant when compared to the ambitious programmes of reshuffling the population which some of the newly independent governments in Africa and Asia have embarked upon. In Tanzania, which is an outstanding case, the overriding objectives behind these incisive measures were to promote rural development and raise the quality of rural life. In practice, however, the gap between ambition and fullfillment sometimes attained yawning proportions, given the shortage of qualified personnel and other resources for implementation combined with administrative ineptitude and political impatience (cf Hydén 1972, Loficie 1978 and Mascarenas 1977). Although this study does not directly relate to development problems in the Third World,* the Tanzania case is important in illustrating some of the general aspects of settlement, time resources, changes in activity structure and intensification. Large portions of the Tanzanian population practised variants of shifting cultivation with fairly long fallow periods prior to resettle-

---

* The present study is not chiefly indended to deal with Third World development problems, although what is presented here is not without implications for that theme. The author hopes to publish some materials from his previous project on 'Time Allocation, Innovation and Agrarian Change' (cf Carlstein 1978a, 1978 b). The Tanzanian materials lend themselves very well to a rewarding analysis of the mobilization of human time resources in relation to settlement policy, rural development and political participation, but only very few aspects of this complex can be brought up here (cf also the end of this chapter).

ment. These local cultivation systems had to undergo substantial modifications and alterations as a result of increased local population size and sharing of day-prism habitats.

Let us initially confine our discussion to a few aspects of *settlement size*. To get at the dimensions of this experiment in rural mobilization, Tanzania in 1970 had a population of 500,000 living in settlements of village size, the rest residing in homesteads and hamlets below some 200 individuals. The average size of the existing villages was in the order of 270 inhabitants, which by Asian standards is very small. By 1975, 9 million people were living in 'planned villages'. This rate of population resettlement makes the rate of urbanization in England or the rest of Europe seem minute compared to the rate of population relocated per time unit in Tanzania during only five years (cf Table 6:1). By around 1975 the average village size was 1260 indviduals, while in some regions (Kigoma, Mara, Mwanza and Shinyanga) it was between 2200 and 2600 persons, the latter being regarded as somewhat above the ideal size, however. In Dodoma, which is a semi-arid region with a particularly fragile environment, some villages were even larger than this.

It is beyond our scope here to discuss the broad stragegy of *ujamaa* or 'planned villages'. But a few general observations can be made about similar programmes of resettlement. Local settlement size has numerous consequences. Some were outlined in Chapter 5 for 'typical' shifting cultivators, where some basic links between settlement structure, activity and time allocation were stressed. If agriculturalists practising long-fallow cultivation are moved together into villages of more than a few hundred, let alone more than a thousand people, competition for land (space-time) within the local day-prism habitat becomes immense, and all the more so if they also have livestock and are expected to expand cash crop cultivation. Yet, many of the Tanzanian regions were dominated by variants of swidden agriculture of the long or short fallow variety. Even in regions with climates conducive to more intensive cultivation (such as the highland regions or those close to the great lakes, e.g. Arusha, Kilimanjaro, Mbeya or West Lake), swiddens were often found parallel with more intensive forms.

When shifting cultivators change sites of their own accord, much careful deliberation goes into the choice of new sites, since their location affects distances to potential fields, to water sources, to roads, to relatives in nearby hamlets, to grazing, and to firewood, for instance. Likewise some locations are less infested with malaria, tse-tse flies or other pests or have other advantages of terrain. Some sites were chosen because of their exceptional fertility (e.g. at a given locus on the catena), thereby allowing nearby gardens to be more continuously cultivated. In fact, traditional settlement systems are

TABLE 6:1

THE INCREASE IN AVERAGE VILLAGE SIZE IN TANZANIA 1970-1975

REGION	Average Number of People per Village						TOTAL POPULAT.
	1970	1971	1972	1973	1974	1975	1975
Arusha	208	238	215	212	231	1532	272,765
Dar es Salaam	—	—	—	—	188	800	40,000
Dodoma	352	973	1339	1128	1427	1634	630,858
Iringa	33	332	329	370	395	1876	804,858
Kigoma	197	206	887	887	906	2343	452,285
Kilimanjaro	300	328	209	206	227	281	4,508
Lindi	247	343	280	287	646	1196	386,664
Mara	487	339	339	339	2105	2232	626,687
Mbeya	362	131	138	145	161	1000	934,800
Morogoro	316	93	206	167	266	521	123,256
Mtwara	372	497	406	423	508	863	667,413
Mwanza	164	147	152	176	267	2371	1,437,095
Pwani	863	773	603	614	702	528	157,641
Rukwa	—	—	—	—	207	900	346,800
Ruvuma	75	144	144	175	349	1201	378,511
Shinyanga	129	82	124	112	138	2540	940,335
Singida	429	255	226	226	447	960	247,834
Tabora	321	227	170	168	184	1708	553,770
Tanga	208	272	318	318	265	348	105,184
West Lake	255	206	202	156	207	377	26,432
TANZANIA	272	345	357	360	511	1260	9,140,299
NUMBER OF VILLAGES	1956	4464	5556	5628	5008	6944	—
POP. TOTAL IN VILLAGES *in thousands*	531	1,545	1,980	2,028	2,560	9,140	9,140

(Source: *Maendeleo Ya Vijiji*, Prime Minister's Office, June 1975.)

often very well adapted to all kinds of local conditions, and the pros and cons of alternative locations are carefully weighed against each other, including their symbolic meanings (some sites being unlucky, haunted or blessed, etc.) The siting of the village affects the relative location to other stations used and hence the *travel time* associated with various daily activities.

The resettlement scheme in Tanzania was hastily implemented and numerous villages were located on arbitrary sites without due consideration of local conditions. Such implementation would lead to a number of built-in costs, one set of such costs being reduced efficiency of travel and consequent waste of human time. On the other hand, the villagization programme assumed the introduction of a host of other time demanding innovations, while some of those designed to save time and effort often failed (cf p. 250-57 below).

The consequences of the drastic introduction of villages were not slow to appear. The country was thrown into a food crisis, conveniently blamed on drought, but mainly attributable to a combination of reduced farming efficiency and non-cooperation by disappointed cultivators (Loficie 1978). The new local settlement structure thus made intensification vital, but the means of such rapid intensification were partly or largely lacking.

Let us examine what the possibilities for adapting to the new situation were: A quick and simple option would be to use land farther and farther away as nearby lands become exhausted (this is widely employed by shifting cultivators traditionally, cf Figure 5:8). *More travel* is generally feasible but obviously time demanding and far from comfortable in the given climate and terrain. Even though travel time costs may rise sharply in the daily and annual time-budget, considerable competition for space-time close to the prism centre remains when villages reach a size of one or two thousand individuals. Sooner or later people start building temporary dwellings outside the villages, and *double residence* becomes a fact, although the villagization scheme was intended to bring people together in daily interaction so that they could 'learn from one another'. A suitable compromise might be for the active farmer categories to commute out from the village for several days at a time, leaving children, aged and other household members behind (cf Figure 3:21). The latter solution saves considerable travel time which can be put into actual cultivation instead, but it reduces the scope for daily interaction among different population categories.

A third response could be to adopt a *faster mode of travel*, e.g. to improve transport by means of bicycles. This widens the day-prism habitat by those having access to cycles (Figure 6:7), thereby alleviating the excessive sharing of space-time near the village. Unlike horses which are time-saving in transport but are land demand-

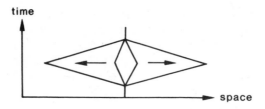

FIGURE 6:7   The widening of prisms when walking is replaced by bicycle riding.

ing since they need fodder, the bicycle requires no special source of energy. On the other hand, in Tanzania bicycles are rarely produced locally, and their adoption brings with it increased reliance on the outside world. Even pressure on local land would increase, since people would have to grow space-time demanding crops to pay for such vehicles, i.e. introduce or grow more cash crops. At the national level, bicycle production has not kept pace with villagization, nor has that of other industrial inputs, such as artificial fertilizer. Moreover, cycles assume good paths, tracks or roads and are less of a time saving innovation in the rainy season when roads are muddy. Hence prisms would only be selectively widened by the introduction of bicycles, both during the year and within the local population.

A fourth possible response may be that of a wider adoption of *high yielding crops*, such as maize, at the cost of lower yielding ones, e.g. millets and sorghum. But the latter crops are more drought resistant and it is the long-term average yield which counts, not just the harvests of propitious years. *Pseudo-intensification* by means of tubers is likely to be a more widespread adaptation, since it is technically simple and implies the saving of both land and time in a situation where many new time demands are imposed as a result of the 'planned village' programme.

A fifth response to enlarged settlement size close at hand is *local intensification* by means of shortened fallow periods. This is a simple and initially even successful response, but one which generally paves the way for problems in only a few years time, when the vegetation cover has not been allowed to regenerate and the fertility of the soil has declined. To stabilize such a system requires a good many changes in cultivation practices. Competition from weeds must be reduced by more scrupulous weeding, the soil must be better prepared and tilled, planting more timely, and so on. Some means of improved fertilization must also be introduced, such as green manuring or the application of animal dung, techniques which all involve changed forms of both time use and land (space-time) use.

Intensification assumes technological changes. These appear more readily in the gradual long-term perspective of decades and generations envisaged by Boserup. In the case of Tanzania, local intensification took place at a fantastic rate. The *de facto* population in a locality may have grown by more than 1000 per cent in a year or two. Under these conditions, local intensification may be anything but a success story. * Such rapid changes towards more frequent and intensive use of nearby lands in many instances leads to land deterioration in the immediate vicinity of a village, especially in view of the joint impact of humans and beasts on soil and vegetation. This has certainly been the case in many places in Tanzania and is a problem that will remain for a long time.

However, there are also ways in which the spatial structure of local intensification can be turned to the advantage of the village population, namely through a pattern of spatial zoning. This is a much neglected topic both in the anthropological and economic versions of intensification theory, but one which ought to be an important ingredient thereof.

SPATIAL ZONES OF INTENSIFICATION

Agriculturalists commonly adapt to excessive sharing of land close to a prism centre and to increased time distances to their fields by using *nearby* land *more intensively* than *distant* land. They arrange land use in zones of decreasing intensity from the village centre, and at this local community level, a discrete spatial pattern of intensification will chrystallize out.

The tendency for distinct zones of land use to emerge concentrically around towns and villages was noticed at an early stage by the German estate manager von Thünen, who applied Ricardo's Economic Rent concept to explain why concentric land use zones appear as a result of differential *transport costs*. His explanation was formulated as an ideal theoretical landscape in what he called 'The Isolated State' (von Thünen 1826). The economic rent of 'a particular piece of land is the return which can be obtained above that which can be got from the land which is at the margin of economic cultivation' (Chisholm 1962:25). When transport costs of getting the commodity to the town are added, it may not pay to cultivate a given crop in a

---

*   Boserup has a chapter on the 'vicious circle of sparse population and primitive techniques', where she points to low population density as an impediment to development. But she does not explore the possible consequences of local concentration and resettlement of low density populations.

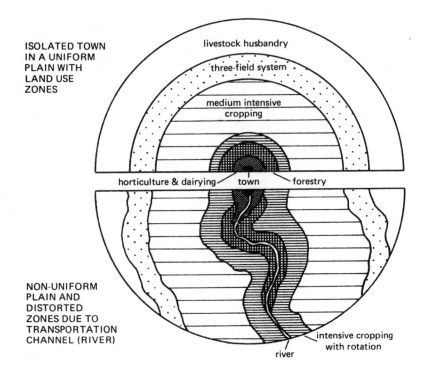

FIGURE 6:8   The concentric zones of land use around a town or city in the 'isolated' state envisaged by the German economist, economic geographer and social reformer Johann Heinrich von Thünen in 1826. Although this idea has been much discussed in the literature on geographical analysis, the fact that each zone away from the city generally meant a decline in land use intensity has perhaps been less emphasized.

particular zone compared to another type of crop. Von Thünen's model was one of marginalist choice by individual farmers selling produce on a monetized market. As was customary in Europe at that time, cross-country transport was by means of horse-drawn carts. Von Thünen's model (Figure 6:8) resulted in several zones centered on the innermost urban zone: the zone closest to town consisting of horticulture and dairying, followed by a silvicultural zone (delivering firewood, a phenomenon peculiar to Germany at the time), and then by a zone of intensive cropping with rotation but no fallow. Outside this zone was one of medium intensive (short fallow) cropping with cultivation 3 years out of 7, followed by the three field zone (cultivation 1 year in 3), and finally came a zone of ranching, i.e. of extensive land use. (Cf Haggett 1972 for a modern summary of von Thünen's land use 'rings'.)

Land use zones have been widely observed in the geographical literature (cf Chisholm 1962) and have been attributed partly to natural conditions, partly to socio-economic factors such as transportation costs. At the *local* level, transportation costs are, as Chisholm rightly points out, more directly a function of *human time costs*, whereas at the regional, national or international level, time costs enter the picture but the real costs are also dependent on the mode of transport, technical conditions, etc. which underly their customary measurement in monetary terms. However, the relationship between land use zoning and the *intensity of cultivation* in the space-time sense has been less widely observed.

Although land use zoning by intensity is widespread in many parts of the world, it tends to be clearer and less distorted in medium intensive agricultural ecotechnologies. In the old Scandinavian mixed farming and agro-pastoral system, for instance, the infields near the stalls (where livestock was fed during the winter) were intensively cultivated and were the only land receiving manure. Beyond the infields was a zone of fodder production which had formerly been close-to-dwelling swiddens. From this zone grass and leaved branches were carried to the stalls and stored for the winter. Commonly, outside this zone was a zone of shifting cultivation proper combined with forest grazing. Although many variations on this theme can be found, the pattern of more intensive and time demanding cultivation close to dwellings (day-prism centres) and less intensive useage farther away stands out clearly (cf p. 134-35).

Looking into the production function of the widespread infield-outfield systems of Europe, Christiansen (1978) points out that:

> ... the outfield production alone is advantageous, when labour is scarce and land not ... /the/ use of outfield alone was possibly the origin of the system. But it is also evident that a need for higher carrying capacity, which implies higher yield per unit of area, makes a combined use of infield and outfield necessary. Population pressure may have been one of the driving forces behind this (E. Boserup 1965). The normal development of the system makes the infield encroach upon the outfield.

The livestock component in this system greatly contributes to the increased yields of the infield as a result of the nitrogen supplements from manuring. Manuring is thus the key to the productivity and prospect for intensification of the infields, especially when animals are stallfed or driven daily to the village for milking and protection (cf Chapter 4 on daily return systems, and mixed farming, p. 221-27).

Even in pure shifting cultivation systems, people commonly have intensive kitchen gardens with more continuous cultivation close to their stations of residence and the tendency towards the emergence of spatial zones of intensity is present in most systems. Sometimes the pattern may be more subtle, as when residence is dispersed, as in the Kofyar case:

> Bush fields in the hills may be used from six to nine years, depending on the length of the preceding fallow period, but the greater pressure on the plains land *near* villages means that *fallows are shortened* from more than fifteen years to less than ten. (Netting 1968:64) /My italics/

Grove, whose results were used by Boserup (1965:19, 59), not only pointed to the overall regional population density and intensity of cultivation but further looked into the situation at the village level in the north of Nigeria, a region well suited for the study of agrarian systems at different levels of intensity. He demonstrated the relationship between intensification and spatial zoning, especially in areas of 'moderate' population densities, i.e. roughly of medium intensity. This region of north Nigeria had an average population density of 12 inhabitants per km^2 (or 30 to the sq. mile).

Taking, among other places, the example of the village of Soba in Zaria province, Nigeria (a place initially described by Prothero) Grove notes that:

> ... the pattern of land use is irregular but several concentric zones can be distinguished... The village walls enclose about a hundred acres where small patches are well manured with household waste. Outside the village walls, within about half to three quarters of a mile, nearly all the land is cultivated with manure from locally owned donkeys, goats and sheep. Fulani /pastoral nomadic herdsmen/ are invited and even paid to kraal their cattle on this land near the houses... there is more cooperation in Soba between pastoralist and agriculturalist, to their mutual benefit /a kind of mixed farming/, than there is in the more sparsely settled lands of western Bornu. Outside the manured land, in a third zone roughly half to one mile from the *gari* /i.e. village/, farms are cleared and cultivated for three or four years and the land reverts to fallow for at least five years... In the outermost zone of Zoba village, more than three or four miles from the *gari*, bush growth predominates... Concentric arrangements of land use, similar to those around the *gari*, are repeated on a smaller scale around the four hamlets of Soba, and they appear to be a common feature of settlements in areas with population densities of about 50-150 to the square mile /i.e. c. 20-60 to the km^2/. (Grove 1961 124-5)

The *spatial structure* of intensification at the local level is readily discerned, a zoning arrangement summed up in idealized form in Figure 6:9 *a-b* (shaded areas in the figure denote cultivated land.)

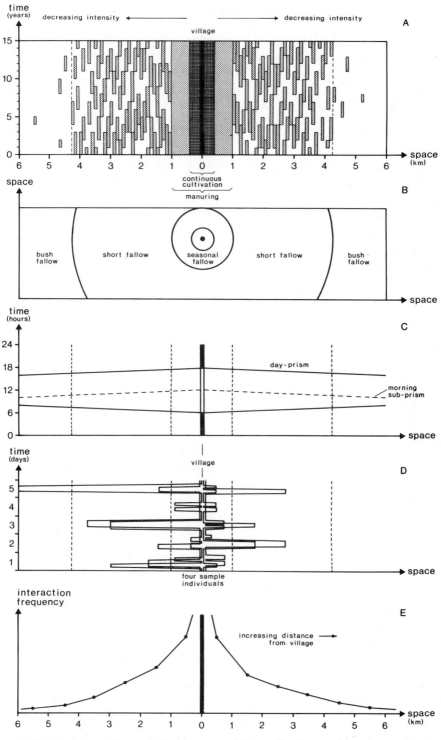

FIGURE 6:9   The spatial structure of intensification and zoning at the local
level. (See text for explanation.)

Southern Europe is another place known for its conspicuous concentric land use patterns, and Chisholm (1962) calls attention to examples from Sicily, Sardinia and Spain. Some of these settlements are better designated as (agro-)towns rather than villages and have populations of thousands rather than hundreds. In the agro-town hinterland of Canicatti, Sicily, for instance, tree cultivation dominates the innermost zone up to a distance of 4 km, beyond which arable land takes the upper hand. This pattern is also characterized by declining intensity since the arable land 'tends to be left in fallow more frequently at greater distances and when so left requires little attention /i.e. time/' (Chisholm 1962:62-65).

It is thus clear in the case of Tanzania that with the excessive concentration of population into fairly large villages, a result of enforced villagization, land use zoning is one of the potential forms of adaptation to the new situation.

Prisms and space-time packing: Reinterpreting movement, distance decay and interaction

The time-geographic path model of society-habitat assumes that phenomena which can be physically delimited can be projected onto the existential/ontological dimensions of space and time. In the geographic disciplines, society as well as nature and earth are by tradition studied with respect to the spatial structure and organization of phenomena, living and non-living. Human geography also shares with the natural sciences such as physics the use of space and time as basic analytical dimensions through which various substantive properties of different entities and systems can be specified.

As geographers began to take a closer look at different social and environmental systems and sub-systems, they made the discovery that numerous recurrent sociocultural processes, especially movements such as freight transport, migration, or commuting, conformed to what physicists would call a gravitational pattern around bodies in space, in our case around regions, cities, towns or other units of settlement. * The affinity between the gravitational theory of New-

* This introduction to the gravity model, distance and interaction is written for the benefit of non-geographers, who are less likely to be familiar with the topic. For a thorough analysis of migration and distance decay, cf Hägerstrand (1957). A more general survey of locational analysis in human geography can be found in Haggett (1965) and Haggett, Cliff and Frey (1977). Cf also Olsson (1965). Most basic textbooks in systematic human geography deal with the topic in one form or another (for instance, Abler, Adams and Gould 1971, Ch. 7).

tonian mechanics and gravity patterns of socio-cultural phenomena was strictly in terms of a spatial morphological analogy. This resulted simply from the choice of dimensions on which given real-world phenomena were projected, i.e. space over time. This is why human geographers refer to this particular patterning as the *gravity model* rather than *gravitational theory*, which is the term employed by physicists.

In general the finding documented for a variety of times and empirical settings was that the statistical frequency of interaction centered on a settlement (village up to a city) or region tended to decrease as a function of distance away from this region. The spatial structure of a migration field, for instance, could be summed up in statistical-mathematical form, as in Figure 6:10, conforming to a Pareto-type formula $y=ax^{-b}$, where y is frequency of interaction, e.g. migrants in a given time period (sometimes also per 1000 inhabitants), x is distance, and b is the gradient or slope.

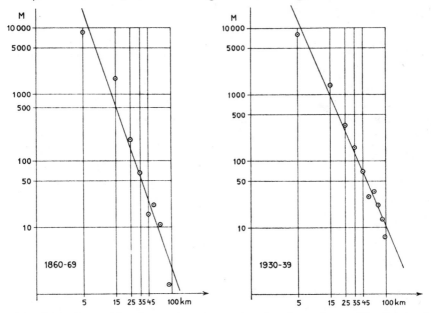

FIGURE 6:10   The figure shows out-migration from the parish of Asby in Östergötland, Sweden, in relation to distance and population in 100 thousands of every zone (from 1 up to 5 km, 5 to 15 km, etc.) The migration frequency is weighed in relation to population distribution outside Asby parish on the assumption that migrants are attracted to other places in relation to the number of people already living there, jobs and relatives available there, and so on, but the frequency of interaction between Asby and its environment is also modified by distance and negatively so. The gradient of the migration field has changed over the years. Between the 1860's and the 1930's the 'quantitatively most important change is not the increase in the number of very long migrations but rather the moderate elongation of the many short ones.' (Hägerstrand 1963).

A similar distance decay profile, constituting an aggregation of numerous movements at the micro-level, is modified in various ways by the overall regional structure of settlement. What may be an ideal pattern of concentric zoning of decreasing intensity of land use around a given settlement, can become a very distorted composite pattern, when several settlements of different size and relative location in the region enter the picture. Yet, at the local level, say of a village, interaction frequency still tends to decrease with increased distance away from it.

There have been many different interpretations of the empirical regularities of 'distance decay' and associated interaction fields. It would be far beyond the scope of the present book to go into this. We shall only here discuss one common interpretation which is related to carrying capacity and the use of time resources, namely the minimization of movement hypothesis.

On the one hand, the spatial patterns of interaction found by geographers looking into the matter bear witness to the general propensity of human populations in given settings not to go father afield in the aquisition of goods and services or in the seeking of jobs, etc., than they have to. Thus, as Christaller (1933) explained, each good or service has a particular range in space, and there is an hierarchical structure of central places (settlements-cum-market-places) with the effect that the goods and services most frequently demanded are also those closest to the consumer. This and other evidence speaks for a minimization of movement hypothesis.

On the other hand, the minimization of movement hypothesis is difficult to specify and calibrate for different conditions. To claim that travel, transport and other kinds of movement and spatial interaction would always be minimized would be to subscribe to a very ambiguous (not to say dubious) law. Given that all individuals pursue various individual and collective, more or less institutionalized, projects, movement minimization is hardly an end in itself, given that movement is a *means* to achieve certain targets. Hence, the extent to which movement would be minimized could only be empirically ascertained by taking the various projects and their locations, inputs and outputs into consideration. This has so far only rarely been done.

Interestingly, the social ecologist Zipf (1949) regarded minimization of movement as only a special case of a general *principle of behaviour*, which involved the *minimization of effort* (presumably time and energy) in all human pursuits. His treatise is an amazing collection of empirical regularities to this effect which, however, are somewhat detached from their structural context, both in terms of society and habitat. He applied the principle of the least effort *(lex parsimoniae)* to the most diverse areas, from language to the distribution of cities. Again, one cannot assume that all effort and activity is minimized in extent. Many people are in fact willing to expend a considerable amount of time and energy on pursuits which they culturally or individually favour. Hence effort minimization is only applicable if projects are regarded as given and means are allowed to vary. This would leave us with a kind of cost minimization hypothesis, but even this formulation is open to objections. Nevertheless, for all the shortcomings of Zipf's original formulation, his extensive monograph is brim full with data that need reinterpretation and reformulation. Perhaps, much of the underlying logic would stand out more clearly if the resources involved were specified more carefully (e.g. human time, space and energy) in relation to projects and positive constraints, on the one hand, and to carrying capacity, packing, capability and coupling constraints, on the other.

In the present study we shall make an attempt to reinterpret the distance and interaction problematics in the limited field of spatial zoning and decreasing intensity of land use with distance from settlement. In so doing, we shall define interaction frequency in a stricter *temporal* sense and not as statistical frequency only. In our case, the frequency *and* duration of travel away from the village or dwelling cluster combine to determine how much time in the local population time-budget is allocated to travel (cf Figures 1:2 right, 3:9, 3:18, 8:1 and 9:3). The temporal frequency pertains to several time scales, since the number of trips or travel occasions can be related to either the day, a series of days (such as a week), or an entire season or a year. We have already mentioned (p. 168-70) how the Siane men went to their gardens every third day or so after the introduction of steel axes (Salisbury 1962), while the women went nearly every day (some 9 days out of 10, cf Figure 9:3). Richards (1939) describes how Bemba women fetch water every other day at two or three miles distance. It is obvious that if the travel associated with given tasks (e.g. gardening, fetching water and firewood, going to school, shopping, etc.) can be carried out at a lower frequency, travel time can be saved (cf Figure 6:12). One means of thus lowering the intensity of land use activities at a distance is obviously to lower the frequency with which this distant place is visited. Planting a crop which needs less frequent attendance farther away may be a solution, for instance. Another way of lowering land use intensity away from the domiciles is to plant crops there, which require less work input in terms of duration. The duration and frequency of visiting fields can be illustrated as in Figure 6:9 d-e.

We can then see some general relationships between five different projections of the same basic problem, that of spatial zoning and of distance decay in agricultural land use intensity (Figure 6:9). This figure depicts both the local space-time budget (A), the conventional zoning map (B), the day-prism (C), sample individual-paths (D), and the distance and frequency of interaction graph (E).

Chisholm interprets the generation of zones of land use explicitly in terms of time resources:

> The basis for the argument ... has been the proposition that human labour is scarce ... Now it is commonly held that this ... assumption ceases to be valid in circumstances where labour is unemployed or under-employed; /and that/ *human time* saved in one process of cultivation adds nothing to the general welfare, there being no productive use to which it can be put. If this were to be so, no kind of zoning would be apparent, because all the territory would be farmed to the same level of *intensity*. (Chisholm 1972:71-2) /Italics added/

In shifting cultivation systems as well as in more intensive systems, one form of spatial zoning or another is the rule rather than the exception. It must also be remembered that the particular pattern of intensity zoning may be generated by time demands at peak seasons.

Spatial zoning and investments to facilitate intensification

Boserup argued that in general, the intensification of an agricultural system involves investments, mainly through a greater application of labour-time. Clearing and stumping land is one such investment when transforming a long fallow system into a short fallow one, let alone the time consuming investments of constructing irrigation terraces or irrigation channels at stages of higher intensity. Other activities necessary to increase or maintain yields as fallows are reduced also involve the mobilization of human time.

Although the natural productivity of land does not vary directly with settlement size and distance from a village, the productivity of labour-time certainly does, when calculated as gross working time that includes travel. If the time spent on travel to distant fields on a near daily basis during the cultivation season is instead diverted to investments in land improvement close to the village, some of the extra time demands induced by large village size can be obviated. This alternative is even more attractive if the village is located on the most fertile land in the area, say land that responds well to manuring. In many instances, of course, a village may be located on land which does not belong to the most fertile segment of the local area, for reasons of defence, accessibility to water or the like. But then pressures are still strong to intensify the land closest to the village for reasons of travel time economy.

FERTILITY IMPROVEMENT AS TIME DEMANDING ACTIVITY AND INTERACTION IN TIME-SPACE

As a cultivation system becomes more intensive in its occupation of space-time, the natural fertility embodied in the vegetation (as in a long fallow system) is greatly reduced. If we disregard particularly fertile sites such as alluvial valleys and inundated plains, fertility has to be maintained or improved through increased input of human time and effort. When vegetation is used as a means of fertilization, a natural factor of utmost importance is rainfall. Regions getting ample rain (for instance, areas with bimodal rainfall around lake Victoria in East Africa) or equatorial regions with nearly continuous precipitation throughout the year (such as the New Guinea region) are favoured in comparison to semi-arid regions with only one rainy season. Mulching or green manuring as a means to improve soil fertility is a more viable alternative in regions where adequate rains ensure sufficient vegetation.

A note on 'unmixed farming' in New Guinea

It is beyond the scope of this book to go into any detailed discussions on the principally interesting cultivation systems of Melanesia-New Guinea. Apart from the many excellent monographs on specific societies and places (such as Pospisil 1963, Rappaport 1968, Waddell 1972 or Christiansen 1975), Brookfield (1962, with Hart 1971) and associates (Brookfield, ed. 1973) have already made impressive contributions to the analysis of these human ecological systems.

New Guinean cultivation systems are often characterized by a lack of seasonality and the fairly continuous rainfall throughout the year means that these systems operate at a higher intensity than those found on the African savanna, for instance. Although there may be perennial fallows in New Guinea, the seasonal fallows can largely be eliminated, not least since root crops have longer maturation periods than cereals, and the latter are less common in the New Guinea setting than in many African savanna or Indonesian outer island settings. Continuous rainfall implies that both planting and harvesting can be carried out on a more continuous basis over the year. The pseudo-intensification process has therefore gone quite far in New Guinea in terms of the elimination of seasonal fallows. But otherwise, there is striking variability between the long fallow swidden agricultural systems such as those of the Tsembaga (Rappaport 1968) and the Siane (Salisbury 1962) and those with very short periods of fallow typical for the Chimbu (Brookfield and Brown 1963), Kapauku (Pospisil 1963) or the Raiapu Enga (Waddell 1972).

Take the case of the Raiapu Enga. In order to reduce perennial fallows and maintain cultivation on a continuous basis, they must put considerable labour-time into green manuring and mounding. Some major properties of their human ecological situation are:

> ... a general pressure of population on resources, a dispersed settlement pattern, an open field/mixed garden dichotomy, and a large dependent pig population. (Waddell 1972:183)

Their average population density is around 96 inhabitants per km^2 and unlike the Siane and the Tsembaga, this population is not concentrated into villages. This dispersion of settlement, also typical for the Chimbu, must be interpreted as a response to the work load and time inputs associated with agriculture at this level of intensity so as to reduce travel.   The pattern is the reverse of that found in Tanzania. So rather than having temporary (double) residence outside the villages, the Chimbu, for instance, have temporary villages (although the sites may be 'permanent') used mainly for ceremonial occasions.

Mulching and mounding is designed to facilitate continuous and (by Boserup's definition) intensive cultivation:

> Basic to an understanding of the mounding system is the fact that it is designed to permit continuous cultivation of the soil /within the open fields of dominant importance/... The cultivation cycle /annually/ is a continuous one in which, at the final harvest, the mound is broken open and the earth piled in a ridge at the perimeter... The weeds which have colonized it, together with unwanted sweet potatoe vines, are thrown to the centre... the vegetative matter incorporated into ten mounds with dimensions averaging 3.1 in diameter and 0.6 m high revealed a mean of 20.2 kg per mound... /there was a/ lack of apparent selectivity in the mulching material. (Waddell 1972:44-5)

Although there is a faint resemblance between many of the New Guinea systems of tubers-mixed-with-pigs and mixed farming, this is only in the sense that pigs contribute to fertility maintenance by rooting up the soil and leaving droppings in the perennial fallows where they are allowed to browse. The real complementarity lies in the enrichment of the human diet, tubers delivering the carbohydrates and pigs the proteins. However, it is interesting that for this complementarity in food consumption to be of benefit, much human time must be spent on keeping the pigs and the tuber gardens *unmixed* prior to the consumption stage, which is achieved by fencing. Fencing to exclude pigs (there are few wild animals to keep out), is an arduous and time demanding activity, even after the introduction of steel axes (cf Salisbury 1962, Pospisil 1963), and is a price to be paid for maintaining the ecologically workable pig-tuber complex.

In true mixed farming systems, where animals deliver manure and often traction power for carting manure and pulling ploughs, etc., the human time spent on animal husbandry (feeding, herding and otherwise catering for them) is partly an indirect activity of fertility maintenance. In New Guinea, mulching is a way of maintaining soil fertility without having to spend the time on animal husbandry for that particular purpose, while fencing is a time-saving way of keeping pigs out of gardens without having to herd them or guard the crops.

Mulching/green manuring/composting is a successful means of keeping the soil fertile in areas infested with pests (such as tsetse) which preclude certain breeds of livestock. Again the Tiv of Nigeria is a good example:

> Tivland is... /generally/... tsetse infested, and as a result animals in sufficient number to provide manure for fields cannot be kept... The government agricultural service tried to introduce green manure crops and compost heaps... Tiv so far /in the 1950's/ have not been made to see that if they plant part of their land with a crop which will be dug under to provide green manure, they will get much better yields... To Tiv such practice represents hard /and for that matter time consuming/ work with no visible gain... Composting has been even less successful... They are downright anti-compost: 'Have *you* ever tried to carry such stuff on your head?' (Bohannan and Bohannan 1968:63)

Mixed farming without traction animals

A people who do carry manure-compost in baskets on their heads are the Kofyar on the Jos Plateau, neighbours of the Tiv. Living in a territory of some 492 km^2 with an average regional population density of 112 persons per km^2 (or 290 to the sq. mile), some sort of mixed farming or cultivation with artificial fertilizing is a virtual necessity. Kofyar cultivators apply vast quantities of dung and compost to their nearby fields to keep them fertile.

> In every area where stone is available the Kofyar build circular enclosures from ten to fifteen feet in diameter and up to eight feet deep. Within such corrals... which are a fixture of every homestead, goats are staked out each day for the entire length of the growing season. Food for the animals in the form of grass and leafy branches is brought every day by the household members. The forage is always more than the goats can eat. It builds up in

layers mixed with goat dung and urine... for nine months of the year every available bit of manure is retained... When the corral is emptied just before the rains, a substantial quantity of fertilizer... can be returned to the land... In one case the compost from a filled corral 15 feet 8 inches in diameter and 7 feet 7 inches deep, with a capacity of 1,452 cubic feet /or 41 m^3/, was distributed in 388 basket loads over .4 acre /i.e. 41 m^3 per ha/. (Netting 1968:61-2)

Goats can provide manure but not traction or carrying power, so the whole structure of Kofyar cultivation is affected by the heavy time inputs and portability constraints of carrying manure from homestead to fields. Since the temporal intensity of land use has reached nearly continuous cultivation with a nine month growing season, the level of necessary time input is high and Kofyar are extremely hard working and are also experts in natural resource management by their terracing and drainage techniques.

In view of this, for a people unfettered by government ambitions of villagization, it is only possible to keep work loads within bounds if settlement is dispersed so that all the inputs (i.a. fertilizer) only has to be carried very short distances. Under a system of villages of the Tanzanian average size (say some 1000-1500 individuals), the Kofyar system of goats providing manure but not traction or porterage would have been prohibitive or induced great time costs in terms of alternative activities, even with spatial zoning.

Another system resembling that of the Kofyar is the system practised by the Wakara on the island of Ukara in Lake Victoria (Ludwig 1968). Just as the Kofyar suffered from impaction, the Wakara had retreated to an island with limited land. In 1965 the average regional population density was 207 inhabitants per km^2, which is roughly twice the density of the Kofyar. Historically, the advantage of living on the island was that of more efficient defence against warlike neighbours. Of total land, 98.6 per cent is used agriculturally, which is a remarkable figure, and out of this land 6.7 per cent is irrigated.[*]

... during the 18th and 19th century, the people of Ukara experienced a long history of land shortage and population pressure. Consequently, after the colonial powers had gradually succeeded in controlling East Africa, many of the Wakara migrated... They then abandoned their advanced husbandry methods and changed again to shifting cultivation which seemed more economic whereever sufficient land was available... (Ludwig 1968:94)/The Wakara are yet another case of deintensification in the literature./

[*] The system is so land and time use intensive that it ought to be placed in Chapter 7, but it is described here because of its affinity with the Kofyar system.

The Wakara practise manuring, fodder growing, erosion control and,

> /they/ turned to permanent farming before the arrival of the Euro-
> peans. They have already used farmyard manure and cultivated
> green manure crops for a long while. Farming ensues with organ-
> ized crop rotations. Animals are kept in stables and provided with
> fodder. The cultivation of fodder is systematically organized. In
> addition, we find irrigation farming yielding several harvests
> anually. Another characteristic may be seen in the numerous
> measures applied in order to check erosion: terracing, ridge
> cultivation, etc. (Ludwig 1968:89)

Not surprisingly, all these measures to maintain yields at very high
levels per areal unit require that the Wakara work extremely hard.
Their work-days involve spending as much as 10 hours per day on
cultivation (cf Chapter 9). Outside their island, they are known to
be 'assiduous and respected workmen', and their whole ecological
system conforms neatly to Boserup's general intensification theory.
Not surprisingly, the Wakara have also undergone pseudo-intensification in recent years:

> In recent time cassava has gained appreciably in importance /in
> relation to the staple millet/... because more people can be fed
> per acre with less effort than with millet. (Ludwig 1968:105)

Since the Wakara have to work so hard for modest returns (subsistence plus a small cash income), many cultivators have preferred
to migrate to Sukumaland on mainland Tanzania, where they get
better returns for their labour. However, the Wakara are exemplary
as cultivators and indicate the extent to which intensification can be
pushed with the appropriate techniques and a high level of time
input.

With the dispersion of population, a common trait of intensive
systems such as the two mentioned, the clear spatial zoning around
settlement units may disappear, as Grove (1961:127) points out
regarding certain parts of northern Nigeria:

> ... in Bindawa, ... the proportion of land under cultivation is much
> greater than at Soba /previously discussed/ and there is little trace
> of the zoning of land use according to distance from the /village/.
> ... Some garden crops are raised within ... Bindawa, and the small
> plots outside them are manured more heavily than those at a dis-
> tance; but there is no encircling zone of bush fallow and wood-
> land. Only about one third of the village people live ... /behind
> the village walls/; and the remainder live in compounds *dispersed*
> singly or in clusters through the hamlets. All the cultivated land
> has people living *near* it... (Grove 1961:127) /Italics added/

He also adds in a footnote that people who formerly lived in dispersed compounds were brought together in settlements near roads, to get access to schools, wells, etc. This was in 1952, years before Tanzania introduced its villagization programme. But Grove also adds that ... 'the distant land may suffer from lack of manure, and trees near the settlements are felled for fuel.' This is as might be expected.

Mixed farming with traction animals: Relaxing the human constraint on portability

Mixed farming was and still is a major form of agriculture in Europe. It was even more dominant prior to the introduction of artificial fertilizer and mechanization. Mixed farming is also widespread in South Asia and China, for instance. But it was in the European setting that geographers, looking into the working of mixed farming systems, noticed some of the many relations between human time allocation and settlement:

> The point that... distance should be measured in terms of the time taken, was made by Müller-Wille, who prepared a map for one village with lines showing zones of equal time required in carting manure to the fields... he concluded that when more than one hour was required the amount of manure declined sharply... (Chisholm 1962:53. Cf Müller-Wille 1936 and Lösch 1954:382.)

The fact that animals such as horses, oxen, bullocks or even camels can be used to cart produce and manure as well as to pull ploughs explains why it was possible in many European and Asian settings to manure fields and practise mixed farming without having to resort to dispersed settlement of the Kofyar type. In parts of these regions, concentrated village settlement with a more intensive social life style was compatible with medium intensive mixed farming, since it did not involve having to move bulk materials by human porterage.

The use of beasts of burden and traction not only relaxes the portability constraint and hence that on mobility of goods (although it introduces new coupling constraints between humans and beasts), but it also allows people to harness an extraneous source of energy for work. This allows certain tasks to be performed faster and with less capability constraint than the mere application of human energy. Although the plough has been used for centuries in northern Africa, its relatively recent introduction into Africa south of the Sahara has often enabled cultivators to prepare larger areas of land than with hoes. An interesting consequence of this is that rather than land preparation being a bottleneck in the annual cycle, weeding of the larger area has taken its place. Thus in hoe cultivation (*horti*culture) gardens often tend to be smaller than the fields of plough cultivation (*agri*culture).

In post-villagization Tanzania, people are now trapped by the situation of large village size without the necessary technical support of domestic animals in large enough numbers for traction of ploughs or carting of manure. This makes the introduction of mixed farming difficult, although it has been made ecologically desirable in order to improve yields per capita. To maintain large settlements with rather limited access to artificial fertilizer also makes it hard to intensify farming successfully. At a regional level, the widespread use of artificial fertilizer assumes an industrial capacity to produce it (or money to import it which would require more cultivation of cash crops, which in turn presupposes more fertilizer). It also demands a good regional transportation network and vehicles in adequate numbers. At the local level, even artificial fertilizer is heavy and bulky and needs to be carted to the individual fields in a country short of carts, wheelbarrows, etc., as well as draught animals (let alone tractors). An alternative strategy might have been to introduce crop rotation schemes   with nitrogen fixing plants (e.g peas and beans) similar to the rotation schemes practised in South-east Asia (cf Friedman 1979: 79-82), but this requires a vast educational undertaking that must be based upon extension agent capability which is so far lacking.

By contrast, it is curious that in Sweden, historically, land and settlement reforms were initiated in the late eighteenth and early nineteenth century in the very opposite direction. Although traction animals (chiefly oxen and horses) were owned by all but the poorest farmers thereby permitting mixed farming, land reforms aimed at the dispersal of cultivators to individual farmsteads. This policy was designed to promote intensification of cultivation and better manuring to raise productivity. Although the policy entailed the fission of pre-existing villages and had considerable social costs in terms of reduced (frequency and duration) of interaction among villagers, it also freed cultivators from the coordination and coupling constraints associated with village life. The widespread synchronization of agricultural and other activities (such as fence repairs and harvesting) tended to impede agricultural innovation, since the majority always had to be convinced of the virtues of new methods before they could be implemented. Technical progress was therefore slow, productivity was low and the state could collect less tax (hence the concern by the government).*

*   This is a very general and superficial statement which overlooks many regional variations and other aspects. To the knowledge of the present author, there is no useful general monograph on Swedish land reforms available in English.

INTENSIFICATION AT THE DOMESTIC UNIT/FARMSTEAD LEVEL

Having covered some major issues of intensification both at the regional and local level, we shall now have a closer look at relations and structures at the scale of the domestic unit-cum-farmstead.

Fragmentation of domains and space-time holdings

The fission of villages into dispersed homesteads, as in the case of Sweden, would have been of little avail as a strategy of intensification if the land holdings of households had remained fragmented and dispersed. Then travel time to and from parcels and the time/energy costs would still have been high and even prohibitive. Since the Middle Ages and onward, the extreme fragmentation of holdings has been common throughout most of Europe due to inheritance rules and subdivision of land so that each heir would receive land of varying quality to suit different purposes (and for other reasons). This system is still common today in many parts of continental Europe. It is also common in other parts of the world where mechanization, land reforms or other factors have not reduced its occurrence.

All studies of the fragmentation of holdings point to the adverse effects with respect to travel, management and fertility maintenance. The holdings distant from the farmstead tend to receive much less manure and upkeep than the closer parcels of land, and if settlement structure was to be optimized according to this function alone, it would imply the consolidation of holdings and dispersal of settlement. However, there are many other influential factors and sources of distortion of such an ideal pattern, for instance the availability of water (cf Blaikie 1971).

For 227 farms scattered over central Sweden, Larsson (1947) found that output per hectare decreased substantially with increasing distance to fields, and did so more for larger farms than smaller ones (cf Chisholm 1962:56 for a summary statement). This is partly a result of the increased time and energy costs of carting manure. Even when traction animals are available, however, they usually move rather slowly and have to be employed for other activities as well — ploughing, for instance. Thus shortage of *animal time may also affect the spatial pattern of cultivation intensity.* Peak season shortage of one resource, be it human time or animal time (or in mechanized agriculture, machine-time) may thus affect the relative intensity with which another resource is used, here space-time (land). Peak season time shortages can thus, by structuring the overall settlement system, 'spill over' and affect activities at non-peak seasons.

Consolidation of holdings may *de facto* be carried out by means of lending, leasing or selling outlying parcels, a measure which can be combined with acquisition of nearby land on similar terms. If acquired parcels are contiguous to those already operated, consolidation proper takes place. But in societies where both the sexes and all siblings inherit land (partible inheritance), strong forces stimulate further fragmentation, especially if the population is growing. Parcels of land will then tend to get smaller, which is an obvious obstacle to mechanization or other improvements. In societies where custom dictates that land should be redistributed at periodic intervals of a generation or so, many of these adverse effects on cultivation can be mitigated, but such traditional systems are now rare and agricultural commercialization has led to their abolition, as in the Pakistani case reported by Barth (1959) for the Pakhtuns (Pathans). A growing land shortage may have the same effect. The Nyakyusa of Tanzania, for example, underwent a transition from communal village and extended family tenure to nuclear family tenure:

> A *redistribution in each generation* operated without much friction so long as land was plentiful and there were few permanent improvements or long-term crops, but as land became scarce and men built more lasting houses and planted coffee /a perennial crop/, then it became impossible. Owners were neither prepared to lose the improvements /i.e. time demanding investments/ nor, having secured fertile land, to vacate it. (Wilson 1963:384; italics added) /Cf Chapter 7 for reasons behind Nyakyusa land shortage./

To take a case at a different scale, the Chinese revolution in 1949 ushered in an enormous consolidation of holdings at the village level rather than the household level and eliminated much of the land fragmentation problem which was serious in pre-revolutionary China (cf Buck 1937).

Consolidation of holdings as a mechanism of time-space packing

The consolidation of fragmented time-space holdings is a mechanism of repacking activities both in the local space-time budget and in the population time-budget. Particularly when residence is dispersed rather than concentrated to a village, the consolidation of holdings may save much time and effort in travel and porterage, and the time thus saved can be turned towards the intensification of land use instead, i.e. towards increasing time inputs per areal unit.

To get a better picture of the importance of consolidation of scattered land holdings, one must place this problem in the context of the great number of movements made to and from fields both in the daily round and cumulatively in the year, bearing in mind also the travel generated for other purposes such as fetching water and firewood, herding livestock, visiting market places, and so on (cf the daily round graphs in Chapter 9). There are also the problems of guarding a great number of scattered plots (from vermin or people) to consider, not to speak of the costs of fencing a large number of small plots as opposed to a few large ones. In many of the European infield-outfield systems, the small parcels were not fenced by themselves but whole sections of infields were fenced together (as in Figure 6:11a). It is also common in many parts of the world that residents of one village may own plots in other villages also, and if these outlying parcels are to be cultivated, the travel time costs incurred will obviously be higher. Such distant plots are then frequently leased, exchanged or externalized in other ways, the total effect being that the ownership pattern does not coincide with the pattern of actual cultivation. Scudder, for instance, mentioned such a mechanism among the Gwembe Tonga:

> While possessions of gardens in another neighbourhood does not jeopardize one's rights to the land in question, the main drawback is the energy /and time/ expended in getting from one garden to another. When the task of cultivating becomes too arduous /and time demanding/, the owner usually allows kin with closer access to take over cultivation. (Scudder 1962:69)

In the Punjab, to switch to another part of the world, one of the main handicaps with fragmentation of holdings was the problem of guarding fields (cf the human indivisibility constraint) and maintaining boundaries intact. Theft and cheating with boundaries gave rise to innumerable quarrels and expensive litigation with neighbours. (This is documented in many of the studies undertaken by the Punjab Board of Economic Enquiry in Lahore, Pakistan.)

The need for consolidation can be illustrated by the case of a typical southern Swedish village of the year 1700 which indicates the extreme level to which fragmentation of holdings may evolve (Figure 6:11a). On the average, each farm has some 80 to 90 parcels (Wester 1960). Subsequent to the land reform in the beginning of the 1800's, all agricultural land in the village had been consolidated into seven large holdings (Figure 6:11b). Since the village was comparatively small, all the homesteads were allowed to remain agglomerated, but in many other villages of similar and larger size, a number of farmsteads (often more than half) would be compelled to move out to the centre of their lands.

Consolidation of holdings and dispersal of farmsteads so that they are closer to the fields can thus be time-saving in at least two ways: firstly, the frequency of travel can be reduced so that actual move-

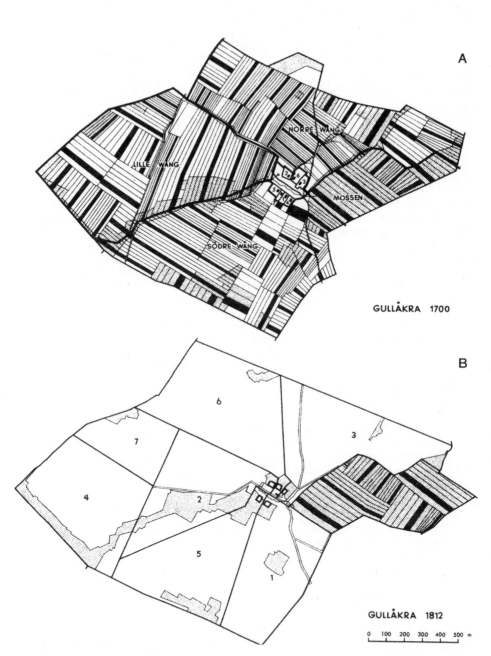

A

NORRE WÅNG

LILLE WÅNG

MOSSEN

SÖDRE WÅNG

GULLÅKRA 1700

B

6

7

3

4

2

5

1

GULLÅKRA 1812

0 100 200 300 400 500 m

FIGURE 6:11 The fragmentation of holdings in a southern Swedish (Scanian) village of the year 1700. Farm number 4 had some 90 scattered parcels (indicated by black), but only the four blocks of parcels ('wång') were fenced.

After consolidation and land reform (bottom map), few small parcels remain, and the whole village has been alotted large pieces of land, except for what was formerly common land (a peat bog, 'mossen') which remained fragmented in theory but was used by all as pasture regardless of boundaries (Wester 1960).

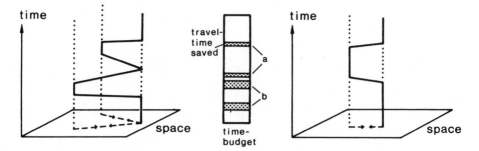

FIGURE 6:12   The time saving effects of consolidation of holdings and dispersal of farmsteads. Compared to a situation with fragmentation of holdings (left), consolidation and dispersal (right) saves time in two ways: a) by the shorter journey from farmstead to field, and b) by reduced frequency of travel in relation to net time inputs in the field itself. (The figure assumes that two plots have been reduced to one. Shaded area in the middle figure indicates the travel time actually *saved* due to changes in travel distance and frequency. Cf Figure 3:9.)

ments per week can become fewer; secondly, the average time distance to the fields becomes shorter (Figure 6:12).

Intensity, households and domain structure

So far, numerous examples have been used to demonstrate how land use intensity varies with distance away from farmsteads or residences. But it is also affected by the *total size of the holdings* of the domestic units. In Sweden, Larson (1947) found that:

> ... those farms with greater average distance to parcels were also larger farms. This would affect the intensity of production, for in general the per hectare level of gross and net incomes falls as farms increase in size... The man with the small business /or farm/ must extract every bit of income he can from his property and therefore tends to push production to the maximum possible level on all his land. (Chisholm 1962:56-7 commenting on Larson)

In other words, cultivators with comparatively less land tend to use their limited space-time resources more intensively than those with larger farms, in practice by applying relatively more of their time per space-time unit. In countries like India with great discrepancies in the size of land holdings per household, this is clearly revealed in farm management surveys and other materials on land use economy. Bharadwaj (1974) made a penetrating statistical analysis of a great number of farm surveys from all over India and found a very strong negative correlation between labour days per acre and size of holding

as well as between yield per acre and size of holding. A similar in-
verse relation between productivity (intensity) and farm size was also
found by Buck (1938) in his extensive survey of pre-revolutionary
China. It is in many respects extraordinary that none of the main
theorists on agricultural intensification (e.g. Boserup, Sahlins or
Harris) have noticed this condition and considered its implications,
namely that, although there is a broad correlation between population density (growth) and agricultural intensification at the regional
level, the latter factor is also an issue of *land distribution*. Particularly in medium and high intensive agriculture, there may be a very
broad range of intensity variation among farms concealed by the
regional average.

Thus Bharadwaj notes in the case of India:

> A basic premise underlying a number of explanations of higher
> degree of labour use on small farms refers to the greater avail-
> ability of family labour relative to land on smaller holdings...
> the small farms in fact tend to use a relatively greater amount of
> labour per acre... total labour days per acre show a tendency to
> decline with an increased size of holding — the inverse relation
> being statistically significant in most cases... (Bharadwaj 1974:19)

Bharadwaj goes on to say that:

> ... there is generally an inverse relation between earners per acre
> and the size of holding which is statistically significant in the
> majority of cases...
> /The table/... indicates that labour days per acre spent on 'all crop
> production' are inversely related to the size of holding generally,
> and significantly so in most cases. Relating labour days per acre
> on individual crops we find that in only a few cases, ... the rela-
> tion turns out to be statistically significant and is inverse. With
> regard to the rest of the crops, except in a few cases, the correla-
> tion coefficient, although not significant, bears a negative sign.
> (Bharadwaj 1974:97, 98)

We may interpret this in terms of a specific relationship between the
household time-budget and its *space-time budget*, as in Figure 6:13
(cf Figure 1:2, 8:1 and 8:7). In Figure 6:13, we have ordered some
sample households by increasing size, whereas their land holdings
are of variable size, reflecting the given domain structure (i.e. distribution of land holdings). The mismatch between household time
resources and land resources is considerable. The large household
with 9 members (say 3 adults) will be able to put in much more
time per space-time unit than the household with 5 members (and
3 adults) due to the disproportionate relations between human time
and the land to absorb it. This is on the assumption that every one

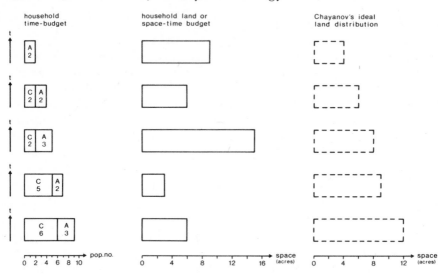

FIGURE 6:13   The relationship between the household time-budget or time resources and the household land resources. Commonly throughout the world, there is little correlation between time resource capacity and the size of land holdings, as in the case of India (described by Bharadwaj 1974). The land will then be used at very uneven levels of intensity (left and middle of figure).

Chayanov assumed equal access to land according to time resources available in households for cultivation (left and right of figure). Chayanov thus assumed a land surplus situation which is anything but a universal condition. (A : adult; C : child.)

works only their own land and does not lease it or exchange their labour-time, which by definition implies 'production for domestic consumtion' or 'production for use'.

In a country like India, for instance, this kind of uneven domain structure and land distribution puts those having a good deal more than average land in a position to extract a maximum return, if they choose to parcel out their land in small enough parcels to people holding little land of their own but having plenty of time for work:

> ... the big cultivators, while aiming to produce a surplus, may yet prefer not to cultivate the land for a number of reasons including the existence of opportunities for making profits or wielding power through non-farming activities...
> ... the landlord ... may choose to parcel out land ... to the very small tenants, who in turn will be compelled by economic necessity to cultivate their plots *intensively*, applying owned inputs *(particularly labour)* far beyond the maximum net return. The /big owner/ by so parcelling out land may be in a position to extract a maximum return... (Bharadwaj 1974:4-5)/Italics added/

By leasing land or exchanging land for labour-time, and of course by redistributing land among people, a more even level of intensity can be achieved among holdings. It is also possible to achieve a greater average regional intensity of land use and occupation by adjusting availability of land to household size and capacity. Thus, although the domain structure and regulatory constraints on land access are far from ecological phenomena, they affect both society and ecology through their impact on differential levels of intensity of land occupation.

Chayanov on intensity and the developmental cycle of domestic groups

Dealing with the balance between the household time-budget and its space-time budget inevitably leads on to the theory of labour intensity proposed by Chayanov (1925, 1966) as well as Sahlins' (1972) interpretations of his work. Chayanov was interested in the factors regulating the 'volume of economic activity'. Here this is interpreted as the volume of human time mobilized mainly for projects of agricultural production and domestic chores. Essentially Chayanov was dealing with peasant production for *use*, i.e. for own consumption rather than exchange.

Chayanov departed from what demographers refer to as the *family cycle* or what anthropologists have termed the *developmental cycle in domestic groups* (Goody 1958). This cycle comprises the stages a household passes through as its members go through their life cycle (some being born, others moving or dying). An important aspect of this is how the household time-budget (cf Figure 8:7) and age-sex-skill composition varies throughout this developmental cycle. In one of his early time-geographic studies, in which Hägerstrand (1963) depicted a sizable rural population as a flow through time-space for a hundred year period, he also showed the domestic cycle for a family on a small Swedish farm (a croft). In this graph (Figure 6:14), both the overall outcome and specific events such as births, deaths, migration and other aspects of population turnover can be discerned (for a further description of a single farm for a hundred year period, cf Hägerstrand 1978a).

Chayanov had access to a massive economic-demographic data base for the so called repartitional communes in late 19th century Russia, and he found a number of empirical regularities. Some of these were epitomized in 'Chayanov's Rule', which says that:

> /The/ intensity of labour in a system of domestic production for use varies inversely with the relative working capacity of the producing unit. (As phrased by Sahlins 1972:89, 91)

FIGURE 6:14

The Family Cycle, an example from Asby, Östergötland province, Sweden.

Individuals represented by life-paths or life-lines, circles indicating birth, crosses death and arrows movements.

'At the age of four years the male follows his family on his first movement. Seventeen years old he leaves his home in order to work as farm-hand on a number of neighbouring farms.

From the age of one year the female follows her family on a series of short movements. Sixteen years old she leaves home in order to work as maid-servant on a number of neighbouring farms.

The couple met when working on the same farm and married in 1886. This step is followed by a movement to a dwelling of their own - a small croft. The first child is born. Already in the next year the family moves to a somewhat bigger croft, in which it remains for a rather long time. Six further children are born.

The man dies at the age of sixty-six. After a few years the widow and an unmarried daughter move to the youngest son. The other brothers and sisters left home for the first time at ages between seventeen and twenty-two. Characteristically they return home once or more.' (Figure and quote from Hägerstrand 1963.)
(Cf Hägerstrand 1978a.)

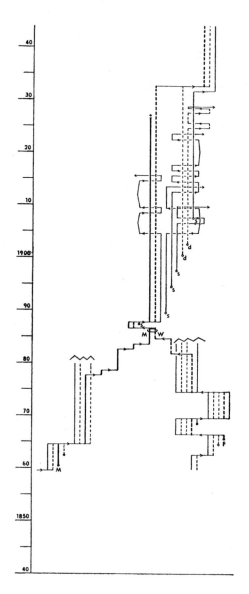

As is evident from Figure 6:14, the household time resources (and its labour force) build up as more children are born into the unit. On the other hand, this is tempered by the fact that children have more severe capability constraints than adults and can only start contributing their time for work gradually with age. Only as they approach adulthood can they make a full contribution, and when small they impose greater time demands on the adults than they can compensate for by the time they supply to agriculture and domestic chores (cf Chapter 9).

Chayanov expressed this process in terms of how the ratio of consumers to workers altered with the developmental cycle. This ratio becomes increasingly unfavourable as small children came to add more to consumption than production, but the ratio reaches an inflection point as children gain the ability to work. Under contemporary Russian conditions, a family producing nine children would in a 26 year period reach this inflection point some 14 years after the first child was born. Prior to that, domestic growth leads to an increasing measure of 'self-exploitation' within the family (cf Chapter 11 on service exchange within groups and sub-populations.)

Another aspect of family growth is the increased need for food and other goods, and the domestic unit is compelled to grow more food. Chayanov thus found a clear correlation between *household size* and the actual *area sown*. Small households cultivated less land than larger ones, and it is interesting to note what determined what:

> ... the connection between family size and the size of agricultural activity should be understood as a dependence of area of land for use on family size rather than conversely. (Chayanov 1966:68)

Here we find a striking parallel between the work of Chayanov and Boserup. The latter analyst viewed population growth at the regional level as the independent variable and determinant of intensity. Chayanov in fact looks upon population growth at the household level as the independent variable of intensity (we return to this topic below).

If we plot Chayanov's position in Figure 6:13, he proceeded on the assumption that cropping could be expanded in direct proportion to family size (and consumption requirements). To broaden the validity of this assumption to cover most peasant and household farming throughout the world would be very dangerous. It is only applicable to systems operating in the context of an abundance of land and comparative equality of land holding and accessibility, a condition most likely to be found in rather sparsely settled regions.

It is also interesting to note that, since Chayanov was working in temperate regions (with severe winters at that), families satisfied their demands for land through *spatial expansion,* i.e. by increasing the area sown. Thus Chayanov did not consider *temporal intensification* (cf

Figure 5:4 and 6:3), the latter being a feasible alternative in the tropical regions which Boserup concentrated her observations upon.

Chayanov assumed a symmetric expansion of the household time-budget and its space-time budget, which is a very strong assumption. As was obvious from the Indian materials on land use and domain structure, cultivators frequently have to intensify their labour time inputs on what land they already hold in order to raise production volume. Alternatively, they are compelled to lease more land, for instance, in exchange for their labour time. Or in monetized societies they must sell their time in exchange for money to buy food with. There are numerous similar short-term regulatory devices for stabiliz-ing the proportionality between human time and space-time resources, or, conversely, for destabilizing them. This is quite apart from long-term land allocation mechanisms such as land inheritance, redistribution or purchase.

In Sahlins' interpretations of Chayanov's rule, he found in the Gwembe Tonga case that:

> ... Chayanov's rule holds — in a general way ... although not in detail... The acreage cultivated/gardener [ratio] mounts in rough relation to the domestic index of consumers/gardener... (Sahlins 1972:106)

But for the Mazulu village discussed, nearly 60 per cent of the land cultivated consisted of *temwa* or swiddens (cf Scudder 1962:38). As we have remarked earlier (above p. 196-97 with respect to parallel cultivation systems, the Gwembe Tonga had ample opportunities to expand their holdings of swidden land with little restraint other than the labour time spent on the venture. But in the traditional cramped valley cultivation system on the inundated river fringe, Chayanov's rule if applied is unlikely to be very accurate, even when implemented on the basis of land actually worked rather than land legally held (people often exchanged plots to suit needs).

Even Chayanov himself admits that the Russian case may not be widely applicable for reasons of nonpartible inheritance and similar conditions of land tenure (domain structure):

> Although we have established by analysis of group averages that the volume of peasant farming depends on family size and compo-sition (... the higher limit is determined by the maximum avail-ability of the family labour force, and the lower by the minimum means of existence for the family), in order to avoid incorrect treatment of our conclusions we ought to stress that at any par-ticular moment the family *is not the sole determinant of the size of a particular farm*, and determines its size only in a general way. (Chayanov 1966:69)

Sahlins' concept of the 'Domestic Mode of Production'

Using Chayanov's Rule, Sahlins (1972) outlined what he regarded as a general condition prevailing in primitive (preindustrial) societies based on hunting-gathering, swidden agriculture or medium intensive agriculture, namely that of *underproduction* by domestic units. Comparing a large set of ethnographic case materials, he examined the intensity of utilization both of human time and land resources. Underproduction (of goods, since he says nothing of services) and underutilization of resources was so typical that he designated this condition as a specific 'domestic' mode of production (the DMP), which had three main characteristics: 1) the underuse of land resources, 2) the underuse of labour power (i.e. time resources for 'work', cf Chapter 9 further on), and 3) that intensity varies among domestic units in such a way that 'an interesting percentage of households chronically fail to provide their own customary livelihood.' (Sahlins 1972:74).

Populations in these kinds of society not only live well within what we have defined as the living possibility boundaries in Chapter 2, but they also produce far fewer goods than are within their 'objective economic possibilities.'

> ... the primitive economies are underproductive. The main run of them, agricultural as well as preagricultural, seem not to realize their own economic possibilities. Labour power is underused, technological means are not fully engaged, natural resources are left untapped. (Sahlins 1972:41)

Underused labour power implies that the average length of the working day is relatively short compared to more intensive kinds of agriculture and that people do not work 'continuously'. /Cf Chapter 9 on variations in work load./ Nor is land used except far below carrying capacity. Sahlins lists a number of swidden agriculturalists, such as the Tsembaga, Lamet, Iban and Lala, as well as peoples with medium intensive agriculture (as here defined), for instance, Gwembe Tonga, Kapauku and Chimbu. He concludes that:

> We have noticed that the domestic mode of production is *discontinuous in time*; here we see that it is also *discontinuous in space*. And the former discontinuity accounts for a certain underuse of labour, the latter implies a persistent underexploitation of /land/ resources. (Sahlins 1972:98)/My italics/

Before entering into a discussion of the specific dimensions of the DMP, the general point should be made that in view of the rather weak data base, methodology and conceptualization of resource use found in some of the literature sampled by Sahlins, it is very difficult task indeed to determine the actual utilization of resources, let alone the 'objective economic possibilities' or many of the general living possibility boundaries as we have defined them.

Sahlins contends that slash-and-burn cultivation lends itself uniquely to 'quantified assessments of economic capacity'. But as we have tried to show, numerous problems crop up when carrying capacity is measured mainly in terms of some regional population maximum, on the assumption that technology will remain constant, that settlement structure does not vary, or that people will go on cultivating in the traditional way unmodified by change in internal or external circumstances. Such assumptions of inflexibility and non-adaptability are hardly tenable and, regardless of whether population growth is a prime mover or not, Boserup was at least right in pointing to the great potential for technological change as a means of human ecological adaptation. (Again technical change is hardly autonomous but is a consequence of social system dynamics.)

Throughout this book we have tried to demonstrate how time and path allocation is intimately related to settlement and activity structure. In this perspective, Sahlins's assumption that 'critical density' in swidden agriculture (for instance) can be assessed in terms of average regional population density — as if human activity was *unfocused* in space — does not invite credibililty. In regions or large territories, a good deal of the land cannot be used in the daily round anyway, due to coupling constraints and prism effects of the kind illustrated in Figures 5:10-12. To reckon production possibility boundaries in terms of regional population densities is not even adequate when the population is dispersed as in the case of the Kofyar. Even then activities take place in dwellings, fields and other locations, serving as stations of return and reunion. Human activity and interaction is *spatially focused*.

If anything is typical, it is the fact that hunter-gatherers, swidden agriculturalists and other cultivators exploit *nearby* resources much *beyond* carrying capacity, and they frequently respond to incipient deterioration of the environment by moving to a new place. As Chisholm put it long ago:

> Shifting cultivation displays in extreme form a principle of widespread importance: while the further lands are unused or underused, the nearer are exploited beyond the limits which they are able to sustain indefinitely. (Chisholm 1962:71)

Taking this proposition one step further, it is precisely because local populations exploit their land in this way that they are in a position to 'underuse' their time and spend less of it on food production with associated travel. Put crudely in the time-space idiom, such populations overoccupy space-time in order to save human time (and spend

it on more desirable activities instead). Such a conclusion is certainly
at odds with Sahlins' tenet that both are underoccupied. On the other
hand, if people were clumsy in their adaptation to space, using unnecessary travel time, this would bring production closer to the
production possibility boundaries, since travel can absorb a great
deal of time in the daily round.

Sahlins' discussion on the role of village size is also debatable but
consistent with his thesis of underproduction. He seems to elevate a
short statement by Carneiro into a general rule:

> ... villages seldom get a chance to increase population to the point
> at which they begin to press hard on the carrying capacity of the
> land... a factor of greater importance has been the ease and fre-
> quency of village fissioning for reasons not related to subsistence.
> (Carneiro 1968:136, quoted in Sahlins 1972:98)

However, the world is full of instances where local populations with
their specific technology do press hard on the land. But the point
is that even villages of relatively small sizes (say in the region of
200 individuals as among the Iban, Lamet and Siane and numerous
others) 'press' hard enough on nearby resources around a village (a
prism centre) for them to shift residence at given intervals, or alternatively they have to spend more time on travel (the men sometimes
delegating this extra travel to the women, as in the Siane case).

Sahlins goes on by further renouncing the role of 'pressure on
land' in relation to other factors:

> In a given cultural formation, 'pressure on land' is not in the first
> instance a function of technology and resources, but rather of the
> producer's *access* to *sufficient* means of livelihood. The latter is
> clearly a specification of production and property, rules of land
> tenure, relations among local groups, and so forth. (1972:49)

This is clearly a theoretical muddle since, firstly, production can
hardly be specified without reference to resource ingredients, such as
land and the landscape, the latter often being a *cultural* landscape,
i.e. a natural landscape already modified or in part produced by
humans (e.g. through the impact on the vegetation or topography).
Secondly, production cannot be specified without reference to technology, which is obviously cultural and not extraneous to society.
But the main point to be made is that if one is to tackle structural
problems this is not meaningfully done by identifying a collection of
factors and then sorting them out into two crude classes, one being
'social' and the other 'techno-environmental' (or even pushing
beyond this by naming one as determinant). It is more a question
of looking into how structure is the outcome of the mutual imposition of constraints defining the limits to possible and impossible

forms (cf below).

We demonstrated earlier on how accessibility to land and the intensity of land use was a function both of time resources/time distance and normative-institutional factors such as land tenure. Thus the spatial fragmentation of holdings as well as the domain structure with its proprietary fragmentation of holdings play a part, but it is because an *institutionally* defined holding has a given *physical* size that a certain level of intensity under a given form of technology is arrived at. Likewise, accessibility is affected by transport technology, as has previously been demonstrated in a number of contexts. This deserves to be mentioned considering Sahlins' statement of above, where 'access to sufficient means of livelihood' is regarded purely in terms of regulatory/institutional constraints. This is curious in view of the fact that in his discussion of hunter-gatherers, Sahlins (1972) incorporated the role of movement, settlement and habitat in his discussion of mobility versus storage. But while he explained why hunter-gatherers shift residence, the topic is ignored in his treatment of swidden agriculturalists. This is in sharp contrast to one of his early essays on Moala in the Fijian islands (1957), where the factor of field-to-dwelling distance was ascribed a structural role that was out of proportion to its real significance. In 'Stone Age Economics', however, this factor is dropped all together. It is also interesting to note that Chayanov, whose theory and method Sahlins (1972) uses, was heavily influenced by von Thünen and his land use zones based on transport and distance cost.

In spite of the virtues of Sahlins' (1972) work, he is also inconsistent with respect to the structural scale at which he tests his underutilization of resources hypothesis. When dealing with *land*, he operates on an average regional level as Boserup does, although he explicitly rejects the idea that 'population pressure' at this level may contribute to intensification. On the other hand, when he discusses the underutilization of *time* for material production within the DMP, he has skipped the local community level (apart from a few comments on the fusion and fission of villages) and moves straight down to the micro-level of household groups.

Any theory propounding the underutilization of land resources is bound to be incompatible with ideas of population pressure on land. Defining the DMP at the household level (but testing it at the regional level), Sahlins concludes that:

> the results /of empirical carrying capacity studies/, although highly variable, are consistent in one respect: the existing population is generally inferior to the calculable maximum, often remarkably so...

However, he goes on by admitting that this does not,

... preclude that localized subgroups (families, lineages, villages), under the given rules of recruitment and land tenure, will not experience 'population pressure'. Such is a structural problem, not posed by technology or /land/ resources *per se*. (Sahlins 1972:43, 46)

But the width of this structural problem (and the extent to which it involves the structure of society in a narrow sense or that of society-habitat as a composite formation) is not specified further. Sahlins does not present any model of land use or carrying capacity with respect to villages or households as units of focal activity in space. Instead the problem of land use intensity is tackled by means of a high level proxy, one resting largely on the imprecise notion of maximum man/land ratios. But it is exactly this kind of aggregate model which has stimulated so many crude formulations of population pressure against which Sahlins rightly objects. Aggregate man-land models have also fostered false population versus habitat dichotomies. So while one can readily subscribe to Sahlins' (1972:49) scepticism of simple arguments of demographic cause and go along with his idea that 'the definition of population pressure and its social effects pass by way of existing structure', his broad statement that '... quite generally among tribal cultivators, the intensity of land use seems a specification of the social-political organization' and little else, is too exclusive. As we demonstrate in this book, there are several additional factors and forces behind land use intensification.

Returning to Chayanov's rule and the intensity of land and time use by domestic units, Sahlins' total rejection of demographic cause is also remarkable in view of the fact that Chayanov's rule is not just a cross-sectional comparison of household size and work level. When looked upon in a time-series or diachronic perspective, Chayanov's rule is nothing but a theory of population pressure. Chayanov never hesitates to regard growth in the household population as the *independent* variable of intensity. He shares this view with Boserup, who similarly regards population growth as the independent variable in her intensification functions. However, to claim that population growth has certain consequences is not tantamount to the reduction of causality to a factor extraneous to society — a thing that seems to be vehemently feared by all sociologists. An increasing population, which is largely what Chayanov envisaged, may be the result of a conscious process or strategy at the household level in response to given institutions, resource demands or socio-environmental predicaments. But more about that in Chapter 9.

In regions, such as India and Bangladesh today, where the general regional population density is very high, where the average land use intensity is similarly high, and where the distribution of land among households is uneven so that some households use land much more intensively than the corresponding regional average, 'the producers' access to sufficient means of livelihood' (including land), in Sahlins' words, is a specification both of population distribution, time resources, land resources and technology, as well as various institutional rules regulating access, and the proprietary distribution of land. We shall return to Sahlins' versus Boserup's ideas of intensification in Chapters 7 and 9 (suggesting some further modifications needed in both cases). However, when dealing with society-habitat and its structure in given places and time periods, it is wise policy to refrain from applying simple forms of causation or determination, for instance with respect to intensification and other topics of resource utilization. The advisable thing is rather to adopt an approach in which a given structure of society-habitat is seen as the result of the interaction of various constraints (positive and negative), between and within multi-level systems and sub-systems.

POPULATION, ACTIVITY AND SETTLEMENT AS A COMPOSITE SYSTEM: A TIME-SPACE STRUCTURALIST PERSPECTIVE

We can now return to the topic of structure as the interaction and aggregate outcome of constraints (positive and negative). We propose to look into this theme through an analysis of the interrelations between the population, activity and settlement systems primarily, but also through the sub-systems contained within them. To aid us in this task, we have already in previous chapters partitioned the composite population-activity-and-settlement system according to level of aggregation, starting with the regional (inter-local) level, and moving downward to the local, household and individual level.

The population system

In the entity referred to here as society-habitat (to emphasize its joint nature), one major sub-system is the population system, the *web* of *interacting* individual-paths, which forms a system in its own right with its own logic, that of *path allocation* in time-space. Much of this was already outlined in Chapter 2, and various aspects thereof have been described in subsequent chapters. Regardless of the particular society and time-space setting chosen, there are some very general capability and coupling constraints affecting path allocation, such as those dictated by human indivisibility and limited speed of travel. Some of these constraints can be posited as exogenous variables (e.g. indivisibility or the fact that there is always *some* limit to speed of travel, although the latter constraint is also a function of the *means* for travel at the disposal of a given population and the kind of terrain and settlement system in which movement takes place). A further factor built into the population system is that path allocation in some forms (i.e. to given time-space locations) always has a displacement effect on other forms of path allocation. This is related to time allocation by the population. The members are both carriers and sources of human time, but no configuration of individual and collective activities is feasible if it is incompatible with the possible configuration of individual-paths. In other words, the activity system is constrained both by path allocation and by the allocation of limited time resources (cf Chapter 8 and onward).

The activity system

The activity system likewise has its own logic. Activities are vehicles for reaching human goals and carrying out various individually and socially defined projects. Some activities are guided by underlying ideas, norms and values, or are more or less instrumental in character, but all institutions, forms of social interaction and even culture as a whole (cf Chapter 12) are expressed as activities. These activities not only demand the human time of specific population categories but they are also to a greater or lesser extent locationally time and place specific. Some activities are fixed or coupled to particular sites in the natural landscape while others are confined to time-space locations defined by convention and contract. But the activities are not randomly located in time-space.

Different activities are interdependent for several reasons: a) they may involve several individuals who have to coordinate their paths in time and space (i.e. are subject to coupling constraints), b) they may have to be combined in a certain sequence in order to lead up to a given kind of output, and c) different activities compete for the time of the same population and thereby displace one another. We have shown above, for instance, how more time spent on travel means less time spent with others in the village, or how increased input of time in agriculture may reduce the volume of ritual activity (e.g. the Siane case in Chapter 5). But we have further illustrated how the daily round is an integrated activity system in which component activities do not appear at random but have a structure. So even activities which belong to different sectors and are aimed at very different goals and projects become interlinked and interdependent although they were not intended to interfere with one another. This interference is simply a material outcome of the structure of the situation.

The settlement system and habitat

The settlement system is neither an isolated system external to society, something 'physical' but not 'social', as is often inadvertently assumed in social science, nor is it the sole system in society, an impression sometimes conveyed in geographic writings. The settlement system, as the human structured habitat, is an integral part of society. It is the joint outcome of interaction and confrontation between different elements and sub-systems, which are all located in space and time. Some features of settlement are epiphenomena of the population and activity systems, while others act as constraints on these two systems, much depending on the time scale chosen. In the daily perspective, the lay-out of the settlement system is fixed and taken for granted, acting as a constraint on many daily activities In the longer time perspective, however, there are changes associated with construction and destruction, with the turnover of buildings, roads, and other facilities or places of natural resource extraction. The structure of stations and transportation channels is thus the result of human action and environmental modification.

To carry out projects and activities, space must be overcome through movement, at the cost of time and energy, not least since people, resources and facilities occupy space, displace one another and become spatially separated by intervening objects and people. Many of the constraints on packing in space-time, regionally, locally or at the household level, have already been described.

While projects and activities may differ in terms of locational and space-time requirements, it is the same settlement system which must house them all. The structure of settlement and the human influenced habitat, the cultural landscape, is the result of demands pulling in different directions. The historically received structure of settlement is thus a compromise outcome within a *multi-purpose* spatial system between a multitude of *sectoral projects* pervading it. Some of these projects have had a dominant impact while others have had to adapt and fit into the remaining slots in a manner consistent with available time and space resources, path allocation constraints and other constraints such as regulatory constraints. Likewise, some population segments have been able to impose their projects on the settlement, population and activity systems to a greater extent than others, who have had to yield and adapt. In this sense, the settlement system is the arena of goals and policies of all kinds. (Cf Hägerstrand 1973b).

The settlement system is the outcome of compromises in yet other respects. By constituting to a great extent embodied human time — physical capital in the form of buildings, roads, fields, irrigation facilities, and the like — it represents a localized physical infrastructure, which it may have taken an enormous amount of human time to construct (cf Chapter 7 on irrigation systems). In many societies, major portions of this physical infrastructure cannot be abandoned without seriously impairing ongoing projects and socio-economic output. In such societies, it is generally less costly to relocate the population in space to fit the existing settlement structure than vice versa. Surviving even political upheavals, a highly built-up artificial habitat is subject to considerable inertia. Reforming the overall settlement system becomes a long-term project taking decades to effectuate. In other societies, such as many of those based on shifting cultivation, there are fewer coupling constraints to localized investments and therefore greater flexibility.

Intersystemic and intrasystemic relations within the composite society-habitat formation

As pointed out in Chapters 1 and 2, our concern in this study is not to find the ultimate determinants, prime movers or causes of cultural evolution, but to delineate living-and-activity possibility boundaries and limits to the manner in which a system or sub-system can be articulated.

Waddell in his excellent time and space utilization study of the Raiapu Enga with respect to society and ecology phrases one of his analytical problems as follows:

It is generally recognized that settlement pattern represents a concrete "... and obvious expression of the relationship between social institutions and the use and allocation of productive resources" (Brown and Brookfield, 1967:119). However, the exact, causal nature of this relationship is seldom the subject of detailed investigation...

Among the Raiapu there are striking /empirical/ regularities in the stability, composition, and distribution of residences. These regularities reflect, on the one hand, social patterns of marriage and divorce frequencies and male-female relationships generally, and on the other, the very close association between residence and garden. It is pertinent to the study that both, in turn, be given detailed consideration in order to segregate the independent from the dependent variables. (Waddell 1972:22)

But the problem is more complex than simply the sorting out of dependent from independent variables because, more basically, it is one of selecting strategic variables or factors and delineating in an analytically useful way different sub-systems, all of which have to be articulated in models. These models in turn must be attuned to the various real-world cases which they are designed to deal with. Thus, for example, in covering the relations mentioned by Waddell, such as residence, marriage and divorce, male-female relationships, or residence-garden relations, * these features may have to be reinterpreted into different analytical categories for use in models, in order to make it clear how these various factors are interrelated. The crucial importance of choice of categories is evident from Sahlins' misleading discussion of population pressure, intensification and the role of 'social' versus 'ecological' factors.

In this study, all too few of the possible relationships can be explored between the elements, sub-systems and systems in the total formation of society-habitat. This applies even more to the totality of *causal relationships*, of necessary and sufficient conditions for given outcomes. What we do have is a *web* of such relations and this is why simple, two variable cause-and-effect relations (of the population-growth-as-the-cause-of-intensification type) must be eschewed. In adopting a systems approach to the dynamics of society-habitat, we thus strive to replace simple causality with some kind of 'structural causality', or a causality at the level of systems or sub-systems rather than elementary variables. Elementary causality is rather built into the models (articulating these sub-systems) to begin with, as a matter of scientific design. This is why so much of this study has been centered around the building up of a *set of interrelated sub-models of human interaction* in space and within the population, since this is a precondition to higher level forms of structural and causal analysis. We thus follow Friedman in his definition of structural causality in terms of a kind of reciprocal causality within and between systems characterized by the *mutual imposition of constraints*:

> Expressed mathematically, it is analogous to mutually limiting functions in systems of equations which impose inequality side-conditions on one another. Here the functions are autonomous, but the range of values which they can take is limited by the other functions. Structurally, it is a case of constraints of the possible combinations of given elements or on variations in their relations ... Within this framework, a contradiction is defined as the *limit of functional compatibility* between structures ... (Friedman 1979:25-27).

In an essay on structural causality in the context of population, kinship, and production, Godelier (1975) departed from a marxist model (infrastructure-superstructure) of pre-industrial, non-European social formations. By defining relations of production according to

---

*   Cf the factors mentioned by Stauder at the end of Chapter 5 above (p. 185).

the structural job done rather than in terms of substantive or cultural categories, structural causation emerges as an ... 'effect of relations of production at a given level of development of productive forces, that is to say the mode of production, on other levels of social organization.' As explained in Chapter 2, we are not basing the present study on the infrastructure-superstructure model, and (although it borders on heresy), there is not even an entity referred to as 'the economy' in the present study.

In lieu of 'economy', 'mode of production' and the like, we take the liberty of installing *activity system* in this pivotal position. By thus incorporating human intentionality and normative action – the trajectories of mind and matter inherent in the concept of *project* – we ensure that our society-habitat formation is vested with a truly social and normative dynamic, and that human individuals and populations are neither relegated to be the staged puppets in a reductionist drama, nor the forlorn subjects to an endless array of negative constraints. However, by not proclaiming activities substantively defined as 'economic' or 'infrastructural' as having structural causal precedence, we leave the matter of determination open and rest content with looking at society-habitat as both the genitor and matrix of a 'struggle' between projects of all kinds,* (while fully realizing that certain kinds of projects have to be carried out, if people are to survive and the system is to reproduce itself.)

Having said this, the stage is set for discussing the struggle between projects that took place in Tanzania in conjunction with its villagization programme, between the modern projects pursued by the state and the traditional projects of the rural people, the arena being the settlement system.

---

* The 'struggle' between projects (cf Hägerstrand 1973b) can be generalized to include also those found in nature, again with an open mind as to whether or not the anthropogenic projects always get the upper hand. As natural hazards research has shown, humans are often the victims of the forces in nature in spite of the constant human efforts to channel the trajectories of other species or forces in the environment into directions which harmonize with human intentions.

STRUCTURE, CONTRADICTION AND INTENSIFICATION: VILLAGIZA-
TION IN TANZANIA REVISITED

Earlier on, we discussed some implications of the villagization programme in Tanzania and the possible adaptations to the new situation.
One of the consequences of increased sharing of the local prism
habitat was land use intensification. But little was mentioned as to
why this form of mobilization policy was introduced or what were
the other structural effects and contradictions of the political mobilization process in rural Tanzania. The centralized state pursued
numerous projects which were to be introduced or imposed in the
local communities of the country.

There are some interesting parallels between this process, and that
of intensification and household resource mobilization envisaged by
Sahlins (1972). Although he dealt with 'the economic intensity of
the political order' at the level of kinship, headmanship, 'big-man
-ship', chiefdom and tribe, these traditional polities in Tanzania must
also be placed within the contemporary framework of the state.

A major point made by Sahlins (1972:82, 130, 135) is that 'political life is a stimulus to production' and that,

> in the archaic /say preindustrial/ societies, *social-political pressure
> must present itself the most feasible strategy of economic de-
> velopment*... /given/ that resources are often not fully turned to
> account... between the actual production and the possibility there
> remains considerable room for maneuver. The great challenge lies
> in the *intensification of labour*; getting people to work more, or
> more people to work. That is to say, the society's economic
> destiny is played out in its *relations of production*, especially
> *political pressures* that can be mounted on the household eco-
> nomy. (Sahlins 1972:82) /Emphasis added/.

Since Sahlins assumed that the domestic mode of production was
inherently underproductive and that intensification for 'ecological'
reasons was virtually ruled out, the remaining forces of intensification
would be politico-economic (an effect of relations of production).
Between themselves, households caught in the DMP not only exhibi-
ted centrifugal tendencies with respect to economic relations but also
tendencies towards *spatial* dispersion. Within households, however,
the tendencies were centripetal (cf principle of return and reunion),
cooperative and towards pooling of goods and services. Within them,
'householding' prevailed:

> ... householding is the highest form of economic sociability: 'from
> each according to his abilities and to each according to his needs'
> — from the adults that with which they are charged by the

division of labour; to them, but also to the elders, the children, the incapacitated, regardless of their contribution, that which they require... (Sahlins 1972:94)

It obviously takes a big political push to extend this altruistic principle beyond the cosy range of the immediate domestic and minimal kinship unit, at least if it is to be widened to a village population of some one thousand individuals or more.

In Tanzania, a customary concept of aid and cooperation among kin was that of *ujamaa*. Seeking a foundation in tradition for a new form of indigenous socialism, an African socialism, the *ujamaa* concept was elevated to the chief symbol and vehicle for the national development programme, engineered by the policy makers working within the TANU party. As laid down in the Arusha declaration of 1967, *ujamaa* should mean living together, working together, owning together, pooling and sharing, an ideology greeted with much sympathy. But to turn this into praxis would amount to much more than checking the 'centrifugal tendencies' to which domestic units might be bent. It would involve a massive mobilization of human time and land resources; to extend the range of familial relations to entire village populations would be a time demanding process at the risk of creating unwieldy units with little cohesion other than spatial, if even that. Historically, policies of concentrating scattered populations have been common, the objective having been to place people under the control of the state, to protect them, to mobilize their resources, to make them pay taxes or simply to make them enter the regional market. In Tanzania, this list was augmented by modern objectives such as providing people with services, technical facilities and new production inputs. The state was to give and take, not merely the latter.

The *pooling* of resources and *sharing* of products and facilities assumes spatial interaction between formerly dispersed units. For good logistic reasons, spatial interaction is facilitated by proximity (as widely observed, while the time and path allocation reasons behind this fact are not). When people are concentrated in space, the sharing of scarce facilities such as schools, roads, agricultural equipment, wells and piped water systems, and other *indivisibilities* can be done at a lower cost and the capacity of such systems can be utilized better (i.e. with higher temporal intensity). But as events were to show, there were great disutilities of public utilities such as pumps breaking down. If they had been the main attraction to people moving into villages, they quickly became symbols of failure and disenchantment, as people had to walk great distances daily for water while waiting for spare parts that take months to arrive. And there is one basic facility which is not shared better when population gets concentrated to the extent assumed in the *ujamaa* programme, namely land.

There is a wide difference, however, between political intention and project implementation, and to translate policy into action requires people who can do the job. We know from Melanesia how 'big-men' assume leadership and set examples, working harder than others and exhorting also them to work more. Such 'big-men' serve as intensifiers of production, according to Sahlins (1972), and kinship networks fulfil similar functions. But to adopt this strategy on a modern national scale takes a horde of mobilizers, and without this capacity to back up the programme, it hardly helps to move people together into villages.

The 'big-men' of modern times would be the agricultural extension agents, the teachers, the medical personnel, the managers of co-operatives, the authorized dealers in goods, the administrators and the local politicians. Through them the state can reach the people for the good of the people. This service-giving minority must *meet* the service-taking majority in concrete path-to-path interaction, but such interaction is very time demanding (cf Chapter 11 on the time allocation problems of service exchange). If overall service-giving capacity is low, due to the relatively small number of people in this category, the state must get the most mileage out of this limited set of service-givers; not on the roads, however, but in actual contacts with the people. As one leading development planner puts it:

> ... the most difficult task /is/ the actual contact /surface/ with farmers. I think we probably all agree that the idea, if anyone still has it, that the extension service can visit all of some *million farmers* is simply impossible ... /experience has shown/ ... how far we are from covering even one quarter of the farmers... So the idea, I think, must be abandoned; and therefore we have to think of *groups*... (Hunter 1976:199) /Italics added/.

In Tanzania, the government did not think merely of groups, but of whole *villages*! But this was to a considerable extent a gut reaction policy. It was based on the assumption that if only people could be brought together spatially on a massive scale, they would be *reached*, they would reach one another, work together and generate development. It was this policy, untempered by considerations of capacity and capability, or the fact that ordinary people had their own traditional (but still) projects, that was to pave the way for a good deal of frustration, revealing underlying contradictions.

Perhaps one of the main contradictions was in the realm of time resource utilization by ordinary cultivators as opposed to that of the cadre of specialist mobilizers. Many of the services given in the village context, with the possible exception of roads, schools and wells, were used only *intermittently* by the cultivator households. But in order to use these services people were concentrated into villages on a *permanent* basis. So for the adult population to enjoy a few hours of services per week, they had to spend many extra hours travelling to fields, perhaps remaining there throughout the day rather than being able to walk back and forth several times per day. The time booster given to the service-givers is probably manifold repaid in travel time by other segments of the population, because of the structural contradiction in the settlement system between space-time consuming cultivation activity, and centralized service activities requiring relatively little space-time. Of course, if it is assumed that ordinary people suffer no time constraints, there is no need to worry about the time resources of the population majority. But most writers on the under-productive use of human time in rural settings fail to realize that travel can absorb a substantial amount of time in settlements where daily travel distances have been increased. They further tend to forget all other travel associated with activities such as fetching firewood or herding livestock. Although the introduction of tubewells, piped water and other facilities may have had considerable time-saving effects in the daily round of some Tanzanian villages, the whole policy of a contact-intensive closely knit village life with fields nearby (as in Figure 6:15 top) may largely have been disrupted by the increased distances to fields resulting from the fusion of settlements.

Some of the problems of village size were aggravated by bad siting of villages in many cases. Inoptimal siting had its roots in the same constraints as many of the other problems, viz. the limited capability and capacity of government agents to deliver needed services. There were simply not enough 'manpower' or specialist time resources to allocate to careful planning, field-work and implementation of the programme. Who had the time to scout around and weigh different factors of location such as soil quality, access to water, proximity to existing or future roads, good plots for actual buildings, presence or absence of tse-tse infestation, distances to other villages, and so on?

Political impatience and desire for many things to happen fast regardless of quality also made things worse, because when a huge task is to be undertaken by a very limited volume of suitable personnel in, say, three years instead of twelve, capacity shortage is increased by a factor of four. Somebody had to suffer the effects of such hasty implementation and at the receiving end were the ordinary rural people. With lowered productivity and increasing food shortages, the urban population and its élite were also affected.

Let us end by briefly returning to Sahlins' ideas on the political factors behind the mobilization of human time and space-time resources. One may readily agree with his point that,

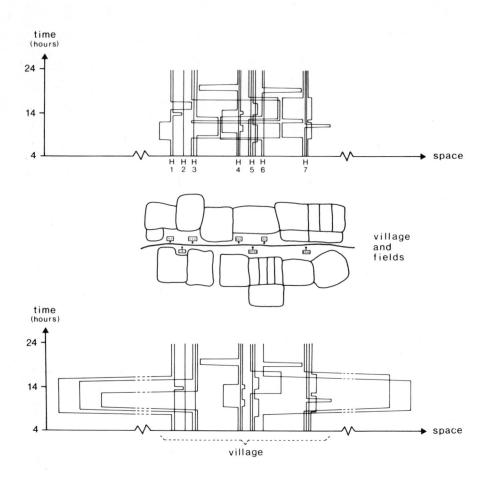

FIGURE 6:15   The structure of a village which is close-knit and contact intensive internally (top), and that of a village which is more open in its contact structure due to factors such as a high proportion of its inhabitants working outside
the village (bottom). /H: household./

Often one effect associated with modernization of a village or the break
down of local self-sufficiency of a village (cf Wilkinson 1973 for the latter
concept), is that villages are opened up so that a greater frequency and duration
of total contacts are taken with individuals residing elsewhere. With industrialization or commercialization a transition from the closed state to an open state is a
common occurrence throughout the world. This balance between external and
internal contact, participation and interaction also affects the maintenance of
local as opposed to inter-local, regional and national culture.

The graph is thus a principal one which is applicable to a diversity of contexts.
In the present context, it serves to show the difference between the close-knit
kind of villages envisaged by the Tanzanian government and the open kind of
village with less internal contact which most likely resulted.

> As the structure is politicized, especially as it is centralized in ruling chiefs /or state government servants for that matter/, the household economy is mobilized for a larger social cause... the primitive headman or chief ... /or village administrator/... incarnates collective finalities; he personifies a public economic principle in opposition to the private ends and petty self-concerns of the household economy...
> Everything depends on the political negation of the centrifugal tendency to which the DMP is naturally inclined... (Sahlins 1972:130-131 on the 'Economic Intensity of the Political Order'.)

At another level of scale, one could look upon local populations as analogous to domestic units and see the state as the political and economic mobilizer. This is the role the present Tanzanian elite and state apparatus has chosen for itself, and although it may be criticized, this role is reasonably justified in the world of today.

But what we have here is not simply a contradiction between forces and relations of production or between classes. If we take the projects of the state, the dominant polity, they are themselves far from mutually consistent. Quite the contrary, the state system is laden with contradiction and the limits of functional compatibility between different projects and sectors have been surpassed by a wide margin. For pre-state society, Sahlins mainly called attention to the contradiction between centrifugal and centripetal forces, that for 'prevailing institutions of hierarchy and alliance' ... 'success is measurable rather by the concentration of population' ... and that,

> My concern here is merely to indicate the *problématique*: each political organization harbours a coefficient of population density, thus in conjunction with ecological givens, a determinate intensity of land use. (Sahlins 1972:131)

This may be, but if we want to understand why, we must look into the various projects pursued by different individuals, groups and organizations so that we discover also the internal inconsistencies and contradictions between projects. Hence, in Tanzania, land use was not primarily intensified because of political exhortations to people to work more and produce more, but because people in the new situation probably had to travel more and spend more time to cope with the ecological exigencies of the new situation, e.g. declining fertility of land. The political organization of modern states does not harbour one coefficient of population density at all, but rather a series of conflicting pulls, depending on the projects chosen and what remains of them when they are implanted into the real world constraint system.

Mobilizing and economizing on human time in the settlement system

Without having reached a very complicated or advanced level of structuralist theorizing, we have nevertheless tried in the preceding chapters to give some examples of the mutual imposition of constraints between various sub-systems within society-habitat. This interdependence (or reciprocal causality) was found at all levels of scale, from the region down to the household.

The mechanisms described, whether simple or composite, show how spatial location and the packing of activities into the settlement system impinge on human time and path allocation, or conversely, how the packing of activities in the population time-budget affects their packing in the space-time budget of a region, locality or household. One major mediating mechanism is that of the prism, and the effects of time and path allocation on the carrying capacity of the local prism habitat has been amply illustrated. Moreover, the artificial or man-influenced habitat, the cultural landscape, is also constrained by the limited time resources for activity, since it takes time to transform or structure the habitat (cf 'structuration', p. 62).

The attempt has also been made to demonstrate that there are several structural conflicts, incompatibilities or contradictions inherent in the *synchronized allocation* of population-time and settlement space-time. Apart from the conflict between stationary and mobile activities in the daily time-budget — a conflict which has to be resolved by economizing on time in space — another typical contradiction is that between the time gains in (high intensity) person-to-person interaction accruing from the spatial concentration of population, and the excessive sharing of the local prism habitat that this might entail. Furthermore, there is the contradiction between storage and mobility, from its simplest forms found in hunting-gathering societies to more complex forms.

There are thus mechanisms for economizing in human time within settlement space by organizing activities into certain time-space daily routines. There are also mechanisms for economizing in settlement space-time (room) so that more or less space-time demanding activities can be accomodated. Since all human activity involves the dual occupation of time and space resources, the *economizing* in these resources is always *synchronized* and subject to coupling constraints. This is one of the themes in the next chapter, where water(-time) is added to the two resources already much discussed.

7 IRRIGATION AGRICULTURE

THE ECOTECHNOLOGY OF IRRIGATED CULTIVATION

Water is a crucial input in all cultivation of plants; without an adequate supply of this basic resource, any cultivation project would fail. In temperate regions with otherwise sufficient moisture, water is essentially a constraint in the timing of agricultural activities in the annual cultivation cycle. It may, for instance, be an impediment to the harvesting of crops, if too frequent or abundant. In the arid or semi-arid tropics, however, water is generally a more decisive constraint on cultivation than the availability of fertile land. For cultivators in these regions, unwatered land is indeed sterile land.

Irrigation, in all its simple and complex forms, is the human technique of manipulating water as an input in cultivation projects. As such it is ancient, dating back thousands of years to the fifth millenium or so B.C. Irrigation systems formed the technological backbone of the earliest large-scale societies, the pristine states of Egypt, Mesopotamia, the Indus Valley and China. Some of these societies also produced an urban revolution and large cities, centuries and millennia before the advent of the industrial and urban revolution in Europe.

Historians of all kinds have placed great emphasis on the role of irrigation systems in the ancient states, although the topic has been highly controversial. A watershed in the analysis of irrigation-based societies was the appearance of Wittfogel's (1957)* treatise on the roots of 'Oriental Despotism', a political system which grew out of the particular form of bureaucratic management thought necessary to initiate and maintain large-scale irrigation systems. Although his theory has been subject to much criticism but also to supporting assertions (cf Harris 1977:153-63), the fact remains that irrigation technology in its advanced applications constituted the first large-scale, region-wide (not to say interregional) technosystem devised by Mankind. As such it is bound to have a number of social

* Wittfogel actually presented his 'hydraulic hypothesis' in the 1930's.

structural concomitants of pervasive character. In the general analysis of technology, irrigation systems are also of great significance, since in the world of today the large-scale kinds of technosystems have diversified and multiplied in number (e.g. communication and transportation systems, energy transfer systems, region-wide computer-cum-information systems, military weapons systems, etc.), systems which through their joint and cumulative impact have given rise to virtually intractable management problems.

By subsuming water under Land rather than treating it as a specific resource, as a *factor of production* (like human time and energy) water has largely been ignored in economic and social theory. This is only one reason for the existing gaps and incongruities between economic and ecological theory. However, in rural and urban, as well as in ancient and modern societies, water is anything but an insignificant ubiquity. In the historical perspective, for instance, Wittfogel (1956:152) points out that,

> The peculiarities of agrohydraulic civilization become apparent as soon as we realize the role that the management of water has played in the subsistence economy of certain agrarian societies.

This is true for regions such as the Nile Valley, Mesopotamia, Indo-Pakistan, Indonesia and China to this day.

Irrigation is one of the available technologies for human interception in the Hydrological Cycle. The latter, of course, is not a cycle in the temporal sense of the term used in this study but rather a *circulation system* consisting of the *storage* and *transfer* of water in nature and society. (It is, in fact, analogous to what was time-geographically defined as a system of stations and transportation channels guiding the 'flow' of paths (human and non-human) in time and space.) Human capability to influence the *precipitation* stage of the hydrological cycle is still very limited even today, while the *runoff* stage is more amenable to manipulation. As far as agriculture is concerned, there are two complementary facets of runoff; the first being the *application of water,* i.e. irrigation proper, the second being the removal of excess and undesirable water, i.e. *drainage.* In temperate climates like that of Europe, drainage has historically been the dominant problem; in the arid or semi-arid tropics, irrigation becomes crucial while drainage remains very important.

Apart from the actual *watering* of land, a second important but often overlooked function of irrigation is *fertilization,* since the water from many watercourses carries with it substantial quantities of silt and minerals which are deposited on the land. (We have noted in earlier chapters the role of fertilization in the intensification process.) Sometimes, in conjunction with inadequate drainage, however, a state of 'malfertilization' may arise in the form of accumulated salt deposits. This is commonly due to a high water table in conjunction with unfavourable rates of evaporation and percolation. In some areas, positive fertility is only maintained because of the regular input of waterborne alluvium and nutrients. The third function of irrigation systems is that of *water control*. By means of weirs, dams, dykes, etc. the negative effects of too rapid runoff and flooding (typical of dry regions with scant vegetation or steep terrain and the like) can be considerably reduced. A complementary aspect of water control relates to the temporal and spatial distribution of water, e.g. the modification of an agriculturally unfavourable timing of watering in the annual cultivation cycle.

There seems to be wide consensus in the literature that in order to enjoy the advantages accruing from the use of irrigation, a number of additional functions-activities become necessary in such systems as opposed to those operating within a pluvial regime:

> Irrigation demands a treatment of soil and water that is not customary in rainfall farming. The typical irrigation peasant has (1) to dig and re-dig ditches and furrows; (2) to terrace the land if it is uneven; (3) to raise moisture if the level of water is below the surface of the fields; and (4) to regulate the flow of water from the source to the goal, directing its ultimate application to the crop. Tasks (1) and (4) are essential to all irrigation farming proper (inundation farming requires damming rather than ditching). Task (3) is also a frequent one, for, except at the time of high floods, the level of water tends to lie below that of the cultivated fields. (Wittfogel 1956:157)

Wittfogel also contends that irrigation systems give rise to the functional necessity of elaborate water management and a class or category of people performing this activity which in turn tends to produce a specific kind of polity, a hypothesis that has largely been refuted (cf i.a. Gray 1963, Hindess and Hirst 1975, Leach 1959).

From a time-geographic structural perspective, a number of interesting time-space mechanisms emerge in systems based on irrigation ecotechnology, although *only a few instances can be furnished in this short chapter*. The reallocation of water in time-space permitted by the construction of irrigation systems and the investment of human time in them, permits both the *intensification* of land use (cf Figure 5:4), for instance through double or multiple cropping, and the *spatial expansion* of cultivated area (cf Figure 6:3 c and d). In fact, although agriculture is feasible without irrigation in semi-arid regions, at least seasonally and with greater risk, it is chiefly through the development of irrigation systems that agriculture can expand into the arid zones of the earth. Irrigation ecotechnology thus relaxes some of the capacity constraints associated with natural precipitation, perhaps chiefly by moderating the temporal distribution of water input and concentrating it to desired phases of the annual cycle. On the other hand, precipitation is generally more spatially dispersed while irrigation implies concentrated sources of water. Hence, irrigation ecotechnology invariably brings about a number of new coupling constraints as well as new authority and regulatory-contractual constraints of water allocation and control. We will thus look into the interaction between capability (e.g. divisibility), coupling and regulatory constraints as they affect the temporal allocation of water and other time-space mechanisms operating in irrigation based societies-habitats. The chapter is then ended by linking these interpretations to the main theme of land use intensi-fication and the capacity for human interaction in space and time.

LOCAL IRRIGATION SYSTEMS

The assumption is frequently encountered that irrigation systems are typically region-wide and large-scale. In many regions, however, local small-scale and even farm-size systems predominate, as in the oases of the Sahara and Arabia, in southeast India or in Ceylon. More-over, different forms of irrigation are quite often combined, e.g. well and canal irrigation, just as irrigation itself is supplemented by natural precipitation. A wide spectrum of watering techniques can thus be found (cf i.a. Manshard 1974:68).

> Consider, for instance, the wide range of differences between irrigation dependent on springs, wells, mountain streams, rivers, marshes, or converted ocean waters. The basic requirement that water must be diverted to the crop growing area would be met, in each case, by a widely differing technology with varying social implications. (Fernea 1970:157)

Taking first among the components entering in different combina-tions into local irrigation systems the *source of water*, this may be a diverted stream, a river, a spring or artesian well, or a tubewell. If the *storage* component is also included, the source may be a reser-voir, pond, tank or even a cistern. The capacity for storage in these containers varies with factors such as evaporation and seepage. As for the *water transfer energy* component, the time and energy costs for water transportation per unit volume of water is generally least in gravity-flow irrigation and when actual water lifting can be avoided. This in turn makes specific demands on the topographical lay-out to the obvious effect that the source of water should be located higher up than the destination. However, although water lifting may involve the expenditure of much human time and energy (as well as that of animals such as oxen or camels), water lifting from wells may be a very worthwhile proposition. With industrial technology, of course, motor pumps or electric pumps have relaxed pre-industrial capability constraints to an almost revolutionary degree in many places, as has the diffusion of tubewells (cf i.a. Nulty 1972). Turning then to the *channel or aquifer* component, ditches and canals of some variant are the common forms, but the transmission losses through seepage and evaporation vary with the general spatial layout and the kind of material in the aquifers as well as the strategy of water allocation in time and space (cf below on watering efficiency). Finally, there is the *field system* component, which may be more or less constructed to suit the other components and operations in the composite system. In most cases some form of bunding, levelling and/or terracing is necessary to keep watering and cultivation efficiency above given minimum levels. Although terracing is commonly associated with irrigation, it is also frequently used for other reasons (e.g. to use hillsides in a situation of land shortage), and in some of the great alluvial plains (e.g. the Indus valley), the slope of the land does not require terracing. Extreme instances of irrigation waterworks in association with monumental terracing are found in China and the Philippines. Given these various components, some cases will now be presented which show the scope for different combinations and their actual and possible operation.

The Sonjo system combining springs, streams and gravity-flow

The Sonjo of northern Tanzania described by Gray (1963) inhabit an arid zone of meagre and unreliable rainfall. They combine cultivation and 'permanent' settlement with goat pastoralism in 'temporary' camps.* From an intensification theoretic viewpoint the Sonjo case is telling. Their fortified settlements for defence against the Masai have arisen as a typical impaction effect (cf p. 176, and 196-206), and cultivation is intensive both for reasons of high demand for space-time in the local prism habitat associated with village size, as well as because of their irrigation system. Their choice of crops with a sizeable portion of tubers, sweet potatoes and cassava, contributes to a higher land use intensity. Irrigation not only facilitates

---

* This combination of fixed agriculture and camp settlements from which domestic animals are herded as well as the kind of habitat is very similar to that of the Karimojong described earlier on (p. 127-30). However, the Karimojong did not irrigate, and according to Dyson-Hudson (1970:121), 'There are problems in damming the rivers for irrigation. The large rivers come down with such force that it is almost impossible to harness them...' Irrigation projects in the area have failed because of such floods.

a compact settlement system but also double cropping. Unlike the Kofyar (p. 223-4), Sonjo farming is 'unmixed' and their goats are not stall-fed to provide manure but are herded; nor are animals used to lift water. These are important factors reducing Sonjo labour inputs comparatively. Sonjo irrigation also maintains the fertility of their fields which are located at the valley bottom and adjacent lower hill-sides. In a typical village, a stream which follows the escarpment,

> ... is jointed by the water from three springs rising from a spring line near the bottom of the escarpment... in mid dry season, the stream supplied about one quarter of the total water supply. On the other hand, during the rainy season /in wet years/, when water for irrigation is normally not required, the stream increases gently in flow and frequently inundates the whole valley bottom. These periodic floods undoubtedly maintain the fertility of the valley soils by depositing alluvium, thus making it possible to plant crops year after year on the same plots. (Gray 1963:50)

The Sonjo practise double cropping, one crop per annum on each of their two kinds of fields. On the valley-bottom land *(hura)* the principal crops are sweet potato, bulrush millet and sorghum, cultivated in the *dry* season exclusively by means of irrigation. The second kind of fields *(magare)* are located on the adjacent sloping land. They are cultivated in the *rainy* season and are both rainfed and irrigated. Should there be too much water in the stream in the rainy season, excess water is channelled to bypass these sloping fields, at the same time as valley land in this season is risky to use since crops there may be damaged by temporary floods. Since the sloping land is not fertilized by alluvium, it is fallowed every other year, and each household consequently holds two sets of *magare* land on which sorghum is the main crop and millet the secondary crop. The system is outlined in Figure 7:1.

In the perennial perspective, crops are produced in abundance but in a year of very low or badly timed rainfall, the loss of crop may be total on the hillslope land. In such situations the water from the springs always safeguards the second crop on valley land *(hura)*. It is not just the amount of water which matters but also its time distribution and reliability, the latter being the great merit of the springs as opposed to the surface streams which vary with the weather.

The Sonjo irrigation system is simple but effective:

> Owing to the circumstances that the sources of water are close to the cultivated fields and that the land is relatively flat, no very difficult works of hydraulic engineering were required in constructing the irrigation systems at any of the villages. (Gray 1963:52)

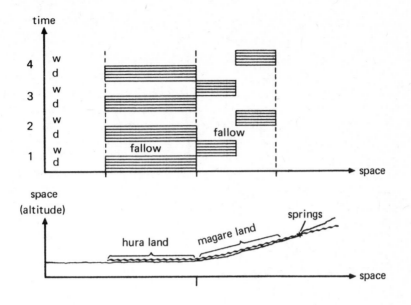

FIGURE 7:1   The Sonjo system of land use with valley and hillslope fields. The temporal intensity in the occupation of land is shown in the top figure. The valley land *(hura)* is used twice as intensively as the *magare* land. (*d*: dry season; *w*: rainy season.)

This point is interesting in that it does not support Wittfogel (whose hypotheses Gray actually sets out to test.) The Sonjo have built only small earth dams reinforced with revetments of small timbers from which small dykes conduct the water into the nearby fields.

> By manipulating the sluices the water may be shunted entirely into one of the channels, or it may be divided so as to flow concurrently through several of them. The sluices are opened by merely removing earth from the dam to allow the desired flow of water. (Gray 1963:53)

The joint utilization by many people of one or a few concentrated sources of water certainly imposes coupling constraints, not only between water and water-demanding plants, but also between people. If the climatic capability constraints on plant growth are to be relaxed, new coupling constraints as well as regulatory-authority constraints tend to be imposed:

> The successful operation of the irrigation system, upon which Sonjo society is dependent, requires the close cooperation of the men day after day. It requires minute planning and continual

supervision. Authority must be unified and loyalties must be un-
divided, for any overt social conflict may disrupt the operation
of the irrigation system and cause irreparable damage to crops.
(Gray 1963:170)

Although cooperation at a general level need not exclude competition, and although overt social conflicts may be disruptive, Gray
perhaps overstates his case on the necessity for absolute unity.
However, the use of a water flow from a concentrated source and the
necessity of dividing it up using various technical and social mechanisms, does impose a number of coupling constraints and regulatory
constraints on water capacity utilization (as will be very evident in
the section on the temporal allocation of water in space further
below.)

Qanat irrigation: The case of Oman

An irrigation system based on a man-made kind of spring is that of
the *qanat*, a famous West Asian (most likely Persian or Afghani)
invention which has diffused widely, the last wave perhaps associated
with the spread of Islam.* The *qanat,* also known as *karez* in Pakistan
and *foggara* in the Sahara, consists of a gallery dug into the sloping
mountain, whereby underground water is brought to the surface
through a gently sloping tunnel. A principal advantage of the *qanat*
is that water does not have to be lifted but can be fed into lower
lying irrigated fields (Figure 7:2).

An exemplary study of the *qanat*-based irrigation technique
and its various social concomitants is that by Wilkinson (1977) on
'Water and tribal settlement in south-east Arabia'. In this region, a
complete irrigation system of fields and a water source is known as
a *falaj*. Providing that the 'mother well' in the *qanat* is suitably
located in the fan of a drainage basin (in the piedmont transition
zone between mountain and desert foreland), a *qanat* will have a
dependable base flow and some of the seasonal variations in water
supply will be smoothed out by transmission delays in the ground.
Again, it is the reliability of water supply which is the important
factor, and as a technology, the *qanat* is in many ways superior to a
number of traditional alternatives. It has been estimated that in
Iran, there are some 40,000 qanats 'comprising more than 270,000

* *Qanat* systems are widespread and most authorities are of the opinion that
the technique is closer to three than two thousand years old. English (1968)
believes that it had its origin in mining projects. One excellent article on this
technique is that by Cressey (1958), another dealing with Iran is found in
Highsmith (1961:23-29).

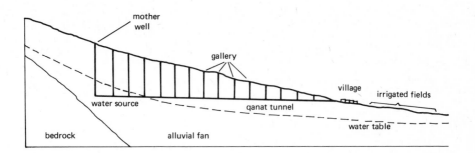

FIGURE 7:2   Cross-section of a *qanat* located in the alluvial fan at the foot-hill. The gallery consists of a number of vertical shafts, the uppermost one being referred to as the 'mother well' and it penetrates a water-bearing formation. The gently sloping tunnel leads the water out to the village and its fields. A *qanat* is a kind of 'horizontal well' from which water need not be pumped or hoisted.

km of underground channels supplying 35 per cent of Iran's water (Rahman 1981). Of course, modern industrial techniques of water supply, e.g. motor pumps applied to tubewells, are often more efficient than *qanat* and have been introduced to replace them in many places today (at least they are more efficient in the short run until they start lowering the general water-table). The fact remains that *qanats* do not normally tap fossil water and hence one and the same unit, once constructed, will have the remarkable life-span of hundreds or thousands of years if properly maintained (Wilkinson 1977).

The size of a settlement gaining its support from a *qanat* varies with the volume and reliability of water discharge as measured in *water-time units*. In Oman where date-palms form the major crop, one liter a second (ls^{-1}) of base flow irrigates around one hectare or 200 palms, while it takes some 40 to 50 palms to support a domestic unit of five persons or so.

> In central Oman the regular discharge of quite a number of the better *qanat* in their present state of repair is probably in the order of 60 to 80 ls^{-1}, but in general it would be unwise to place the average figure of the better *qanat* in the *bajada* /piedmont/ zone at much more than 40 ls^{-1}... This means that a reasonably sized qanat settlement ... covers some 40 ha of permanently cultivated land and has a population of about 1,000 souls ... while the larger settlements may typically be considered to support about 2,500 inhabitants... (Wilkinson 1977:92)

In line with the previous discussion of spatial zones of land use intensity, it is interesting to note that the principles outlined in Figure 6:9 also 'hold water' in the Omani context (cf Wilkinson 1977:66-71).

Another principle borne out in this region is the complementarity between different forms of irrigation. Wells thus form an important supplement to *qanat* irrigation, yet,

> It is obvious that a system of irrigation as remarkable as *qanat* must impose certain characteristics on its dependent communities. The interests of the inhabitants of an isolated village who survive in the desert by a thin life-line of underground water are evidently going to produce a quite different social organization from that, say, of the one-well, one-garden land-owning system. (Wilkinson 1977:97)

We shall return to this theme in the discussion below of the temporal allocation of water.

Tank irrigation: The case of South India and Ceylon

Tank irrigated cultivation has been of major importance in South India and Ceylon for centuries. Epstein notes in her interesting study of modern canal irrigation as an innovation in Mysore that,

> Irrigation is by no means a recent innovation; ... Mysore State, with an area of 29,489 square miles, has more than 30,000 tanks. These tanks or artificial reservoirs ... vary in size from small ponds to extensive lakes... Mysore State thus had many sources of irrigation well before the establishment of *Pax Britannica* in 1799; but these were dispersed and only very few were sufficiently large to facilitate the development of exchange economies. (Epstein 1962:4).

Like so often in history, it was thus profitable for a new regime to renovate and extend an existing irrigation system, a process which neither began nor ended with the British in India.

However, as Leach (1961) points out in his penetrating study of land and water tenure in the village of Pul Eliya in dry-zone Ceylon, the British colonial administrators often did not appreciate that it was water tenure which took precedence over land tenure, and that reforms in the latter sphere disregarding water rights were doomed to partial or total failure.

Technically, a tank irrigation system is constructed as follows:

> A village tank is created by damming up a natural stream and building a long earthwork wall /a so called 'bund'/ to hold the water up behind it... Very roughly, the full tank covers much the same area of ground as the land below it which it is capable of

irrigating. Clearly, the location of tanks must conform to the natural lay of the ground. Although, in a generally flat terrain, the villager has some choice about how he constructs minor works, the site of the larger tanks is pre-determined from the start. (Leach 1961:17-8)

The villages themselves are often located at the lower end of the bund where seepage makes the earth moist and permits moisture demanding crops in the kitchen gardens. Otherwise the main field system is located more or less immediately below the tank and furrows lead the water to the bunded parcels found there, the water being regulated by means of sluices. A village with its tank and its associated fields constitutes a small-scale socio-environmental formation with a remarkable capability to reproduce itself over the centuries (Leach 1961, Djurfelt and Lindberg 1975).

The great advantage with tank irrigation ecotechnology is that water can be collected during the monsoon rains and can then be distributed more evenly both over the first cropping period and often enough also over a second such period. As for maintenance, the repairs of tanks and furrows can be slotted into the slack seasons (cf Leach 1967). Should water be particularly scarce for the second crop, the cropped area can be reduced. In Pul Eliya, for instance, this was effectuated by bunding off individual parcels so that their size was sufficiently reduced. As in all irrigated cultivation systems, a number of mechanisms are needed to adjust cultivation to fluctuations in water supply, thereby ensuring a certain flexibility.

Well irrigation: Feeling the capability constraints on water-lifting

In many arid regions, wells form the main source of irrigation water. In some districts of the Punjab (before the partition of India), for instance, 40 per cent or so of the land could be well irrigated, even when canal irrigation was represented in the region (Punjab Board of Economic Inquiry, 1934, dealing with Gujranwala district.) In Oman to take another example, Wilkinson (1977) reports that wells predominated in the coastal regions where *qanat* were infeasible. Wells have the advantage that they can be individually owned or shared by only one or a few households, thereby requiring less coordination among people (i.e. fewer coupling constraints). They are flexible also in terms of timing by serving as a store of water which can be tapped when crops optimally should receive water, providing they do not run dry. Like any source of water they have a limited flow capacity, but this may very often be in excess of the real operational bottleneck, namely the capacity constraints on water lifting.

For all its other merits, the use of a well assumes water-lifting, and the lower the water-table, the more work in the form of time and energy expended there is per unit of water lifted. The gravity-flow technique has the tremendous time and energy saving advantage that human activity is not needed in the actual porterage or transport of water. Lifting water from wells can be done by means of human effort, of course, as is frequently the case when water is drawn for domestic consumption and for watering animals, i.e. when relatively small quantities of water are involved. But water-lifting by human effort at the scale of plant cultivation imposes heavy demands on human time and energy. Throughout Asia and elsewhere, a number of traditional techniques have thus developed to harness animals to supply the bulk of the water-lifting energy and time. The Persian wheel is one such technique suitable when the water-table is fairly high, while hoists using a leather bag sunk into the water is another. Once the water has been hauled up, however, both these systems rely on water flowing by gravity to its final destination.

The Boka wells in Pakistan (leather bag hoists) and the Persian wheels are interesting in that they illustrate the need for synchronized inputs of human time, animal time and water-time. Animal-time (e.g. bullock-time) may thus be another limited time resource in agriculture. In a village in the Multan District, for instance, water had to be drawn continuously at some stages of winter crop cultivation:

> In winter when canal water is not available and a vast area under rabi /winter season/ crops has to be irrigated the wells have to be worked continuously for twenty-four hours. Four pairs of bullocks are required to keep the well going day and night, each pair working for three hours at a time and six hours out of 24. (Punjab Board of Economic Inquiry 1938:79) /Figure 7:3/

The amount of land that can be watered in this way varies with the kind of soil. Soft soils absorb more water and hence take longer time to irrigate. On the other hand they retain moisture longer and can therefore 'withstand any delay in /later/ irrigation longer'. The acreage that can be watered varies, but to water around 0.7 acre (or 0.3 ha) takes about 30 hours by this technique and the first watering when land is very dry might double this figure. To lift water in this fashion is obviously a rather labour-time intensive enterprise (for humans to some extent but much more so for animals). Again, the temporal organization of water use depends on how many people share the water-time from a single well or who share the draught animals.

time

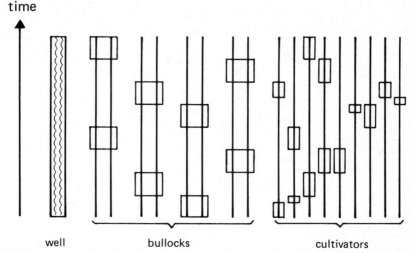

FIGURE 7:3   The synchronized use of water, bullocks and people over time. The well and its Persian wheel is used continuously while four pairs of bullocks take turns at the wells. On the human side, time inputs can be more flexible as long as someone is always taking care of the animals at the well.

THE TEMPORAL ALLOCATION OF WATER IN SPACE

Dimensions of the irrigation allocation problem

Irrigated cultivation not only involves the application of water to land, but in a wider perspective it assumes the synchronized and synchorized utilization of water-time, space-time ('land'), human time and animal-time. (Domestic beasts are often used to lift water, plough the fields or carry out other activities.) Aggregate time-budgets for these resources can be depicted as in Figure 7:4. The resource demanding activities which can potentially be packed into these budgets as well as the associated paths are all located in the time-space of a given region (Figure 7:4 right).

The carrying capacity of these resource time-budgets is no doubt limited, but the activity volume that can be packed into these budgets is not a straightforward function of the proportionality between them (similar to what economists call factor proportions in their production functions). Activity output rather depends on how portions of potential capacity can be mobilized* which in turn is an *interaction process* involving the synchronization and synchorization of paths in

* Cf Chapter 8 on potential human time supply and its mobilization (and the drains on time in the actual process of mobilization, e.g. in the form of travel time.)

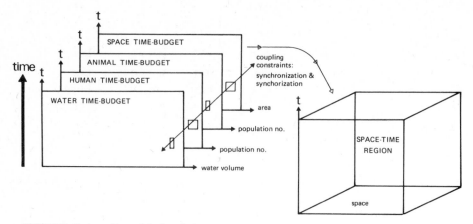

FIGURE 7:4   Four kinds of time-budgets dealt with in this study, each having its limited carrying capacity. Since a vast number of projects assume input from more than one such resource time-budget, the actual activity output depends on how inputs from each kind of time-budget can be mobilized, synchronized and synchorized and how paths can form bundles. Of course, the time-space region and its space-time budget contains the other three kinds of resources, humans, animals and water.

time and space. The overall potential capacity can be more or less *efficiently* mobilized, depending not only on physical and biotic constraints but also on regulatory, normative, contractual and other constraints inherent in the social system. Moreover, potential capacity need not be fixed over the entire period. This is, in fact, rarely the case: people may die or migrate in the period which affects the aggregate time-budget, for example, or basic water supply may fluctuate over the period. Even so, the problem remains of how potential supply capacity can be mobilized and used in cultivation projects.

In arid regions, where water may be a more limiting factor than the other three resources mentioned, water supply sets the major outer boundary around living-and-activity (production) possibilities. We must, however, go beyond the general *availability* of water (i.e. the total water supply at the source) in order to see the real constraining role of water resources. Since irrigation is the activity of combining water with land, thus coupling it to the plants growing there, water has to be made *accessible* in given time-space locations. It is the mobilized and accessible water which determines how much land can be cultivated for how long and with what intensity.

Water can be more or less *efficiently* mobilized in an irrigation system. Low watering efficiency (high mobilization losses) diminishes the extent to which both capacity in the space-time budget and in the human time-budget can be used for cultivation activities. Similarly, if animal-time employed to lifting water is the scarcest factor this constrains the total volume of the water time-budget, which in turn restricts the volume of cultivation in the space time-budget (in turn affecting the volume of cultivated fodder and further the number of animals lifting water that can be fed, etc.).

By interrelating the different resources and the associated constraints, we lay bare what Friedman (cf p. 59 above) referred to as the 'hierarchy of constraints which at each succeeding level determine the limiting conditions of internal variation and development... etc.'. By aggregating the constraints and analysing how they interact and circumscribe possible solutions — living and activity possibilities — the overall system structure and performance is revealed. By choosing here a formal, deductive approach (albeit with a great number of inductive elements) , every actual system is visualized as a 'realized possibility' in line with Godelier's suggestion (cf p. 58-59).

What we thus have at hand are systems in which human time, animal-time, settlement space-time and water-time are subject to synchronized allocation, mobilization and economizing as process and activity over time. Many of the existing economic, operational research or similar models* tend to neglect both indivisibility constraints and coupling constraints, however. Hence many solutions appear feasible which in reality may not be so; and if the indivisibility properties of humans, beasts, equipment and constructions are disregarded, the logic of how they can or should be coupled to each other into specific configurations at specific times and places is also lost, as is much of our understanding of project implementation and social performance.

The divisibility of water

One 'advantage' (i.e. non-constraint) in the mobilization of water is that it is *perfectly divisible* down to the very smallest 'denominations' that human hands and vessels can handle. Viewed in terms of

---

*   The author is aware of the existence of a number of quantitative techniques and algorithms which can be applied in this field, providing one's objective is to find 'optimal solutions' (for whom and with respect to what?) or that one wants to explore constraint systems. However, their meaningful application assumes that the analyst is able to handle them, and even more so that he can identify the essential dimensions of the system. It is the latter task to which time-geographic analysis addresses itself, i.e. by formulation of suitable models of real-world systems so that the bridge between mathematics and reality can be built without, as has often been the case up till now, forfeiting the latter.

For a discussion of the conflicting aspects of indivisibility constraints and assumptions about the form that functions must take in order to ascertain 'global' rather than 'local' optima, cf Macmillan (1978).

paths, there are virtually no constraints on divisibility, but this should not cloud the fact that the pieces of equipment or the people and animals used to manipulate water are *in*divisible. The flexibilities rendered by divisibility are also offset by the fact that water is so *bulky*, i.e. given the relatively large volumes demanded, constraints on (trans)portability have a strong impact instead. The only redeeming feature in the transfer problem is that water is a low-friction substance and gravitational energy can thus be harnessed. This is impossible for solid substances, say, manure which may also be a necessary input but one that has to be carried or carted (cf Chapter 6). Hence there are great capacity gains from the fact that irrigation water can be transported through gravity-flow, just as there are great losses from the fact that water is so divisible that it seeps through the soil also in the wrong places or evaporates all too swiftly in high temperatures.

Water demand: Volume, timing and spatial location

It is useful to start with the *demand* side, before specifying the mechanisms of supply distribution. Different cultigens require water in certain minimum *volumes* at given *frequencies,* both in turn varying with the weather. Most plants thus need more water more often when the temperatures are high, as in summer, while they manage on less the cooler parts of the year. Water requirements also vary over the growth cycle, as well as with the type of crop, e.g. tree crops, cereals, and root crops.

> A FAO report for Saudi Arabia ... recommends irrigating palms on light soils every 6-8 days in summer and 15-20 days in winter (slightly less frequently on heavier soils); it also speaks in terms of a total annual water duty of 9-12 feet (3-4 m)... it should be remembered that it is almost impossible to over-water a palm ... So while the date-palm can survive (without fruiting) fairly bad conditions, there is virtually no limit to the amount of water it can take... here is a wide variation in the practices of irrigating palms in Oman, both with respect to frequencies and volume of water application... (Wilkinson 1977:94)

Many plants need water at shorter intervals than a fortnight and even less, and the irrigation cycles need to be adjusted accordingly. A major problem of meeting demand arises when there is a wide variety of cultigens needing water at different intervals, because it is then difficult to find a compromise solution as regards *timing* and the *efficient use* of water. Geertz emphasizes the importance of timing in his discussion of Javanese rice cultivation:

Timing is also important; paddy should be planted in a well-soaked field with little standing water and then the depths of the water increased gradually up to six to twelve inches as the plant grows and flowers, after which it should be gradually drawn off until at harvest the field is dry. Further, the water should not be allowed to stagnate but, as much as possible, kept gently flowing, and periodic drainings are generally advisable for purposes of weeding and fertilizing... (Geertz 1963:31)

Regardless of whether the crop species cultivated are annuals or perennials, water supply, to which we will now turn, must within certain margins match the time-space structure of water demand.

Water supply and its distribution

Since water for irrigation emanates from a spatially concentrated source in contrast to rain which comes dispersed, it follows that irrigation water has to be distributed by overt allocational activity:

... it is only in so far as we understand the nature of the compromises involved when ... /an irrigation-based/... community produces its own particular solution to the problem of *how to divide an irrigation supply* fairly amongst a *large number of shareholders*, while at the same time trying to make *efficient use* of the water and ensuring that the irrigation system is properly *maintained*, that we will be able to comprehend some of the fundamental aspects of social organization in an Omani village. (Wilkinson 1977:100) /Emphasis added/

Generally speaking, water can be distributed along three dimensions: 1) in space, 2) in time, and 3) among people who hold legal shares in the existing sources of water.

Let us initially assume that we are dealing with a *non-fluctuating source* of water, i.e. one which supplies water at a constant flow rate or volume per time unit. This assumption will be relaxed later on with due discussion of the implications.

Starting with the *spatial allocation* aspects, one basic problem in all irrigation systems is how the water can and should be distributed over a number of spatially dispersed fields/holdings. Disregarding possible natural floods, in small-scale irrigation systems, the water is designed to be led from its source in an orderly fashion through a network of bifurcating channels of increasingly fine size. A constant flow of water may simply be divided up in space by a number of weirs with outlets of certain proportions. Every field in the system thereby gets its water simultaneously. This seems to be the system in

Pul Eliya in Ceylon mentioned earlier.*

However, a continuous flow of water to every nook and corner synchronously creates a number of problems of watering efficiency:

> Splitting can be carried out either by means of sluices, in which case the total flow is directed down each channel in turn, or by means of a weir which divides the flow itself. The latter solution... is ... unsuited to relatively small discharges for the obvious reason that spreading the water *reduces its velocity* and so increases losses in transmission... (Wilkinson 1977:104) /Italics added/

The loss in velocity is due to the fact that the combined outlets of the channel network exceed the capacity of the inlet. Low velocity means that much water is lost by seepage and evaporation in the channels but also that there may not be enough water to soak the fields thoroughly. The more shallow the water is in the fields, the greater is the share lost through evaporation. It is thus principally better to soak a field thoroughly less often than to 'sprinkle' it every day.   When outlet capacity grossly exceeds that of the inlet, it also means that the *watering intensity decreases with distance* from the source (cp Chapter 6 on distance and intensity of interaction), the fields near to the source receiving a disproportionately large share (Fernea 1970:161). However, it may only be when the given water supply is used over the entire area *at the same time* that supply capacity is inadequate or that watering efficiency is reduced beyond tolerance. If, on the other hand, channels are used in rotation and fields are watered *alternately in time,* velocity can be kept up and transmission and evaporation costs kept down. There are thus good economic reasons behind the *temporal allocation of water* in space which explains why the system is so widespread:

> In most systems /in Oman, but also generally/ ... water is distributed to the gardens in a set order, and each plot of land /or holding share/ has *a set period of time allocation* to it...
> /Wilkinson 1977:101/ /Emphasis added/

It is thus entirely valid to speak of the time allocation of water or, synonymously, *the temporal allocation of water.* Many local societies have developed complex systems for the *time-sharing of water* within irrigation cycles operating at certain periodicities.

---

*   Since much of the literature referred to speaks of Ceylon in a historical perspective, I have retained the old name of the island rather than using Sri Lanka.

Time-shares of water

If the base flow occurs at a constant rate, there is no need to measure water volumetrically, e.g. by first filling some cistern to a given level. Volume can be measured in water-time instead, while each holder of water rights receives a given *time-share* in the total supply. However, the water has to be distributed in some orderly fashion, not only for technical reasons of watering efficiency and the need for predictability in cultivation, but also to ensure that each shareholder gets his proper legal share ('legal efficiency'). Hence, in most systems,

> ... the basic principle is that the individual shareholder's water is delivered to him at an *appointed place* and at an *appointed time* by means of the network of distributory channels... (Wilkinson 1977:102) /Italics added/

When water is allocated in time-portions over space, we have a situation of the kind depicted in Figure 7:5, where twelve shareholders each get their water ration one month of the year (the year divided into two cultivation seasons). However, this low frequency of watering is way off the mark; some cultivators get their water in the beginning of a season, others in the middle or at the end, if at all that season. It is obvious that the frequency of water demand by the plants grown must be allowed to dictate the irrigation rhytm. Water cannot, in other words be distributed in such big quanta or at such a low frequency that plants dry up and wither in the meantime. Nor should water be applied so often that there is not enough to soak the soil, on the other hand.

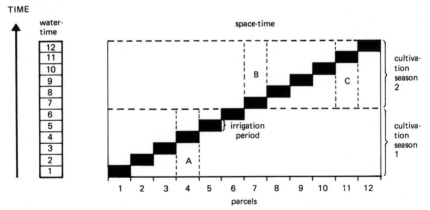

FIGURE 7:5   The time sharing of water among 12 shareholders with equal size water-time and space-time shares. Of course, this distribution of water is not realistic since the frequency is far too low. One month per year and nothing the other months is out of step with plant water demands.

The various time/space relations between different components such as watering periods (i.e. time shares of water), irrigation cycles, watering intensity and watering efficiency work out as follows:

> Suppose eight shareholders build a *falaj* /i.e. irrigation works/ and share the land and water among themselves dividing the land on either side of the channel into eight plots and allocating the whole discharge for a complete day in turn: then it will be seen that the *dawran* /irrigation schedule and cycle/ will be eight days and each day will constitute a share of the *falaj* water. (Wilkinson 1977:101)

This is shown in Figure 7:6, and assuming each shareholder has the same size plot, the *watering intensity*, defined in this case as water volume per hectare and per (portion of the) year, is also equal among all shareholders. There is *direct proportionality* between water-time and space-time held. Putting this ideal into practice ensures to all a uniform and adequate water supply.

It may be, however, that time-shares are equal in size, while the land holdings vary (Figure 7:7). This clearly affects watering intensity; some plots get water above optimum intensity for the crops grown there, others get water below the optimum or minimum and cultivation will suffer.

It is possible to attain a uniform and adequate watering intensity although parcels differ in size, if time shares have been adjusted so there is direct proportionality between water-time shares and space-time shares. In this situation, of course, the water has to be measured with greater temporal precision than full days.

Taking the fourth case, it can easily be shown what happens to watering intensity in each plot if both time shares in water and plot sizes vary so that there is no proportionality between water-time and space-time (as in Figure 7:9). Then variations in watering intensity tend to be even greater; some holdings will be overwatered while others will suffer drought. The total local cultivation system will then be producing below its potential capacity.

Irrigation cycles and complex time-sharing of water

In practice, the administration of water is compounded by the fact that there are *many more shareholders* than there are whole or half days in the irrigation cycle, and for reasons of demand frequency, the cycle has to be kept short (especially in the hot season), say between 5 and 15 days or so. Consequently, the water-time shares generally have to be stated in smaller units than half days.

In the villages of Oman so intriguingly described by Wilkinson (1977), the standard irrigation cycle (the *dawran*) runs around two weeks and is first divided into main shares *(khabura)* of one day. These are split into subunits of half-days (night and day portions called *bada*). The latter unit consists of 24 *suds* of half an hour each. A *suds* is the most practical unit for allocating water, and time shares are often expressed in these units only. Sometimes shares are also specified in smaller denominations, but the smaller the time units, the more difficult are they to apply in the actual administration of water.

In another outstanding study of local irrigation systems, Eldblom (1977) working in the Libyan oasis of Ghadames found that the main irrigation cycle there *(dh·ami)* lasted 13 days (on the average), i.e. for the major artesian well *(Ain el Fras)* in the village centre. In the original time sharing schedule, the flow was split into 12 shares of 24 hours each. Later a thirteenth day was added, the water of which was auctioned off to cover expenses for taxes imposed on the community. It is a common feature in many systems to have a small share of each irrigation cycle or some specified days per annum sold to raise taxes or money for the administration of irrigation. Sometimes a specific share is given directly to the officers in charge of water measurement, distribution and control. Since continuous flow irrigation must be taken care of day and night, this administrative occupation is often of the shift-work variety. The approximately 12 hours of the night (depending on the time of the year), were divided into 24 *dermissa* (c. half an hour), each consisting of 160 *haboub*. One *haboub* of a bit more than 11 seconds was further divided into 24 *khirat* of about half a second each! Needless to say, such incredibly small time units could only be used for legal purposes, i.e. to record the water shares in the so called *zjedd*, the official water register. As a local irrigation-based society like Ghadames evolves with each generation and water rights are inherited according to Muslim law (a sister getting half the share of her brother), shares tend to get specified in increasingly small fractions. If the population grows, water and land shares get fragmented even faster than with a relatively stationary population. In Ghadames, it was thus necessary to translate and reshuffle the legal shares into a practicable system of water-time shares once per year in the so called Annual Proceedings *(zjoumla)*. Even these were mildly elastic since there was generally a lag between the moment when someone's water time was up and the time when sluices were closed and the water shunted elsewhere. Since delays tend to cumulate in such closely interdependent systems as a local irrigation system, the *dermissa* had to be adjusted (shortened) to keep the whole irrigation cycle at its proper legal length.

In Ghadames in summer, the irrigation cycle was shortened exactly into half, so that each shareholding unit got half its allocation of water but twice as often. The irrigation cycle thus lasted 6½ days.

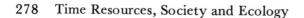
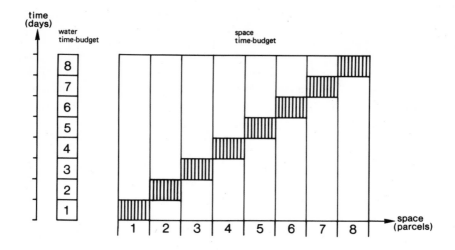

FIGURE 7:6   Every shareholder has equal size shares of water-time and they also have the same size parcels. Hence watering intensity is uniform and there is direct proportionality between water-time and space-time.

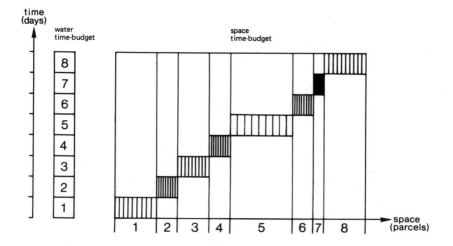

FIGURE 7:7   Every shareholder has equal water-time shares but they have parcels of different size. Water-time is no longer proportional to space-time in parcels and there are differences in watering intensity between parcels.

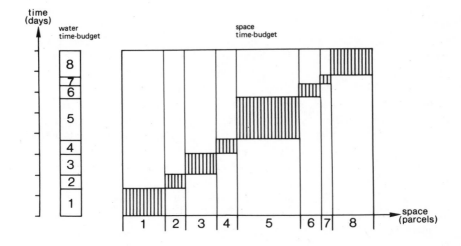

FIGURE 7:8   Every shareholder has different size water-time shares as well as different size space-time shares. There is proportionality between water-time and space-time, however, so watering intensity is uniform.

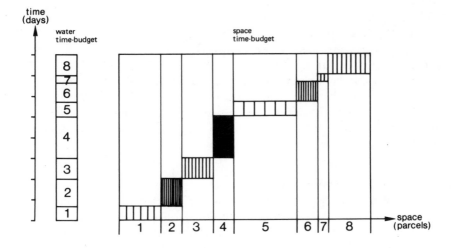

FIGURE 7:9   Each shareholder has different size water-time shares as well as different size space-time shares. However, there is no longer a link between water tenure and land tenure, nor proportionality between water-time and space-time, so the variations in watering intensity tend to be great among parcels.

This harmonized better with plant water demand but has the further advantage of evening out the discrepancies between getting a time share of water at night and in day-time. Half an hour's share at night is obviously more efficient, since evaporation is much less then. If a shareholder with a given position in the irrigation schedule always gets his water at night, he is favoured. This injustice can be eliminated in at least two standard ways. One method is to make nocturnal shares shorter to compensate for the greater evaporation losses by day. Another is simply to see to it that there is an odd number of half-days in the irrigation cycle (e.g. an odd number of *badas* in the *dawran* in the Omani case). Those who get their water by night in the first irrigation cycle thereby receive it day-time in the following cycle. Consequently, although a source may supply water at a constant rate, flexibilities of temporal accounting and administration are still often built into the system of water time-sharing so that differences between night and day-time shares or seasonal fluctuations in the length of the day and night can be compensated for.

The management of fluctuations in water supply

Complexities of water time-sharing are not only due to village size and a large number of shareholders or disproportionalities between water-time and space-time shares. Complications are also caused by temporal variations in water supply, particularly seasonal fluctuations. Earlier on, the rather strong assumption was made that the rate of water supply was constant. In many cases, this is not really so, and hence this contingency is built into the mechanisms of water management in ways which affect both watering efficiency and intensity as well as capacity utilization in the irrigation system as a whole. Wilkinson mentions, for example, that,

> Where there is considerable variation in base flow some of the channels of this primary network are opened only when water is plentiful...: obviously on the land watered by such channels it will be alfalfa or vegetables /i.e. annuals/ rather than tree crops /i.e. perennials/ that will be grown. (Wilkinson 1977:99)

This is a typical example of the synchronized economizing of water-time and space-time. The cultivated area can be expanded for brief periods when water is abundant only if crops of a short life span or vegetation period are chosen (Figure 7:10). Water supply capacity, the scarcest resource, can then be better used, i.e. more cultivation activity can be packed into the limited water time-budget.

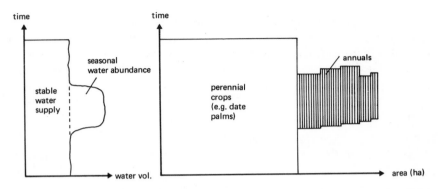

FIGURE 7:10   Example of the synchronized economizing of water-time and space-time. In the times of peak water supply, annuals of quick maturation are sown and harvested before water supply returns to its low flow.

The Sonjo described earlier on have other institutional and technical mechanisms for adapting to seasonal or temporary water shortages. When water supply is good, the cycle takes around 14 days, each 24-hour day being divided into four periods of about six hours each. This is the time-share allocated to every primary shareholder. The Sonjo system is based on *several categories of shareholders.* The first four days in a cycle are given to the seventeen village elders (wenamiji) who have the ultimate control of water by virtue of their hereditary rights and seats in the village council (Gray 1963:58-61). The second category (the eighteen *'minor' wenamiji*) also have hereditary rights in water shares and get the next four and a half days of water, but they have no political control over water. The third category consists of about twenty to twenty-five elders *(wakiama)* who have no hereditary rights to water shares whatsoever. They bargain individually with the village council for *temporary* rights (a month or so at a time), paying for these privileges with goats. However, if the cycle is to be kept within 14 days, the number of these temporary beneficiaries must be kept down (Figure 7:11). The fourth and final category consists of all the remaining cultivators. They are clients of the first three (primary shareholder) categories, whose 6-hour shares of water are usually far in excess of what they need:

> An area which is irrigated in one period /i.e. six hours/ may be owned by two or three different men. These plots are usually located fairly close together, though not necessarily adjacent to one another. It is planned in this way for the most efficient use

of water; thus there will be less waste than if the water were dispersed to different corners of the valley during a single morning or afternoon... /The man with primary rights/... who exercises the most authority, soaks his own plots thoroughly first. When his needs are satisfied the water is diverted in turn to the fields of his companions /clients/, who may not find time to soak their fields as thoroughly as they might wish... (Gray 1963:57)

When there is a shortage of water or when the crops need water more often, the irrigation cycle is shortened and some of the *wakiama* with only temporary water rights are reduced to client status. Anyone caught stealing water has to pay a goat as a fine.

The situation described so far is one of fluctuation towards shortage of water. On the other hand, in seasons of abundant supply, the whole time-sharing system is not applied, fields are partly or mostly rainfed and some of the excess water is simply led to bypass the fields.

Timing, coupling constraints and synchronized economizing

In the systems described so far, the water was distributed to the gardens of the shareholders in a *fixed order or sequence* in time. It was also delivered in a *fixed spatial order* to ensure a greater measure of contiguity between one garden and the next so as to reduce water losses and raise watering efficiency. One problem in the Pul Eliya type system was that if everyone got their water at the same time, not only did this affect watering efficiency, but it also created peak (simultaneous) demand on for human time for activities like ploughing or sowing. In this respect, distributing water among shareholders *with a time lag* means that total time resources in a village will not be taxed as simultaneously.

For the individual shareholder, however, getting the water on a fixed day in the irrigation cycle is not a very flexible arrangement. The time-table of watering easily becomes a dominating factor in the entire *temporal organization* of the community, just as communications time-tables, working hours, opening hours, TV hours and the like form dominant elements in the temporal organization of modern societies. One reason for the fixity in deliveries of water time-shares is that water rights are directly coupled to land rights. However, this constraint is not necessarily represented in all irrigation societies.

In the very long-term, inter-generational time perspective, a local irrigation society evolves under the influence of a certain extent of demographic randomness, both with respect to family size and even more so in terms of the sex ratio. In conjunction with inheritance

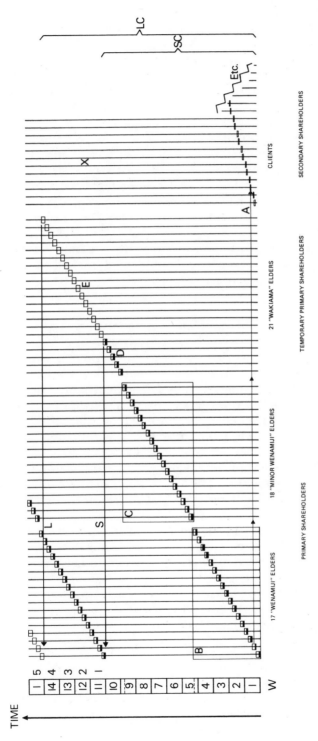

FIGURE 7:11   The hierarchy of water-time sharing among the members of a Sonjo village. The primary shareholders get their water first and then redistribute what they do not want among their clients (arrow A). When there is plenty of water, the 21 'wakiama' elders get primary shares (E) which they also can distribute part of among clients. The system then operates under the long cycle (LC) of 14 days. When there is a shortage of water and the cycle (SC) is shortened to 10 days, a majority of 'wakiama' elders must get their water as clients instead. Only the 'wenamiji' elders have guaranteed shares (B, C) in each cycle. Just how the clients get their water in times of shortage (X) is not indicated in the source material. Only a few clients are shown.

rules, water-land proportionality may show a diverging trend which would affect watering intensity and efficiency. Strategies of marriage and alliance, of buying and selling shares, of lending and borrowing water-shares, and other mechanisms may either amplify this diverging trend in land-water proportionality, ultimately leading to the separation of land and water ownership. Alternatively, a given proportionality between water-time shares and space-time shares may be maintained more or less at a given level; those who have too little water-time may buy it 'second hand' from those with more than they want. In that way, although space-time and water-time ownership may evolve along diverging paths, the proportionality between water-time *actually applied* and space-time shares is kept within an interval which is much less detrimental to watering intensity and efficiency.

In systems where water-rights and land-rights are not closely linked up, there is generally greater flexibility in the allocation of human time to cultivation activities. In Oman, this was the main advantage (i.e. non-constraint) of the system used in Izki village:

> Under the .../Izki/... system, water was completely divorced from land and a shareholder was deemed to hold his share above ... the point where the main channel first divides: he may therefore draw his water wherever he wants from the main network... The advantages are obvious. In the standard /Omani/ system a man draws his water for a *fixed period* on a *fixed day* at a *fixed point*, and the only changes he can make ... are either to acquire some ... /additional shares/... (and even then this water may be delivered at an *inconvenient time*), or to make special arrangements with his neighbours. In the Izki system a man acquires a share of water on whatever day he wants for the period he wants, and he can draw it at the place he wants... (Wilkinson 1977:116) /Italics added/

This is the situation described in Figure 7:12. Not only can water-time then be better synchronized with the input of human time, but also, in principle, with the particular crops and crop mixes chosen. A mismatch between water demands by crops and timing of water input is a chronic problem in irrigation systems where farmers are free to chose whatever crops they see fit, particularly in the large-scale irrigation systems to be dealt with shortly.

> On the other hand there are two disadvantages in such a system; first, there can be a tremendous *waste of water* in the channels, because the supply is always being switched from point to point, and second, there are considerable *time delays* in getting the water from one place to another. (Wilkinson 1977:116-7)

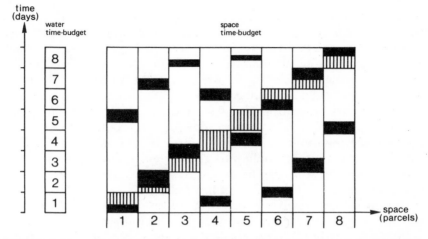

FIGURE 7:12   The time sharing of water in a flexible system of the Izki kind. Since individual holders get their share when it suits their personal time scheduls water is seldom received one immediate neighbour after the other, the net effect being a saving of human time perhaps but greater losses of water. This is another example of the synchronized economizing of resources, human time being the resource economized on. The system is compared to the non-flexible system of Figure 7:6.

Moreover, the value of the water in terms of the water actually mobilized and received at the *point of use* varies with the distance from the source, and transmission losses tend to be high in these flexible kinds of water time-sharing systems. Hence the price to be paid for using the capacity of the human time-budget better is that the capacity of the water time-budget is used less well. This also means under-using capacity in the settlement space-time budget, because if more water is wasted, less land can be cultivated.

In the Omani systems (typical for those West Asian and Sahelian systems where perennials dominate or have a large share), the spatial packing of trees (e.g. palms) and consequently the minimum plot size necessary per 100 trees has chiefly been geared towards the economizing on water:

> The basic reason why palms are packed more densely than the ideal is scarcity of water. Whilst it is true that the average yield per palm may fall below its best in the more closely planted groves, it should be remembered that the land and water needed for the ideal spacing is virtually twice that required for the normal spacing, and so the yield per unit of *irrigated* area is less than with closer spacing. There will however be a marked decline in quality, even if not necessarily in total quantity, in an overcrowded grove. (Wilkinson 1977:93) /Emphasis added/

Land may or may not be scarce enough to have a major impact on the synchronized use of water-time and human time. Sometimes the area in which a suitable set of fields can be laid out is relatively small, e.g. for topographical reasons. Then this may lead to a certain overwatering which should, in principle, allow for greater flexibility in how water is time-shared. Moreover, if the predominant crops are *perennials* that have to grow up for several years before they yield, the irrigated portion of the space time-budget cannot be diminished beyond a given limit at times of water shortage. Hence a certain over-watering must be built into the system to begin with so that bad seasons and years can be bridged over. It is better to use annuals as regulators of water capacity utilization.

We may thus conclude this section on local irrigation systems and the temporal allocation of water by returning to Figure 7:4 and pointing out that both water, land and human time is allocated in space and that the capacity utilization of each is interdependent in a number of ways. Furthermore, the economizing in each resource is subject to a variety of coupling and synchronization constraints which in turn determine the extent and with what effect capacity can be used, and consequently what level of agricultural output can be attained. Output is thus not simply a function of given *factor proportions* at the beginning of a budgeting period, but is rather a function of the very *mechanisms and processes of allocation.*

REGIONAL IRRIGATION SYSTEMS AND LARGE SCALE COORDINATION

Most of the features of timing, time allocation of water, coupling constraints, coordination, watering intensity and efficiency, and capacity utilization remain in local systems that are part of larger systems of irrigation. However, when local irrigation systems are linked together by a series of canals flowing from a main river such as the Tigris, the Euphrates, the Indus, or the other Punjabi rivers, a number of additional constraints enter the picture which are not found in purely local systems. On the other hand, some of the fluctuations in water supply and other limitations associated with local systems are relaxed when the latter are 'plugged into' a larger irrigation grid, so to speak.

The synchronization of rainfall and riverine regimes

As Spate and Learmonth (1967) point out regarding the Indus valley, many of the traditional systems found there depended simply on inundation in the seasons when water was plentiful. Although this gave some possibilities for water control, many of the smaller systems

of tank irrigation were inadequate in dry years, since their main function was to redistribute the water in the rainy season so as to prolong
the cultivation period. However, in years when there was too little
rainfall, it would not even last for one crop in the rainy season.
Of course, even cultivators operating in the great alluvial valleys have
problems with the variable discharge of river water both within and
between years. In Mesopotamia, for instance, Fernea (1970:158-9)
reports that there are great variations in the water volume from one
year to the next. A riverine population of cultivators must also live
with the fact of spatial variations over time, i.e. that portions of the
river channels alter their course as a result of both the depositing of
silt and the scouring action of the river at selective points. Hence,

> their river may supply flood waters one year and a dearth of
> water the next. *Local measures can do little about drought,* but
> local farmers can excert a simple measure against floods which
> will have the desirable secondary effect of helping to maintain a
> longer supply of irrigation water in an average supply year, after
> the flood peak has passed. (Fernea 1970:159) /Italics added./

A basic strategy for relaxing the constraints of cultivation associated
with an arid or semi-arid climate is thus to get a more even distribution of water both in time (over the year or the two or more seasons
of the year) and in space, the latter by leading the excess river water
into areas of insufficient water. Hence, irrigation contributes to
spatial expansion as well as temporal intensification of agricultural
land use. Extending the watering seasons or spreading the water more
evenly over time can be done in several ways, e.g. by dams,
and barrages, i.e. by water storage but also by the shunting of water
between tributaries having different times of peak discharge. Here
the natural conditions vary a good deal among the major river basins
like the Nile, the Euphrates-Tigris and the Indus-Punjabi rivers. The
long course of the Nile stores the water in the river itself (i.e. prior to
the Assuan Dam in particular) so that the lower Nile valley was
inundated in the optimal cultivation seasons.

In Mesopotamia, there is no corresponding great distance between
the headwaters and the alluvial plains, so the April peak flood comes
too late for the winter crops and too early for the summer crops
(Fernea 1970:7-8). Since the optimal cultivation seasons are the
spring and the autumn months, i.e. March to mid-May and late-
September to November, irrigation at these seasons is supplemented
by rainfall which is certainly not the case in the dry and hot season
from May to October. We may compare this also to the Indus valley
which is more favoured than Mesopotamia, especially the Punjab,
where there is very 'timely' complementarity between rainfall and

irrigation. Because of its arid climate, Sind (the lower Indus region) is much more dependent on irrigation than the Punjab. So while the summer monsoon brings rain to the Indus plain, in Mesopotamia, by contrast,

> The water supply radically declines and in many places barely meets the demands of human and animal thirst. Cultivation, totally dependent on canal water, shrinks to one third its winter proportions, and the problems of water distribution and division plague tribe and government alike. (Fernea 1970:10)

In the Punjab, the summer season *kharif* crop is both irrigated and watered by monsoon rains and is harvested in September to November. Then the land is still not too dry for the planting of the winter *rabi* crop, mainly wheat, and the winter crops are similarly rainfed as well as irrigated. The winter rain is also stored as snow in the Himalayas and begin to melt in spring so that the Indus river has its peak by the middle of March while water is still useful for irrigation of the winter crops harvested in the hot and dry months of April and May. On the whole, the river regime in the Punjabi portion of the Indus plain is much better synchronized with seasonal cultivation requirements than in Mesopotamia, and the Punjab has historically been the granary of the Indian subcontinent.

Local capability constraints versus regional coupling constraints

The main advantage with large-scale regional irrigation systems is that they can even out water supply in a larger region and cancel out local fluctuations, although, as mentioned above, the regional system is also subject to seasonality. However, the relaxation of some local capacity constraints is attained at the cost of increased coordination and therefore increased coupling constraints within the larger system. The allocational conflicts, say between local flexibility in timing and overall watering efficiency tend to increase as do the regulatory and authority constraints associated with water management. In ancient as well as modern large-scale irrigation systems, water management is immensely important for watering efficiency and the level of agricultural output attained (cf Bottrall 1978 for a discussion of contemporary systems). It would have been a most rewarding task here to reinterpret the problems of regional irrigation system structure and management on the basis of a time-space constraints approach (capability/capacity, coupling and regulatory/authority constraints). This task would, however, carry us too far afield in this limited study and we must rest content here with the laconic statement that the regional irrigation systems have to await a volume of their own.

IRRIGATION AND INTENSIFICATION

The elimination of seasonal fallows

Given that irrigation is a technique of water management designed to supplement natural rainfall, it is obvious that the temporal distribution of water in space is a crucial determinant of the *temporal intensity of cropping.*

In Boserup's (1965) intensification scheme, irrigation is the chief mechanism to eliminate *seasonal* fallows so that land can be at least double cropped and possibly also multi-cropped (cf Figure 5:4 above and also footnote on p. 193):

> ... the use of irrigation techniques and other capital investment is related to the fallow system. Irrigation facilities and other land improvements, for instance terracing, are never used with long fallow and rarely with short fallow. But the introduction of multi-cropping often depends upon the creation of irrigation facilities and in dry regions the same may be true even of annual cropping. Short-fallow systems, on the other hand, may be practicable without irrigation even in dry regions... (Boserup 1965:26)

In the preceding sections, we demonstrated that when irrigation is applied to relax the capability constraints imposed by the natural climate, new coupling constraints of coordination, timing and temporal allocation of water are introduced. So are a number of legal,

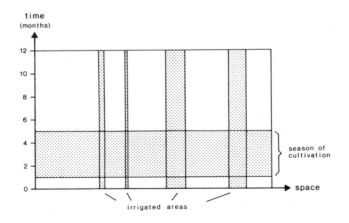

FIGURE 7:13   The introduction of irrigation makes it possible to pack a regional space-time budget more densely and intensively with cultivation activities. The seasonal fallows can be 'colonized' in those areas that are irrigated.

contractual, authority and other regulatory constraints, not only within a given locality, but in the case of larger (say river basin) irrigation systems also among localities.

It is also clear, however, that the preconditions for the elimination of fallows *without irrigation* are also present in humid regions where rainfall is sufficiently evenly spread throughout the annual cycle. This is the case in the New Guinea region, for instance, where numerous very intensive systems of cultivation can be found which do not rely on irrigation, and where drainage may in fact be the main problem. In Boserup's version of intensification theory this possibility is not discussed; it is at best hinted at in the statement above that ... 'multi-cropping *often* depends on the creation of irrigation facilities...' /Italics added/. So while Boserup is right in claiming that population density (and hence, land use intensity) cannot simply be ascribed to factors of the natural environment, the presence of a natural (productivity) factor such as fairly continuous and adequate rainfall does explain how higher intensities may be achieved without the extra human time inputs which Boserup thought were necessary to offset declining output per person-hour in intensive systems.

Not only may seasonal fallows be reduced by irrigation ecotechnology, but the fertility maintenance problem may also be solved to a great extent by the nutrient substances brought with the water. Since fertility improvement by means of manuring may be a rather time-demanding alternative (as discussed in Chapter 6), the fertility factor in irrigation ought in some cases affect labour-time requirements in a downward direction, adding another question-mark to Boserup's main thesis of declining labour-time productivity with increasing intensity of land use. (Cf Chapter 9 on productivity.)

However, irrigation sometimes also leads to water-logging and salination. In his excellent study of irrigation in Mesopotamia, Fernea (1970:159-60) mentions that salinity was one of the major problems with the simple kind of irrigation where the river banks (being higher than surrounding land due to silting) were simply breached and water allowed to inundate nearby lands. The areas close to the river trunks then became spoilt by salt deposits after some years. One major benefit from ditching and canalization was thus, not only that it permitted a larger area to be irrigated, but also that it gave a better control over the irrigation process, including the drainage end of it.

Given the spatial expansion of cultivated acreage permitted by irrigation, it mattered less that a minor portion thereof was kept in fallow for the purpose of drainage, excess water being sluiced into these fields. Although there may thus be a strong tendency for fallows to be eliminated with irrigation, they are not likely to be entirely abolished, since they may still perform certain functions.

Temporal intensification versus spatial expansion

In dealing with the historical changes in Indonesian agriculture, the case of Java demonstrates the great scope for intensification that follows from the refinement of irrigation techniques and associated cultivation methods, according to Geertz (1963). The ecological stability of the wet-rice terrace and its responsiveness to further techno-organizational perfection made Geertz characterize the great population growth and intensification process in Java as one of 'involution' rather than evolution. The capacity of these terraces to 'respond to loving care is amazing', and this affected the choice of expansion at the intensive versus the extensive margin (to use the terms from economics):

> ... there is another introversive implication of the technical per-
> fection of traditional wet-rice cultivation. Because productivity is
> so dependent on the quality of water regulation, labour applied to
> the improvement of such regulation can often have a greater
> marginal productivity than the same labour applied in construct-
> ing new, but less adequately managed, terraces and new works to
> support them. Under premodern conditions, gradual perfection
> of irrigation techniques is perhaps the major way to raise produc-
> tivity not only per hectare but per man... (Geertz 1963:34)

Again we are in a sense dealing with indivisibles. The expansion of an irrigation system may be a major undertaking requiring not only large labour inputs but often tedious negotiations and less easily managed social arrangements to mobilize an adequate volume of labour-time. Since such requirements are far above the time supply of an individual farm household and may even be well above that of an entire local village, it follows that piecemeal refinement of an on-going system is better within the reach of smaller units of action at their suitable times of action, e.g. the slack season. This works to-wards intensification, while it does not follow as Wittfogel (1957) and Harris (1977) claim, that expansion necessarily demands an army of workers organized under the military pressures of an oriental despot.

However, although both Boserup and Geertz stress the intensifi-cation aspects, the spatial expansion aspects of the construction of irrigation works are nevertheless of major importance. Agriculture would simply not be of much significance in the lower Indus basin area (Sind province), for instance, were it not for irrigation. One may futher bear in mind that Java is by no means an area where dry farming is inpracticable. 'Only' about 40 per cent of Java's acreage is irrigated, the rest is rainfed.

TOWARDS A COMPOSITE SPACE-TIME MODEL OF RURAL LAND RE-
SOURCE MOBILIZATION

We are now in a position to sum up some of the major aspects of
intensification in the use of land from the vantage point of a time-
geographic interpretation of land as space-time. In doing so we naturally base this on contributions by studies referred to in preceding
chapters and by leading scholars in the field such as Boserup, Brookfield, Geertz, Sahlins and others already mentioned. However, the
discussion will be confined to intensification with respect to space-
time. The aspects of human time and labour-time intensification
have to be left to Chapter 9 (and later chapters), since a number
of additional dimensions of human time allocation must first be
explored.

In this book, we have chosen to regard intensification theory as
part of the broader theory on resource mobilization in human
societies-habitats. By emphasizing resource *mobilization* rather than
the somewhat bland notion of resource use, we have sharpened our
focus on the *dynamics* and *activity dimensions* of the general problem area.

Resource mobilization also pertains to the active occupation and
capacity utilization of the time-based resources that we have identified in this study. There are upper theoretical and practical limits
to which these resources can be mobilized, limits that are of great
principal significance in all ecological, economic and socio-cultural
forms of analysis. There are thus constraints on the volume of
activities that can be packed into a human time-budget or a regional
space-time budget, for instance. What is the innovative element in
the time-geographic concept of carrying capacity is, of course, that
it defines relations between *activities* (hence processes) and resources
rather than between populations and *food* resources. The traditional
carrying capacity concept may be useful in biological ecology where
it specifies relations between a population and its habitat (generally
in terms of maxima on population density). In human ecology, however, i.e. when directly applied to human populations, this rather
static concept is not fruitful enough, and it fails to do justice to the
kind of mobility and interaction processes in which humans are
involved.

Even when analysing the dynamics of man-land change as shifts
in man-land ratios and population densities, the traditional concept
of carrying capacity (which in effect is also employed by Boserup
albeit not the term as such) proves to be an analytical handicap.
We may look into the dynamics of population growth (or increasing
population density) and relate this to various features of land use,
and it is certainly possible in this manner to arrive at numerous

indicative correlations. Boserup, for instance, does this with a great measure of brilliance, as does indeed Harris (1977). However, by constantly referring the problem back to one of simple balance between a catchall category such as population and the land resource base, the umberella formulation of the problem still becomes 'vulgarized'. This is so because society is something more than a population aggregate; it is a set of socially, culturally and spatially interacting individuals, and models of aggregate population numbers relative to aggregate resources do not tell us enough about flows, interaction, structuration, process and evolution. A main reason why Boserup's and Harris' theories of intensification are still so useful is that these theories contain many interesting component models and arguments, although the way these theories are rounded off in general can be challenged on numerous grounds.

The Boserup model again

Boserup was certainly right in rejecting ideas that rural population density is simply a function of soil quality, climate or other factors in the basic natural environment, or that the kind of cultivation systems found in operation are explicable in terms of the natural environment alone.* The positive correlation between population growth and intensification has been verified in a large number of cases. Moreover, her dynamic view of land use turned the tables on a good many established ideas of agriculture in several disciplines; indeed her basic model induced something of a paradigm shift (shifting cultivation theory, so to speak).

However, her theory was also wide open to criticism on crucial points, a major one being that population growth is the main (or only) stimulus to intensification. But there are other points of criticism, e.g. those listed in the excellent review Grigg (1976) made of responses to Boserup's theses given by various other scholars:

It might be thought that Mrs Boserup's critics have left her theory in tatters. Her central argument, that intensification reduces labour productivity, remains unproven. Few would agree that an increase in the frequency of cropping is the only response to population pressure; the extensive margin can be extended, higher yielding crops adopted, and methods that increase yields introduced independently of increases in the frequency of cropping. Population pressure may be relieved by emigration or the control of

---

*   One sometimes comes across statements by non-geographers implying that these kinds of ideas are still current in geography, while even in this discipline environmental determinism has been dead as a dodo for about as long a time.

numbers. Intensification can also occur without population pressure, under the stimulus of urban growth or the development of trade. .... But whilst it is difficult to accept that population pressure is the *only* cause of agrarian change or that the increased frequency of cropping is the *only* response to population pressure, the thesis is still a remarkably fruitful interpretation of agrarian change ... (Grigg 1976:79)

Perhaps the major reason why the Boserup model was so fruitful and central was that Mrs Boserup started off with a realistic interpretation of land use *over time,* her frequency of cropping concept. Using this concept, radically different systems of agriculture could be organized along one and the same continuum, i.e. that of population density-*cum*-land use intensity. Once this idea was applied, the task of explaining shifts along that continuum had to be faced which is what Boserup did with a good measure of success. And, although some of her initial answers may not stand up to further scrutiny, a good many of the questions raised were of utmost importance. *

*   Her model was also unorthodox in the economic disciplines by being soundly empirical, lacking many of the arm-chair qualities of competing models in agricultural economics. She was well familiar with a great number of detailed empirical cases and by ordering them into a logical framework, certain broad traits in agricultural change and evolution were laid bare, as were various preconditions of why certain agricultural innovations were adopted but not others.

Her theory was also misunderstood by many as a direct analysis of the historically abnormal conditions of rapid population growth occurring in the Less Developed Countries of today. She rather wanted to paint the economic historical background and show that there was scope for the *absorbtion of labour* in agriculture and that population growth did not have to be diverted into urban forms of employment, as was the dominant assumption among many economists in the 1950's and 60's. It was only a minority of economists at the time who emphasized agricultural development rather than one sided industrial, among them Schultz (1964) and Clark and Haswell (1964). In lectures Boserup delivered at Lund University in 1966 and 1967, she started by saying that 'Economists usually make the assumption that a reduction of the agricultural population is a precondition for agricultural development...' She also attacked the assumption ... 'of strictly limited land resources with a too dense population, so that some must be removed in order to raise the productivity of the remainder.' Intensification of land use would create rural employment.

Although her theory is one of preindustrial societies, and although today's Third World countries are atypical of more historical cases by being a part in a 'modern' world, there are still important policy implications of her thesis regarding the creation of rural employment. Some have made the narrow-minded interpretation that she proposes that population growth will in today's world lead to economic development automatically and that governments are free to sit with their arms crossed. Her idea is rather that ... 'with suitable government policies ... and with a proper balance between the use of modern and of labour intensive methods, agricultural plans can be designed which secure the absorbtion of the increase in the rural labour force, except for rare cases of very poor natural conditions for agriculture or extremely high population densities.'

Generalizing the preexisting models

In this book, a major attempt has been made to generalize the existing models of rural and agrarian intensification proposed by Boserup and others. Part of this consisted in combining them, of course, and another part was to pay greater attention to the specification of levels at which intensification of resource use took place, e.g. the household/farmstead level, the local community level and the 'average' regional level. These levels in turn were like Chinese boxes, the larger-scale one containing the smaller ones, and it is important to clarify the interrelations between these levels of structure.

The attempt has also been made to place the various theories and models of intensification within a general model of human interaction. Interaction underlies all social, cultural, economic and ecological articulation of the composite formation we have called society-habitat. Chosing to operate within a structuralist paradigm, it was natural to stress not just interactional aspects but the *capacity for interaction.* It is believed in this book that it is by emphasizing these capacity constraints in particular that we get the instruments for sorting out the feasible from the infeasible solutions and become able to see an actual system as a 'realized possibility' of a general 'possible system'. Moreover, by emphasizing the resource as well as the capacity dimension plus selecting general resources such as human time and settlement space-time (as well as energy and water for that matter), we are placed in a strategic position to add a real economics dimension to socio-cultural and ecological analysis. These resources pervade every sector of society-habitat, not just that defined as related to material means satisfaction or the sector which is monetized. Our farmework is thus thoroughly economic and has a broader applicability to different sectors than the conventional pecuniary framework. (It is obviously beyond the scope of this book to interrelate the ideas presented to the use of money as a regulatory constraints system, although the exercise is far from infeasible. It is only time consuming to work the field out.)

At any rate, we have throughout this book examplified a great number of interaction mechanisms which impinge on the land intensification theme. But before closing the topic, two points will be made:

Firstly, some new concept must replace the rather crude notion of 'population pressure'. In saying this we do not deny the relevance of the population pressure syndrome, but discard a cloudy notion that needs to be redefined and dissolved into more tangible and analytically useful components.

The second point of clarification necessary at the present stage of this book concerns the manner in which demand for space-time is 'encapsulated' in the form of holdings and domains, and the role of domains, territoriality and regulatory constraints in the articulation of projects, or, if one so wishes, in the struggle between projects.

We will address the first of these problems by formulating a a demand model of space-time intensification, starting for convenience of exposition with the *ceteris paribus* assumption that supply of space-time is fixed in volume and only varies with the observation period we may choose.

Injections and withdrawals of demand for space-time

If we take a bounded region for a determinate period of time, say a region of 100 km^2 for 100 years, we receive a regional space-time budget with a limited carrying capacity of 10.000 km^{2}years. The annual carrying capacity is one hundredth of that volume. Assume further that this capacity is not fully mobilized and that each individual in the regional population gets an equal share of space-time, the total level of demand thus being exactly proportional to the number of inhabitants. Assume, finally, that the region is isolated and has no exchange of people, materials or energy with other regions.

We may readily conceive a number of shifts in this initial level of space-time demand, and we could speak of 'injections' and 'withdrawals' of demand. Net population growth would on our previous assumptions serve as an injection of demand, while population decline would constitute a withdrawal, as would net emigration. By contrast, net immigration would be an injection of demand. In other words, sorting out the population along the chief four components of births, deaths, immigration and emigration gives a clearer picture of underlying factors than gross population growth or decline.

A region may be subject to several injections of demand. Wilson (1963:376) mentions that the reasons behind Nyakyusa and Xhosa land shortage were, 1) large increases in their own population, 2) competition with immigrant whites for the land, 3) the introduction of the plough which enables each family to cultivate a larger area than formerly, and 4) the use of land for market crops. In fact, any innovation or element of social change can be assessed in terms of its land (space-time) demands (cf Carlstein 1978b). Some innovations or changes constitute withdrawals, others injections and still others may on the whole have neutral or negligible effects.

There are thus a series of demand *injections* such as:
- population growth;
- taxation in the form of products from the land;
- immigrant settlers occupying land;
- imposition of land reservations (e.g. for wildlife), and so on.

As for *withdrawals* of demand for space-time, these may consist of:
— emigration of population;
— import of land demanding products, e.g. food imports;
— introduction of land saving crops;
— release of land formerly controlled and left 'unused'; etc.

If in a given time period, injections of demand are greater than withdrawals, some kind of expansion in the occupation of space-time within the regional space-time budget must take place. Spatial expansion or temporal intensification are the two forms available, and which form will dominate depends on a number of circumstances. Moreover, as we have shown in preceding chapters, there may be an increase in local demand for space-time within daily prism range although there is no corresponding pressure at the regional level.

In discussing demand for space-time, it should be remembered that this may be induced by necessities other than food production. Even in preindustrial societies there are other products based on land, such as timber, charcoal, firewood and fibers. Apart from the land used for actual settlements, some land may also be reserved for other purposes related to religion or defence, for example.

We may also interpret the effects on levels of intensity from increases in productivity of land in terms of our injections and withdrawal of demand model. If more productive crops are chosen, or if yields are raised by the application of artificial fertilizer or simply by the input of more human time, this may have land saving or withdrawal effects on demand. Of course, it is part of Boserup's model that increased inputs of human labour-time per areal unit will raise the productivity of land, but her main thesis is that there are no gains in productivity of labour-time following the intensification in land use, only losses. Moreover, she seems to deemphasize the gains in productivity from other inputs than the greater application of labour-time, e.g. gains from technological improvements.

The point is that there are a good many flows or 'populations' of paths (cf p. 8) flowing within as well as in and out of time-space regions. Products of all kinds cross the borders circumscribing the space-time budget of the region. As will be discussed in Chapter 10, both prior to the English industrial revolution and in conjunction with it, England tended to relieve itself of certain space-time demanding forms of primary production and imported corresponding commodities instead, e.g. wood products, grain and iron. Prior to the industrial revolution, it took much forest to smelt iron. This industry was therefore better suited to a wooded country like Sweden, but even there the need for 'iron' forest conflicted with the practice of swidden agriculture. Such commondity flows affect the levels of land use intensity within bounded time-space regions as does the flow of people between regions.

Projects, positive constraints and political factors

Having outlined the above demand model as a kind of summary of modifications needed to the previously existing models on intensification, we may now turn to one of our earlier points of departure in this book (cf Chapter 2, p. 47). We stated that the concept of project was fundamental, since it incorporated the dimension of intentionality and goal seeking in human affairs. It was also made clear that projects were socio-culturally defined; they reflected socially and culturally transmitted interests and pursuits and they gave directionality to the human individual-paths, as it were. However, they also acted as allocative mechanisms, since efforts to realize some goals before others and to choose certain courses and paths of action to the exclusion of others gave projects a function of allocative mechanisms. Hence the term *positive* constraint, i.e. constraint by determination of acting units to do some things and not others.

However, we also stated that projects consisted of sequences of activities and that activities demanded resources such as human time, space-time, water-time and energy. It is thus in projects that we have to seek the roots in demand for space-time and the subsequent intensification in the use of space-time that this may lead to. Hence, it is foreign to us here to see the demand for space-time as ultimately only a demographic phenomenon, and it is also obvious that if we want to be able to link the demands for space-time with other processes in society-habitat, we must consider the intentionality aspects of projects as well as the social rationality behind the effectuation of projects.

Projects are thus subject to both positive and negative constraints. We have focused greatly on the negative constraints in this book for the simple reason that these constraints which circumscribe the living and action possibilities have been neglected in social science so far, in the opinion expressed here. It was thought strategic to concentrate on the very general constraints affecting human individual and social life. From there we could move on to other classes of constraints. The constraint framework, of course, was motivated by the desire to get away from empiricism and adopt a structuralist perspective in which actual systems are regarded as realized possibilities. Such knowledge is more applicable to new situations than that which is tied to specific empirical cases, although we have not refrained from using empirical illustrations.

Another thing about projects is that the positive constraints and goals of one actor or action unit, by virtue of competing for resources with other units become the authority or coupling constraints of those units. For this reason, people try to shelter the resource ingredients they need in projects. Institutionalizing holdings is one method of doing this, a method maintained either by contract (agreement) or authority. Hence, in the assessment of space-time demand and its impact on intensification, we must go beyond studying the average aggregate level of population pressure (cf further Chapter 9).

Just as the distribution of *demand* pressure among localities may affect the general level of land use intensity, so does *the distribution of demand among holders of land*. Pressure of demand is institutionalized into systems of holdings and domains, and one of the major points in the discussion of Chayanov's theory (p. 232-38) was to show that there are distributional factors affecting the demand for land as well. There is nothing sensational in this perhaps, but again it shows that aggregate figures and perspectives may be misleading. Although Geertz (1963), for example, points out that the colonialization of Java by the Dutch meant that the crops of Java were brought into the world economy but not its people, and that the indigenous population suffered from new injections of space-time demand (for export crops), one still gets the impression that it was cheifly population growth which led to the intensification and involution of Javanese agriculture. (The fact that labour was taxed and hence in part withdrawn from household benefit may also have to do with population growth, a theme of household expansion to which we will return in Chapter 9.) The picture is further conveyed that the distribution of land in Java was so even that intensity levels among holdings was the same. White, who carried out time and land use studies in Java in the 1970's, was of a somewhat different opinion. He pointed out that there is not only pressure of people on land resources, but that this entire syndrome is thoroughly compounded by the 'pressure of people on people':

> Many village studies also point to the existence of large numbers of landless households and of uneven distribution of holdings among those who do own land. We should note ... that this feature of village life in Java is not an automatic consequence of high population:land ratios, in the sense that equal distribution of land holdings is quite compatible with situations of acute land scarcity even if it does not usually occur. Indeed, attempts to relate inequalities in landholdings to overall land:man ratios may meet with surprising results. Maurer, comparing four villages in Bantul (Yogyakarta) found that the proportion of landless households was highest in the villages with the best overall land:man ratios... Thus, although the economic structure of any Javanese village will reflect a combination of these forces, it is important to distinguish the effects of absolute resources scarcity ('pressure of people on resources') from the effects of differential access to those resources ('pressure of people on people'). (White 1976)

Hence, in any formation of society-habitat there are regulatory and other contractual, institutional constraints which intervene together with capability/capacity constraints and coupling constraints. The reason why we have placed most emphasis on the latter two forms of constraint in this study is that these have been comparatively neglected in social analysis and that their implications have been anything but well explored. Yet, they are essential in any holochronic analysis incorporating the fact that all systems associated with society-habitat are *systems in motion* and *systems of 'flows' in time and space*. Just as we have put theoretical and methodological brackets around pecuniary flows in the present study, we have done so also with respect to much of the institutional substance around which sociology and social anthropology have centered their efforts. While this book does interrelate society, ecology and resources, the choice of *human time as a general resource for activity* leads on to an *action, activity and praxis perspective* which is also a key to the examination of 'institutions' as conventionally defined, although much of this analysis is left aside here for forthcoming publications in which the problems of how action/activity builds up institutions and how institutions affect action/activity will be examined at some length. Let us, for the time being, move on to the specific problems of action and time resources in pre-industrial societies within the general context of intensification theory.

8 TIME ALLOCATION AND THE CARRYING CAPACITY OF A POPULATION TIME-BUDGET

INTRODUCTION

The main theme in the preceding chapters has been path allocation, interaction and the carrying capacity of terrestrial space-time. This was helpful in understanding some fundamental features of many different societies, some having less intensive ecotechnologies with respect to space and time resources, such as hunter-gatherers, other societies making very intensive use of these resources, such as irrigation agriculturalists. Although a good many aspects were brought up, only minor attention was paid to the use of human time resources and the allocative mechanisms behind activity performance.

In the remaining chapters, time resources will be the main focus of inquiry, and the spectrum of ecotechnologies and societies covered will be even broader, ranging from hunting-gathering via more intensive rural ecotechnologies to some of the salient features of industrial and post-industrial societies. Not that it is possible to do justice to them all, and the latter two types are particularly badly treated in view of their tremendous complexity. Nevertheless, some attempt is made to sketch a broad outline if only to pave the way for further investigation along the lines of social time-space theory.

Before comparing the mobilization of time resources for various activities, especially 'work', in pre-industrial societies (Chapter 9), there are some necessary preliminaries that must be put forward — a set of basic concepts and sub-models by which to link data and theory in subsequent chapters. As a first step, the determinants of population time *supply* are outlined with respect to demographic factors, capability constraints and population categories. The next topic is population time *demand*, the structure of which must be analysed in terms of distribution among population categories and with respect to temporal and spatial location of demand. Another important dimension is whether time demand relates to individual or collective (cooperative or joint) activities. In the latter case there are obvious coupling constraints to reckon with.

Any time allocation theory is only interesting when time resource supply is related to time demand or time requirements. The possibilities of meeting various demands by supply and the time allocation mechanisms involved are general topics explored throughout the rest of this book. In the present chapter, the topic is introduced at the aggregate population system level. From there we move down to the individual level and then upwards again. Consistency between the different levels of aggregation is an important prerequisite for solid understanding and fruitful construction of theory.

TIME ALLOCATION AND PACKING AT THE AGGREGATE LEVEL

For an entire population as well as for sub-population units such as villages, households or even individuals, time allocation can be looked upon as a packing problem. Quanta of activities demanding time are to be packed into a container with a limited carrying capacity, and in this case the container is the human time-budget. Not only land areas and human beings can be thought of as resources having a time-budget with a limited 'budget-space' (the latter defined in Hägerstrand 1973d), but other 'populations' (in the ecotechnological sense, cf. p. 7) or resources have a time allocation and time distribution dimension as well (cf. Chapter 7). The sizes of these respective time-budgets form a major production possibility boundary (one form of living and activity possibility boundary) affecting activity performance and system structuration.

Looking at time allocation as a packing process in a population time-budget is a formulation similar to that of investigations of full employment in an economy. The potential labour force is largely taken as given, and the level of employment is allowed to co-vary with other forces in the economy. In this case, population time supply is initially taken as fixed, and the main problem is to establish the extent to which this capacity is used and can be utilised under alternative conditions and forms of organization. Of course, the foremost concern here is not necessarily to maximize some specific form of output or employment in the production sector (however the latter is defined). Human time is a multi-purpose resource used for all kinds of activities such as recreation, personal maintenance, social intercourse, political participation, religious worship, and the various activities associated with production of goods and services. The important thing to remember is that a multitude of substantively different activities are cast in a mould of structural interdependence by virtue of competing for the same ultimate resource, human time.

An initial reservation is due in using the familiar terms demand and supply: In this study, time *demand* is just a simpler term for time *requirements*, while time *supply* is used as short hand for time *resources*. What is not being proposed is a micro-economic price theory of human time, where demand and supply is mediated by money in a market. There are a number of philosophical reasons for this. For one thing, there are too many contradictions of temporal and spatial logic in micro-economic price theory or other special formulations in this field, such as utility theory. Underlying equilibrium or marginalist assumptions are also inappropriate in tackling human time allocation as a real time process, and these assumptions do not incorporate the underlying time-space logic and logistics of human interaction. Moreover, human time is not a storable resource, it is spent the moment it is received (only products of human time are storable) which again upsets traditional distinctions of temporal logic such as that between stocks and flows. Without denying the obvious fact that in many societies, financial transactions are of deep structural significance, the fact remains that a great volume of inter-action, transaction, service-exchange and other forms of intercourse is not directly mediated by money, nor is it accountable for in financial terms. To capture all these fundamental aspects of human interaction in society-habitat, one must go much further than a market mechanism approach. (Cf Carlstein and Thrift 1978 for a few previous comments on this.)   Although this may be objectionable to some economists aware of the new applications of economic reasoning proposed by scholars such as Becker (1965), Ghez and Becker (1975) and others, it seems more than likely that once some general resource allocation theories that are more physically (space and time-wise) realistic emerge, the need for revision of economic theory in and outside economics proper will be felt very strongly.

The aggregate time supply of a human population

Time resources are a strategic determinant of the volume of pro-jects and activities that a human population can perform.   The aggregate potential time supply is a function of two basic variables, population *size* and the *time period* observed (as shown already in Figure 1:2). Such a population time-budget can be depicted in a rather concrete graphic way (Cf Carlstein et al. 1968:18, Hägerstrand 1972a). *Demographically,* the size and shape of this time-budget is a function of the four standard factors of births, deaths, immigration and emigration (or in- and out-commuting) for the region where the population resides. In the daily or other very short term perspectives, births and deaths can generally be ignored while import and export of human time by commuting may be important to consider. The exactitude of accounting should be attuned to the scale of the problem faced, as long as there is an awareness of underlying processes. To depict such a population time-budget for many years in a row is difficult due to population turnover (cf Hägerstrand 1978b).

Describing the aggregate potential time supply as a simple function of the above variables (Figure 8:1 a and b) is a very general statement on time resources for all kinds of activity. Every individual is treated as equivalent and interchangeable: children and adults, women and men, specialists and non-specialists. The fact that individuals differ in their abilities and capability constraints and that certain activities require the time of specific categories must also be incorporated in the specification of time supply. Age, sex and skill and other qualities split up the population into categories, the size of which is decisive for the volume of different activities that can be performed. Although many capabilities change with time, as a function of progression in the life-cycle, accumulation of experience and learning, many capability constraints are essentially unaltered in a daily and even an annual perspective. To change the technical capability of a population in a developing country, for instance, is a process that takes years, not days.

In terms of our model, the interesting feature about capability constraints is that they lead to a vertical cleavage in the population time-budget, along the individuals rather than across. Capabilities are carried by individuals rather than being tied to the population as an aggregate. Figure 8:2 shows a simple case where the population is divided into those able to do activity A and those not. Such constraints thus put a maximum on the time volume which can be supplied to this activity, say the number of farmers in a village able to handle a plough.

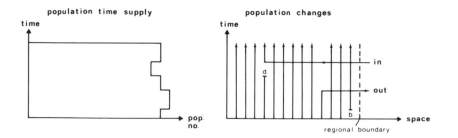

FIGURE 8:1   The relationship between the population time-budget (left) and population changes due to demographic variables such as birth (b), death (d), immigration (in-commuting) and emigration (out-commuting). The population time-budget is defined in relation to a given region.

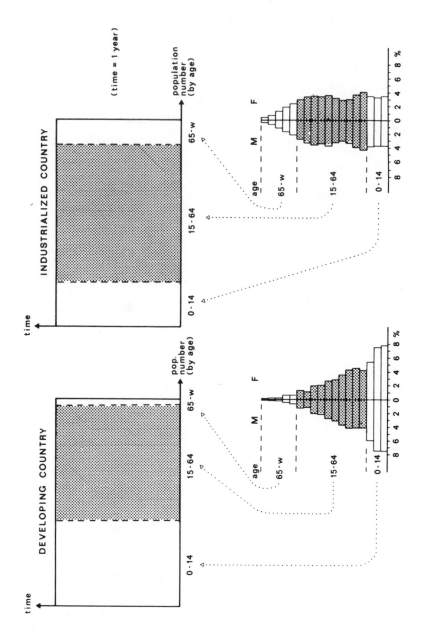

FIGURE 8:3   Age structure and potential time supply. The active ages are indicated by the dotted surface, the left age pyramid being more active (India) and the right representing a stagnant population (in this case Sweden).

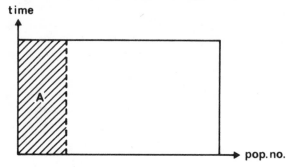

FIGURE 8:2   The maximum volume of time that can be supplied to a given activity A as a function of the size of the category capable of performing this activity. Capability constraints may thus cause a vertical cleavage in time supply.

The age structure of the population is one important factor in time supply and capability distribution. Many activities cannot be performed by children or (no longer) by the aged and are thus left to the active age categories. The volume of time they can supply can be shown as in Figure 8:3, where the rate of population growth is also reflected in the age pyramids, one case being for a developing country (India) and the other a modern industrial nation (Sweden). The distribution of capabilities may also be a function of sex, as in Figure 8:4.

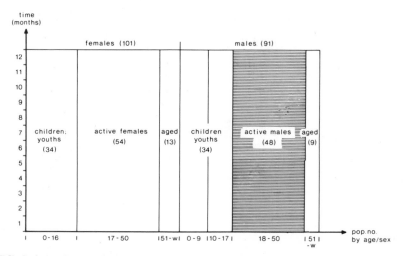

FIGURE 8:4   Sex and age categories in the population also imply a vertical kind of segmentation of the population system according to capability. The active males may be the carriers of some specific skills, which other categories may not have acquired. (The figure describes a Siane village in New Guinea and is based on Salisbury, 1962. The male category described may be those capable of doing advanced axe work, for example.)

It is important to note that these capabilities are not simply a function of biologically based variables such as age and sex. They are just as much a function of social participation and transmission of culture. Linton noted long ago that,

> In all social systems certain basic physiological and psychological factors have to be taken into account. Thus the process of human reproduction imposes certain *limits* upon the way in which a self-perpetuating group of individuals can be organized. Similarly, the differing *capacities* of persons and sexes impose certain *limits* upon the *possible* patterns of organization... age and sex differences ... are linked with different potentialities for social function and all societies accord them recognition in their formal patterns of organization...
>
> Actually, age and sex categories are probably more important for understanding the operation of most societies than are family systems... /an institution having/... received a tremendous amount of attention from social scientists... The age-sex categories and their derivatives are the *building blocks* of society...
>
> The current neglect of this aspect of social organization is no doubt in part due to its deceptive appearance of simplicity. The existence of age-sex categories is so obvious that their importance to social structure is likely to be overlooked. However, the neglect has also been due, in part, to the *lack of techniques for graphic presentation* of the social structures derived from a combination of these categories with other culturally established status determinants... (Linton 1942:871-73)/My italics/

It is remarkable how well Linton's ideas fit into the present format.

Biologically inherited, culturally transmitted and socially ascribed capabilities and capacities are carried by and tied to individuals and thereby contribute to a similar vertical segmentation of the population system and the population time-budget. To the extent that individuals of a specific status, for instance, are mobilized as a task force to do certain jobs, the time they can supply to their activities is of critical importance to social performance. Within the present kind of conceptual model, all kinds of normative, institutional and regulatory constraints which sort out the population into categories can be included (together with biological and technical types of constraints). The caste system of India is a case in point. Capability constraints following caste division impose maxima on the time that can be mobilized for specific activities, so caste specialization divides the population into segments as in the Rajpur village described by Elder (1970) shown in Figure 8:5 (cf also Chapter 9).

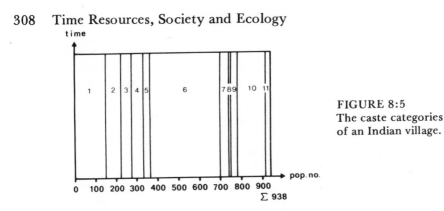

FIGURE 8:5
The caste categories
of an Indian village.

Unmobilized time supply and the fragmentation of time holdings

It must be emphasized that the time supply arrived at by aggregation across the population is the maximum *potential* time supply. It is the *unmobilized* time available within the overall region, not that accessible at the time-space locations where demand is to be met.

In practice, the time resources of a population are as spatially dispersed as the people who are the carriers of this resource. Initially, the aggregate time supply can be dispersed into sub-population time-budgets at the local level within the region (Figure 8:6), and the sub-populations at this level can in turn be broken down into domestic units (Figure 8:7). Not only are human time resources spatially dispersed; the *holding* of time resources is also fragmented and generally under the decentralized control of households in a way reminiscent of the fragmentation of land holdings in a region.

Since population time is a fragmented resource, it must be mobilized in space and made physically accessible, a process implying a drain on this very resource. For most purposes, potential time supply can be regarded as available at the stations of residence and day-prism centres. Were all time demand to have its locus there — each household meeting its own demand (cf Sahlins' idea of domestic production for own use) — potential time supply would be equivalent to accessible and mobilized time supply. But since many inputs other than time are not located at the dwelling station, and since households both demand time from and supply time to other units, time supply must be mobilized. Moreover, many projects are on a scale requiring greater time volumes than domestic units can furnish.

It follows that projects which may seem feasible in an aggregate perspective may not be possible to implement when an assessment is made from the micro-level of households in local settings. As will be shown below, time allocation is more complex than a matching of aggregate time supply with aggregate demand when these are stated in rough 'man-hour' volumes, even when we focus on relatively small sub-populations such as those working in a factory or a high-school.

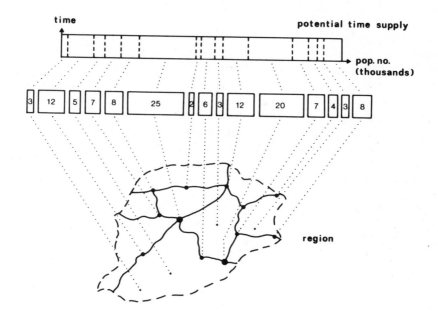

FIGURE 8:6   The disaggregation of potential time supply at the regional level into time-budgets of sub-populations at the local level.

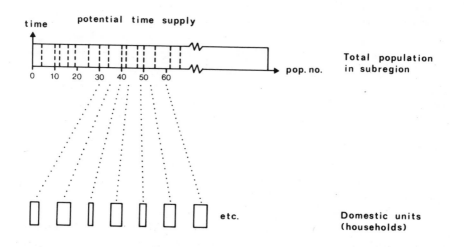

FIGURE 8:7   The fragmentation of time holding, i.e. the disaggregation of time supply at the local level into time holdings by domestic units. The household time-budget in turn can be regarded as the aggregate time supply of the individuals in the domestic unit.

Time demand by human projects and activities

Although a motley collection of goals and interests underlie the various projects pursued in society, all individual and collective projects are implemented through time demanding activities. These multitudinous time requirements do not add up to an amorphous aggregate, however, but are expressed within a habitat and societal activity system of specific structure. Time demands thus have a non-random distribution in the population, in time and in space.

Most time demand is of a fairly routine nature. Were it not, the lack of predictability would militate against coordination and cooperation, reducing the overall standard of activity performance. Assessments of time requirements by different actors are facilitated by previous experiences of similar situations and by the social communication of intentions. Some projects need no prior announcement. Physiologically necessary activities such as eating and sleeping recur on an every-day basis, as do many other activities and meetings of households or other units. The principle of return to certain stations (cf Chapter 3) also lends predictability to future activities and fortifies expectations as to where activities are to take place. Expectations regarding who is supposed to do what, to whom and when are also tied to various population categories in a fairly regular manner. Numerous production projects (such as cultivation) or cooperative social activities are timed by seasonally or diurnally recurring constraints and possibilities. Predictability in timing is reinforced by cycles in nature as well as cycles of social convention, such as market cycles or ritual activity cycles, or by norms, agreements, contracts and regulations. Order in the natural habitat and in social organization thus facilitates realistic assessments of time demand. In addition, predictability is indirectly supported by an understanding of things which *cannot* happen, many possibilities being ruled out by the physical and biotic constraints under which society operates. The level of awareness of these living possibility boundaries may vary, but some understanding is generally there.

Under more irregular circumstances, as during natural catastrophies or social upheavals, activities are not so predictable, however. Rapid social structural changes are problematic for the same reasons. In modern societies this is partly compensated for through planning and making plans and intentions known to other units. In this way the future is colonized by designs in order to channel the system into certain courses. Centres of authority, purchasing power and government may also make direct claims on human time resources to make sure they are used and distributed in line with their intentions. The intentions of some thus become the constraints of others, depending on how they can be politically upheld.

In general, human time demands are not a function of capability and capacity constraints in the same straightforward manner as time supply is, since demand is also a matter of wants and designs. The more stable time demand incurred by physiologically necessary activities such as sleeping and eating constitutes only a part of total time demand. In other spheres of living, aspiration can go far beyond capability and many time requirements may be unrealistically in excess of supply, bearing in mind, however, that aspirations are to a considerable extent socioculturally conditioned.

It is not just the demand an individual makes on his own time that may cause this discrepancy. More complicated are the social demands projected onto specific individuals and categories. The same individual may be affected by demand from a number of other persons at the same time. Some people may be demanded far in excess of their capacity, while others are largely or even totally outside the focus of demand.

It is obvious that time allocation is an *interaction process* within a population system. In order to map the close connection between demand for time and the capacity to satisfy it, we have adopted the methodological device of mapping demand onto those individuals (categories) exposed to demand rather than on those imposing this demand, e.g. onto the employees rather than the employers, or if we are investigating the effects of a school reform, time demands are mapped onto the potential teachers, students and others to do the job, not the politicians who decided it. We thus distinguish between the *sources* and the *subjects* of demand.

For a person doing something for himself to satisfy his own demand, the source and the subject obviously coincides. But if yet another person's time is required to fulfil the first one's ambition, only the source is the same. The main theoretical advantage of this whole exercise is that it facilitates a better physical assessment of demand and the extent to which it can be met by supply.

The distribution of time demand

As suggested by Hägerstrand (1972a, 1978b), it is possible to map the distribution of time demand graphically in a population system. Using this approach, primary schooling for the young generation imposes time demands distributed as in Figure 8:8, for example. School work follows a fairly rigid and prearranged schedule in most parts of the world, it seems, both with respect to temporal location in the day, week and year and in terms of age/sex categories affected. In countries where military service is organized on a conscript basis, an older age category is exposed to this kind of time demand, and generally only males are included. For Muslims and Christians, religious activities in the form of prayers and visits to mosques or churches affect a much broader age category of the population, but also for a much shorter time per day, week and year than primary education does (cf Figure 8:9). Demands for domestic chores such as fetching water may in some societies be focused on young women or all the women who are able, while in other societies, as in some of the villages in India, it may affect a specific caste and mainly the men (cf Figure 8:5).

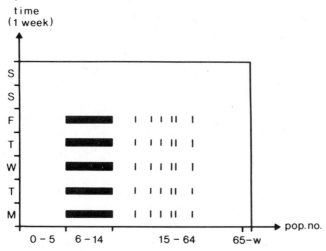

FIGURE 8:8   The time demand for school attendance by pupils (left) and teachers (right).

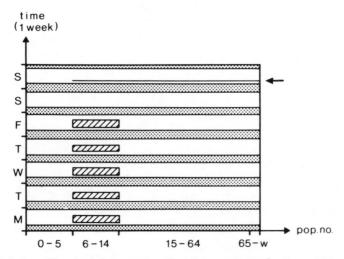

FIGURE 8:9   The time demand for church attendance (indicated by arrow) in relation to time demand for sleep and school attendance by pupils.

Most of the time demands associated with personal maintenance are uniformly distributed across the entire population because they are physiologically necessary, mainly eating and sleeping. These activities are *reflexive* in that the source and the subject of demand coincides and they are *horizontally* distributed in the population time-budget, as in Figure 8:10. Time demand for eating and sleeping are necessary for survival and hence have high priority in relation to other activities which they interrupt.

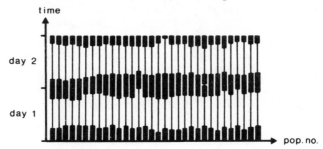

FIGURE 8:10   The time demand for sleep and its typical horizontal distribution in the population. Sleep cannot be delegated to specialists, nor can eating.

Another interesting aspect of sleeping and eating is that these two activities *cannot be delegated* or transferred to other categories in the population or to specialists. There are other activities which also cannot be delegated to others such as resting while one is ill or a pregnancy (the latter only affecting certain age/sex categories, to begin with). If it had been possible in human history to delegate sleep to other members of society, it would be more than likely that a special class of people would sleep on a full time basis so that the powerful could seek pleasure and practice various pursuits on a round the clock scale. But the range of delegation for sleep is narrow in the extreme and this means that packing of sleep in the population time-budget can never look as in Figure 8:2, but has to be something like Figure 8:10.

Other activities can be delegated within much broader categories of the population, for instance among the personnel within a firm or household, or within an age/sex category. Some constraints on range of delegation or transfer are dictated more by social norms (such as only men doing military activities or only women doing cooking) than by biophysical capability constraints.

The fact that sex roles, for instance, may be so ridgid that many activities and the associated time demand cannot be met by the time resources of all adults is a factor of inflexibility in the packing of activities in population time-budgets (cf Chapter 9).

for transfer and delegation exists in that rather than going to another person, one may ask that person to come over. The tendency to transfer the time costs of travel to somebody else is common among categories which are especially subject to pressure of demand. This will be discussed in greater detail in Chapters 10 and 11, but it has already been touched upon in Chapters 5 and 6 on settlement reforms designed to make better use of a limited set of service-givers, while ordinary people affected by nucleation of settlement had to spend more time travelling to fields. The mechanisms of transferring travel time onto others are many and varied and are often subtle (such as the way shops are arranged in modern societies to save the time for movement of the personnel and shift it onto the customer).

Time demand distribution of a simple production project

Activities generating time demand are in many cases parts of larger projects. Cross-cultural, geographical and historical studies show that the same fundamental projects can often be carried out in many different ways and still result in similar or identical products. Even in one locality, the same technique and procedure can be varied, which is a source of flexibility in the packing of activities in a population time-budget so that capacity can be used better.

But there are also constraints in the projects themselves, not least because they require many other inputs than human time. There are certain technical and reactional constraints (reaction in the chemical sense) on how substance can be combined to form new states. The input-output relations in projects also generate start- and deadlines for component activities, which are part of a project hierarchy and order among activities. If some component activities are delayed, this may have repercussions on the entire project.

The time it takes to complete a project can be referred to as the *gestation period*. Projects involving biotic organisms often have a more rigorous and inflexible programming once they are triggered because of intrinsic growth rates of a biochemical nature. To produce a human child is a project involving a nine month gestation period, while growing cereals may take between three and six months, for instance. This is not to say that cultivation projects are uninfluenced by external conditions such as temperature, light and rainfall, or that growth rates are uninfluenced by competition for nutrients, and so on.

The overwhelming majority of human projects do not place a continuous demand on human time throughout their gestation periods, but only do so intermittently. Projects move towards completion in steps or quanta, and activities being part of very different projects thus become interwoven. Aggregate as well as the disaggregate time demand is affected by this interaction.

In specifying time demand, it is inadequate to do so simply in terms of man-hour volumes for a given time period, since the same volume of demand may be distributed in a great many ways within the population in time and in space, depending on the specific circumstances.

We can use a simple production project to illustrate this, for example, the production of salt among the Baruya of New Guinea, a process described by Godelier (1971). Typical for a project of this kind are the constraints on the *sequence* of component activities. In this case, the grass used as raw material first had to be cut, then transported, dried and burnt, and from the ashes produced salt could be extracted through a further stage. This technically simple project would yield, in this particular case, fifteen bars of salt (c. 25 kg).

The aggregate time input was 21 man-days, according to Godelier, and one man-day amounted to 9 hours. Leaving the spatial locational aspects aside, this time volume can be distributed in a number of different ways. In principle, it can either affect 1 person for 21 days, if the project is entirely individual, or 21 people for 1 day (or even more people for less than a day), or 7 individuals for 3 days. There are many other possible ways that yield a total of 21 person-days (Figure 8:11). The actual distribution is case D, which also takes the sexual distribution and the gestation period into account (to the extent that the latter can be inferred from Godelier's data).

In practice, the project could hardly have been carried out by 21 people in a single day because of a number of other constraints. For one thing, the salt grass could not be harvested before the dry season had started and then the grass had to dry for another one or two weeks. The filtering of ashes could probably have been distributed over a shorter time period, but the evaporation of salt bars had to take place in an oven at a maximum temperature of $65^{\circ}C$ or so. The process would therefore take at least five days in the kind of oven used. Since the ashes were storable, it might have been feasible to choose the time for filtering and evaporation within broader margins and there appears to be no strong technical dead-line when the project had to be completed within the year. But the activity of tending the oven so that the water evaporated, requiring one person for five days, could not be done by five persons in one day, although the time demand volume is the same. Again, given constraints such as the capacity and indivisibility of the oven, the process had to take something like five real days to complete. Hägerstrand (1974a) has called the mapping of different activities in a project onto the population and other inputs (such as tools and land) the *score* of the project, drawing on an analogy from the notation system of music.

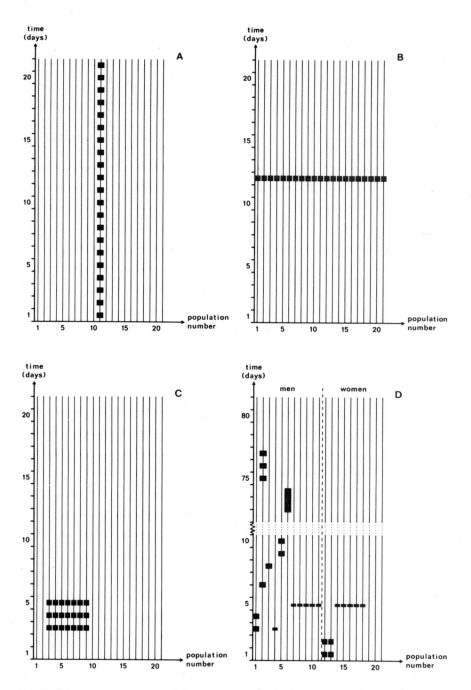

FIGURE 8:11   Some possible distributions of time demand (21 person-days) for a project. The lower right part of the figure shows the (approximate) time demand for producing c. 25 kg salt in New Guinea (data from Godelier 1971).

Matching time demand in the activity system with time supply in the population system

In time allocation analysis, it is helpful to sort out projects and activities into a social sub-system of their own, the *activity system*, as Hägerstrand (1972a) has suggested (cf also p. 244-47). The activity system is the source of demand while the *population system* is the source of supply. In *ex post* studies of actual time use, these two systems are not distinguished. In an *ex ante* perspective, however, they converge on one another as time passes. Prospectively, the activity system (the various activity sub-systems of individuals, groups and organizations) consists of intended or planned activities more or less organized into projects. These are as yet unconstrained by the allocation of people's time and paths or by the allocation of other inputs into projects, such as land, materials, tools, buildings or other facilities (cf the synchronized economizing of resources).

If, for example, we were to add up the projected time demand for a multitude of human projects in a time-space region, we may arrive at an aggregate time demand profile such as that in Figure 8:12a, depicting the volume and temporal location of demand *(ex ante)*. While demand fluctuates over time, population time supply is non-fluctuating (if we disregard births, deaths and migration), as in Figure 8:12b. Given the continuity and indivisibility of each individual aggregate time supply cannot attain the distribution of case *C*, although the total volume is equal to that of case *A*. Case *C* assumes divisible people and storable, discontinuous time supply.

Looking at the period in retrospect, demand had to be adjusted to supply, as in Figure 8:12d. Excessive demand was severed off as time passed, and some initially intended activities were simply not performed. Correspondingly, the slots of time not subject to specific demand were automatically filled with some kind of activities, since all time must be spent the instant it is received as 'income'. In this way, the population time-budget is always fully packed with activities, although the density or intensity of *specific* activities varies.

It can be safely assumed that in no real world society is there a perfect match between given time supply and projected time demand as time unfolds. In some societies, the demand for time may be excessive for a number of reasons. There are the previously mentioned problems of planning and predictability and the influences of a fluctuating environment (such as erratic rainfall altering cultivation programmes). The actions of many other individuals that one collaborates with or otherwise interacts with are also difficult to foresee. Since any excessive demand is automatically cut down with the passing of time, intended output may suffer. Still some choice can be exercised, since planned activities are not random-

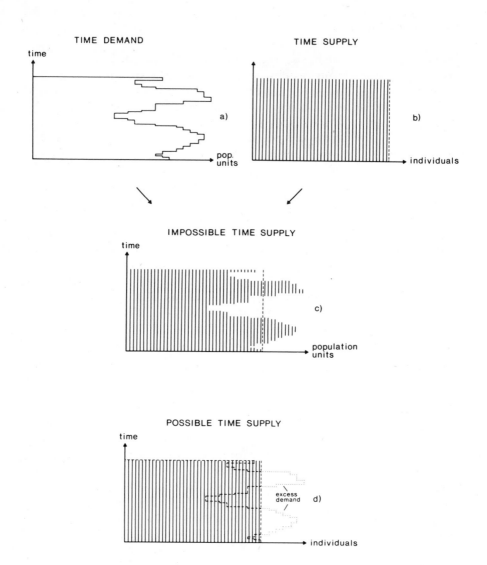

FIGURE 8:12   Possible and impossible aggregate time supply in relation to demand. The person-hour volumes in case a) and b) are identical but have different temporal distributions. It is impossible, however, to distribute time supply as in case c).

ly excluded. Were this the case, too many projects would be hit by missing activity links in the implementation chain. What rather happens is that demand is selectively reduced and other activities are simply postponed to fill time at some later suitable occasion.

The *excess time supply* situation implies that the carrying capacity for projects and activities is not fully utilized. The institutionalized demand for time may be low for various reasons. The general work load may be low, for instance, if the region is so rich in resources and facilities that a satisfactory subsistence and living can be achieved at a low level of cost. The already mentioned case of time affluence due to low basic requirements is one in point described by Sahlins (cf also Chapter 9). There may also be excess time supply because of *a lack of other resources with which human time can be combined,* such as land, water, fuel, tools or other facilities. These inputs may either be missing in the region or else in short supply. Moreover, the political will or structural conditions to share existing resources may not be there in cases where resources are concentrated in the hands of limited sub-populations. Typical cases would be those of season-ally more or less 'idle' peasants lacking access to land, or in urban regions, city-dwellers being cooped up in local environments so devoid of resources that they have 'leisure problems'.

Packing localized activities in a population time-budget

Even when the volume of time demanded appears to coincide with that supplied at an aggregated level, the picture may look rather different at a disaggregated level. This is made clearer if we move down to the micro-level and examine how the spatial and temporal location of activities together with capability (indivisibility) and coupling (coordination) constraints affect time allocation.

The general principle can be illustrated by taking only two individuals and their time-budget (Figure 8:13a). Time demand is assumed to be distributed in time-space as in Figure 8:13b. If this demand is added up, it exhibits an uneven distribution (right side of figure) and does not fit well into the population time-budget. Aggregate time demand is excessive, even before travel time requirements and path allocation have entered the picture. In effect, Figure 8:13b (right) depicts demand assuming that travel is instantaneous.

When allowance is made for the fact that movement incurs drains on time, that some activities are performed jointly, and that some activities cannot be performed in a piecemeal fashion to be effective, the net result is that the *real excess demand* is increased (Figure 8:13c, right). The combined effect of the volume of demand and its time-space location is that up to four individuals must supply their

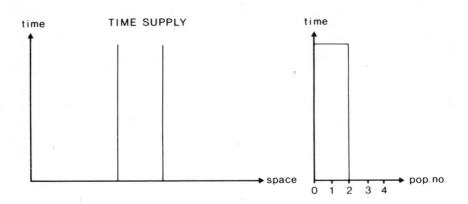

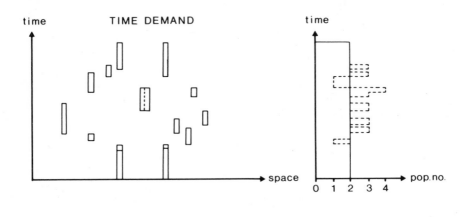

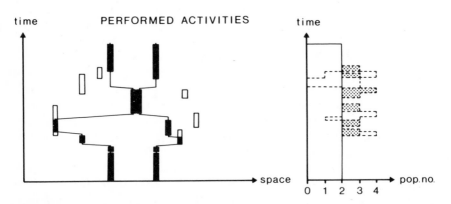

FIGURE 8:13   Time supply and demand in a two person case. In the middle figure (right) the aggregate excess demand is shown as that outside the two person time-budget, prior to actual time and path allocation. In the bottom figure (right) the real excess demand in addition to that shown above is indicated by dotted surface. Travel time is now included.

time if the activity programme depicted is to be feasible.

Often enough, however, there is some flexibility in terms of volume and/or location of demand so that the time of given individuals can be more readily supplied. However, as far as joint or collective activities are concerned, they do impose coupling constraints which means that there are limits to flexibility (as we shall discuss at great depth in Chapter 11). In this chapter we will begin by illustrating some important mechanisms of time allocation and coordination at an elementary level.

TIME ALLOCATION AT THE INDIVIDUAL LEVEL

In the preceding sections we started at the aggregate level and worked our way downwards. It is also possible to begin with the individual and move towards groups and sub-populations in order to reach the population system level thereby illustrating the effects of path allocation, capability and coupling constraints.

Some of the approaches to time allocation in economic anthropology by scholars such as Herskovits (1952), Firth (1967), Salisbury (1962) and Barth (1967 a and b) have departed from the classic ideas of scarcity and choice applied to individual actors. From this point they go on to the consequences of various allocations and the alternatives open to actors in particular situations. A typical formulation of the choice and allocation model is that of Salisbury:

> The one resource used in all activities is the time of the participants. At all moments an individual has to choose whether or not he will enter a situation where a specific activity would be appropriate. At all moments the cost of doing one activity is the activities of other kinds which must be foregone. (Salisbury 1962)

Although both economic anthropology and sociology have had a vigorous debate on formal versus substantive definitions of what constitutes the economic sector of society, few of the participants in this vivid interchange of ideas have denied the fact that choices made have allocative consequences, so that the cost of a particular choice is the opportunities thereby forgone. The debate has focused more on whether the allocation of scarce means to more or less satiable wants is an adequate base for defining the 'economic' sector of society, which it is not.

On the other hand, if one departs from the specific resource of human time, the really important matter is that *all activities* take time regardless of which sector they belong to. The need to define so sharply what constitutes the economic substance of society (in order to build a discipline around it) has had the opportunity cost of displacing a golden opportunity for more holistic analysis in social science. General time allocation analysis has the advantage of integrating otherwise unrelated activities, since all activities use the same basic resource — human time. Why be so excessively 'economic' when society is, so to speak, at stake?

If we apply the conventional scarcity and choice model to time allocation at the individual level, each person has a time 'income' to be used for activities which is equivalent to the observation period, say a 24-hour period. Adding the postulate that the individual has a number of *alternative* activities open to him and that the potential time 'expenditure' on them would amount to *more* than the time 'income', there is a relation of *scarcity* which forces the individual to choose (Figure 8:14). (As noted in Chapter 1, scarcity is not an absolute precondition for choice, but let us assume this for the moment.)

A methodological assumption made here is that activities are discrete and non-overlapping quanta of action, each making certain demands on time resources. The activities cannot be done all at once,

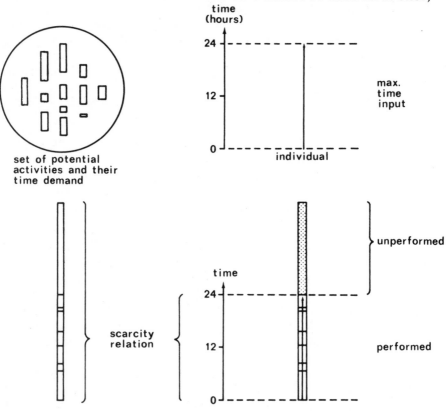

FIGURE 8:14   The individual and excess time demand. The scarcity relation forces the individual to leave some activities unperformed.

but must be performed in a temporal order or sequence. The situation depicted in Figure 8:15 is therefore impossible by definition.

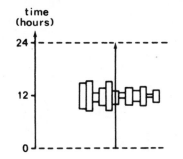

FIGURE 8:15   It is impossible for the individual to perform numerous time demanding activities at the same time.

The time *cost* of one activity is the time which is taken away from alternative activities. If an individual's time is allocated to activities A, B and C, as in Figure 8:16, the *opportunity cost* may be activities D and E, for example, which are foregone and displaced in the process, and rather than specifying which exact activities are being displaced, the more general approach can be used where opportunity costs are simply stated in terms of time resource quantities; each activity is in fact assigned a 'time price'. This is a quite defensible procedure as long as one is not tricked into believing that human time has the same qualities as money (cf further below).

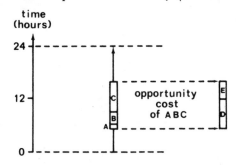

FIGURE 8:16   The opportunity time costs of three activities in terms of other activities forgone.

The main danger with the individual scarcity and choice model is the tendency to get stuck in the behaviouristic mould at the extreme micro-level of the individual. Society is not a collection of Robinson Crusoes each trapped on his own little island. It is a system of interacting individuals (etc.), and many opportunity costs are not just individual but affect other members of society as well. Choices and allocations of some behavioural units feed back onto other units

whose environment is shaped by them. So while individuals can serve
as a point of origin, it is imperative to go beyond single individuals to
groups, sub-populations and the entire population.

In order to be more realistic, let us leave aside the particular issues
of scarcity, choice and the isolated individual and move on to time
allocation as it relates to path allocation and interaction among
individuals. A strategic assumption here is that people do not simply
allocate their time to individual activities but also to *collective* ones.
The latter are indicated by lines crossing through the activities in
Figure 8:17. For individual A to carry out his programme, he may
impose a demand on the time of B, C and D. It is also easy to
imagine the reverse case where the time demands of B, C and D are
aimed at A, or some similar combination.

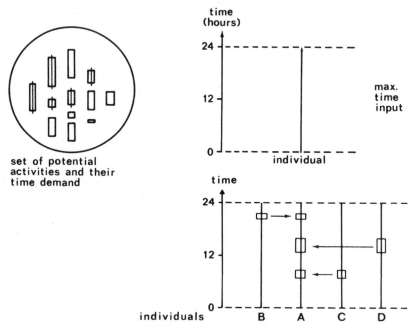

FIGURE 8:17   The performance of collective activities (indicated by lines
through them in upper left figure) involves the time resources of other persons
also. Time allocation cannot be realistically studied for individuals in isolation.

Unless we are dealing with some form of telecommunication, such
as telephone calls, collective activities outside peoples' base station
must involve travel for some members. Who moves to whose places,
and how long does this take? Had it not taken time to travel, individuals could have carried out more of their stationary activities than
is the case when these have to be interspersed with travel (Figure
8:18). The effects of relative spatial location on travel time are

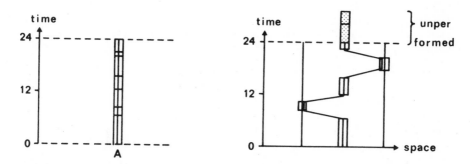

FIGURE 8:18   The displacement effect of travel time requirements on other (stationary) activities.

crucial but are a neglected aspect in most anthroplogical approaches to time allocation.

The effects of temporal location on time demand and supply are also strategic. Suppose, for instance, that individuals B, C and D all want to use the time of A at his home station (Figure 8:19). They meet A in turn and the cooperating parties are accessible to one

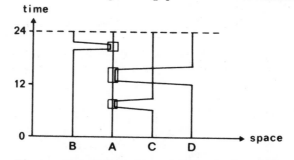

FIGURE 8:19   The meeting of time demand (from B, C and D) sequentially (successively) rather than all at the same time, which may be infeasible, since the activities are mutually exclusive or incompatible.

another. But if for some reason B, C and D were only able to meet A at the *same* time of the day, A would not be able to supply his time to meet the demands of three different activities simultaneously. It would not help that A has 24 hours to supply in one day and that the demands from the others are considerably less. He will still be short of time because of this temporal coincidence of demand, which gives rise to a time allocation conflict. The latter can only be resolved by rejecting cooperation with two of the others at the same time. Had individual A somehow been able to store his time from previous days, he could have enlarged his supply at peaks of demand, but unfortunately human time cannot be saved for a rainy day. The travel solution might also have been different, as long as the various demands were met one after the other (Figure 8:20)

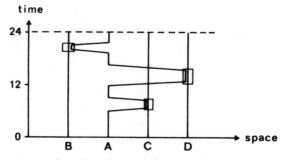

FIGURE 8:20   A meeting time demand from B, C and D at their respective stations, thus assuming the time costs of travel as well.

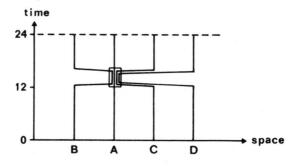

FIGURE 8:21   Person A meeting the demands from B, C and D at the same time at A:s station, on the assumption that these meetings can be combined.

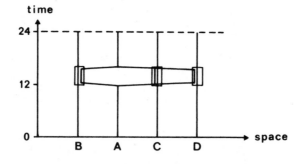

FIGURE 8:22   This is an impossible solution. Demands on A coincide in time, but since A is indivisible, he cannot be in three places at the same time. (Such coincidence of time demand on one person is anything but rare in society.)

We could also make the assumption that the activities for which B, C and D were making demands on A were compatible and could all be done at the same time. Individual A could then invite the others over (if they could make it at the same time of the day), as in Figure 8:21, while A, on the other hand, would not be able to visit the three others simultaneously at their places, since that would run counter to the indivisibility constraint (Figure 8:22).

When an individual is engaged in a joint activity with somebody else, he is generally not accessible to some third person at the same time. This creates what can be referred to as an *accessibility gap*. The allocation of time thus has two sides, one being the *occupation* of time, the other the *displacement effect* and 'social' opportunity cost this has for the other members of society (Cf Chapter 11).

Collective activities assume that potential collaborators are available and accessible in the environment. But this condition can not be taken for granted. There are numerous situational determinants of time supply, and when a suitable or desired counterpart is not accessible, demand is not matched by supply and must either be reduced or else postponed. Both travel and waiting are means of making demand and supply coincide.

These examples of time allocation at the individual level indicate that the allocation of time by a given individual not only incurs opportunity costs to her or him but also that when activities are carried out jointly with others, there are collective opportunity costs. Such collective opportunity costs or displacement effects will be analysed in detail in Chapter 11, but from instances given so far, it is apparent that individuals are affected by each others choices, and what may be freedom of choice for one person may often be a constraint on choice for another. Every social system has its different modes of 'trading in degrees of freedom' with respect to time allocation, as Hägerstrand (1972a) has put it and there are many mechanisms which regulate the way individuals demand and occupy the time of each other. So what is necessary to make time allocation theory capable of explaining social structure and activity performance is an *interaction theory of human time allocation*, not an isolated individual choice kind of approach.

A remark on time, money and economic anthropology

Many economic anthropologists have been disenchanted with the application of modern economic theory to primitive and preindustrial societies. Sahlins, for one, puts this forcefully in an early essay of his:

> It is incredible that anthropology — having at hand comparative materials from the whole range of human history and the entire gamut of cultural arrangements — should adopt, in dealing with primitive economics, a theoretical outlook relevant to historically recent, evolutionary advanced systems of production and exchange ... primitive economic behaviour is largely an aspect of kinship behaviour and is therefore organized by means completely

different from capitalist production and market transactions. The inference is clear: The whole of modern economic theory, resting on the assumption of the market and its concomitants, fails to apply to primitive economies. (Sahlins 1960) *

To some economic anthropologists, particularly Belshaw (1954) and Salisbury (1962), time was an appealing recourse in the analysis of societies which were not at all or little monetized and where the conventional tool-kit and theory of economics was difficult to apply. To Belshaw money was too particular an institution: 'We must therefore find some other more universal element which may be subject to measurement. Such a universal element is time...' The analysis of time was not only a method of turning anthropology more *quantitative* but also of introducing *commensurability* between products, activities and processes of various kinds. Just as miscellaneous goods

---

*   There are, of course, a good many *transitional cases* between the kinship structured economy of primitive or preindustrial societies and the more or less market-oriented urban-industrial societies of today. Moreover, the economic theory relating to socialist economies such as the U.S.S.R. and China cannot be all about the market, and many models used in econometrics are of the general systems and operational research variety and can be applied to societies other than market dominated ones. To cover the whole spectrum of societies from the primitive to the contemporary, the economic branches of all social sciences and economics proper have to merge, which entails revision and redesign in all these disciplines. While economic anthropologists have raised objections to the application of conventional economics to 'non-market' societies or sectors, it must also be borne in mind that numerous scholars in human ecology, sociology and other disciplines have cast doubt on the adequacy of modern economics in dealing with many problems found in the advanced industrial societies themselves. The extent to which present-day economic science has been able to cope with issues of structural unemployment, technological transformation, welfare and quality of life, social inequality and class, the environment, and so on has been the subject of much debate.

In view of these considerations, it is sometimes surprising to find that even those economic anthropologists who are rightly critical of the wholesale application of modern economic theory to pre-industrial, pre-capitalist or contemporary developing regions of the world, are far less scrupulous in their choice of basic concepts. They readily apply the whole set of basic categories which have evolved in economics, such as 'land, labour and capital', 'goods and services', 'production and consumption', etc. This is as true for scholars with a 'liberal' approach as those with a marxist outlook. Any deviation from this established system of categories is bound to be regarded with suspicion, since it is constantly assumed that the mentioned categories are the only ones (categorical monopoly...?). (Cf further arguments in Postscript to this volume.)

If the objective is to further the integration of theory rather than its sectorization, the gaps between systems termed 'economic', 'social', 'technological', 'political' or 'ecological' have to be bridged. In doing so, rather than focusing so much attention on *products* — commodities — it might be better to begin with *projects* and their associated activities and trajectories.

and services become commensurable in terms of money in economic theory, time would permit the scholarly penetration of the 'subsistence' sector and non-monetized societies.

By applying time as a yardstick in measuring relations between actors and activities, it would no longer be necessary to have the 'price' of an activity stated as opportunity cost (or displacement effect) of *specific* alternative activities (as in Figure 8:16). It would be much simpler to express activity 'prices' in time costs directly, given the acknowledged fact of limited time 'income' per capita. The time for activities can then be calculated, added, subtracted, accounted and budgeted. Human time can be treated as a kind of money. This approach ought, one might reason, vastly improve the scope for applying existing economic theory to non-monetized societies and sectors.

At this juncture a good many theoretical pitfalls present themselves, if implicit assumptions remain unquestioned. To those having the ambition to employ the established apparatus of economics outside its original target zone, the time of the population may implicitly be looked upon as some kind of money to be allocated among alternative ends. Human time can certainly be allocated, but the implication is not that time and money have similar properties.

In assuming that time is money (or even analogous to money), the inference would be that human time – like money – is storable, divisible, exchangeable, (trans)portable, additive and accessible or mobilized (i.e. liquid). Money can serve as a store of wealth or purchasing power and has these various qualities, whereas human time has not. It is not storable, for instance, since it is 'expended' the instant it is received as 'income'. Products may be storable and memories also, but not the time that was put into them. Money is (trans)portable, while time is expended in transport. If a person leaves home with a given sum of money, this sum is generally intact upon arrival at the shop or destination, but someone who has an hour in which to do shopping will not have an hour left as he or she reaches the shop, and allowance must also be made for the return trip. The transferability of money between times and places is without counterpart in the transfer of human time. The latter is expended while transferred.

Money can serve as a medium for the exchange of 'goods and services' (cf p. 30), and being a 'store of wealth' it permits the double coincidence-of-wants constraint to be relaxed. But in the exchange of time demanding personal services requiring face-to-face contact, time cannot be used as a medium of exchange and both parties (service-giver and service-taker, cf Chapter 11) have to be simultaneously engaged. Time is not storable and the carriers of human time are indivisible, which affects the divisibility properties of human time negatively, while money is divisible (down to the smallest denomination).

As for accounting, human time for collective activities cannot be subject to the same accounting framework as money can. Human time is neither 'stock' nor 'flow' in the conventional sense and the budgeting of time requires a different framework of calculation, due to path allocation and situational determinants of time supply. (This problem refers to the matching of supply and demand for collective activities discussed in Chapter 11 and is better left aside here.) The operators in a time allocation framework of accounting must therefore be different to be realistic.

In saying this, we are not implying that economic anthropologists and others have been confusing time and money, since that would be to do grave injustice to the wealth of penetrating studies available. The point is rather that, since time allocation theory is still in its infancy, it is important to make the differences in the properties of human time and money explicit, not least in order to understand how human time is actually related to money as a means of transaction. Needless to say, while human time is a real resource, as we see it here, money is a derived resource or a means of controlling real resources and their allocation. However, the way in which money is used to control and allocate human time is outside the scope of the present study. This is not because the role of money is unimportant — far from it — but rather because thousands of scholars have taken money, markets and all its associated features as their point of departure. In this study, we depart from time and space which 'were there' long before money complicated things.

9 TIME RESOURCES IN PREINDUSTRIAL SOCIETIES

INTRODUCTION

Having furnished some models of human path allocation, time allocation and packing in a population time-budget, we are now in a position to outline how time resources are *mobilized* for various activities and projects in pre-industrial societies. To this end, we have chosen a comparative perspective where societies with different ecotechnology are used as examples, starting with hunting-gathering and moving on to irrigation agriculture. This is also the approximate cultural evolutionary sequence in which most intensification theorists have organized their ideal types, although we have altered this order somewhat with respect to pastoralism so as to improve the correspondence between ecotechnology and increasing intensity of land use (space-time occupation).

The major objective of this chapter is not to arrive at a series of conclusive statements on intensification and its relation to ecotechnology, let alone cultural evolution, since there is not enough ethnographic material available to sustain such conclusions. Although there are numerous indications, scholars in this field must still face a good many methodological issues, collect additional data and piece together existing indirect data before further headway can be made. In this chapter, we can only make a contribution in this general direction.

In view of the scant attention customarily given to time resources in ethnographic, geographic and social scientific enquiry at large, the coverage that can be given below is perforce very patchy. In scope as well as content, local community studies ('village studies') make up a very motley collection and to the extent that they include a real time use dimension, they are biased towards 'labour' and 'work' rather than total activity systems. (Intensification theory similarly tends to revolve around 'work' activities.) 'Work' is quite an elusive category, however, and the definitional boundary between 'work' and 'non-work' ('leisure', as some call it) is extremely fluid. It varies from one observer to another with negative implications for cross-cultural, cross-spatial and cross-temporal comparisons. For all the virtues of work-cum-labour analysis, the standpoint advocated here is that it is best replaced by general activity and time allocation analysis, embracing all kinds of activities. Only by adopting this basic strategy will it be possible to develop activity analysis in a genuinely fruitful and powerful way.

The topic of time resources in pre-industrial societies is so broad, unfortunately, that we are compelled to neglect many important aspects, such as the annual cycle. Again, seasonal variations in time inputs not only pertain to activities included in cultivation projects, for instance, but equally to other 'non-work' activities, e.g. those contained in the annual ritual cycle. It would also have been desirable to pay greater attention to domestic units and their *coordinated time allocation* in the daily round. * In fact, the complexities of *bundle formation among paths* and associated collective activities in pre-industrial societies are also neglected in this chapter. (We will to some extent make up for this in the context of industrial societies in Volume II where a good many principal features of time allocation and bundle formation are looked into.) What we will do here is to interpret a few of the best time use studies available in order to outline the structure of time utilization in pre-industrial societies.

The stress in this chapter will be placed on the levels shown in Figure 8:7, viz. the local and domestic unit levels, leaving aside larger population systems. Given the work-bias in most studies drawn upon, a good deal of information remains to be desired concerning activities like education, ritual, political and administrative activities, child care, illness, social intercourse, visiting, play, recreation, shopping or exchanging services. Better specification of the mobility component would also have been valuable. Nevertheless, only future studies can rectify these imbalances and it is hoped that a number of fruitful ideas will emerge from this study, both on the design of future investigations and on the reinterpretation of existing sources. The broad aim now is to furnish some general ideas and case material on the *mobilization of human time resources* for all kinds of activity. Intensification, both in land use and time use, is a vital part of this theme.

---

*   In the not too distant future, I hope to be able to present some comparative material on the annual cycle as well as on the daily round of activities in different societies, mainly as the basis for a theory assessing the impact of various innovations that contribute to structural transformation (cf Carlstein 1978a).

METHODOLOGY AND DIMENSIONS OF TIME MOBILIZATION

Anthropological and sociological studies of rural communities have generally relied heavily on *verbal* description. While this mode of expression is a flexible and direct way of conveying information on local living conditions, it has the well-known drawback of not interrelating things in very precise proportions. It therefore needs to be supplemented by a *quantitative* mode of description, especially with regard to human activities.

> The manner in which individuals spend their time is a basic dimension of ethnographic description. Under such headings as 'the daily round', 'the annual cycle', or 'the division of labour by sex', most ethnographies eventually describe the broad out-lines of time allocation in the community. This information is then used by theorists to construct comparative generalizations. (Johnson 1975:301)

But activity analysis specified in terms of time resources is fraught with methodological pitfalls which must be overcome if we are to make substantial headway in the general and comparative analysis of different societies and habitats. Several social science disciplines have shown a certain amount of interest in time use analysis for quite a few years (cf Erixon 1938, Richards 1939, Sorokin and Berger 1939, Foster 1948 and Tax 1953). In spite of a growing interest in studying human activity with time use precision (cf Szalai et al. 1972), methodology – the bridge between data and theory – has developed rather sluggishly except in sociology which has relied on statistical methods to a great extent.* Other disciplines have tended to lag behind, although there are 'islands' of this kind of work found in the vast literature. Sahlins (1972) made a fine effort to mobilize available time use data in anthropology, for instance, but the collection he comes up with is very heterogeneous. In anthropol-ogy, like human geography, there is still little consensus as to how activity-cum-time-use data should be collected and presented, if at all, nor how they should be combined with population data. Most village studies, for instance, show a remarkable lack of elementary demographic data, and time use figures must be combined with population data to be of any real use. Even Johnson (1975) who strongly advocates time use study in anthropology omits the demo-

---

* The statistical time-budget tradition in sociology has been fruitful in many respects and has boosted the general interest in time allocation aspects, although this does not imply that these aspects have entered the mainstreams of the disci-pline. However, method has perhaps outgrown theory, and much work remains to interrelate time-budgets to society, economy, ecology, habitat and technol-ogy. (Cf Carlstein, Parkes and Thrift 1978b, especially the Afterword.)

graphic data for the population studied by him.

On the other hand, any plea for better method without linking it to an appropriate theory is largely misplaced. This has been one of the weaknesses with some of the sociological work done. The statistical paraphernalia was there for the taking and the data on time use could be reshuffled in all kinds of ways which still failed to do justice to the innate logic of the systems investigated. So this chicken-or-egg paradox must be properly resolved by the parallel development of data, method and theory.

The household daily round as a minimal activity system

Just as the daily round of an individual may be looked upon as the minimal unit in social-environmental analysis, it must be placed in its immediate larger context which is that of the household. In his study of rural systems in Java, White pointed out that:

> ... rural development programmes ... require a better understanding of the existing 'raw realities' of labour utilization, and this understanding must be based on intensive research in which the *basic units* of observation are neither individuals nor 'occupations' but *whole households*. (White 1976:284) /Italics added/

This viewpoint has much to commend it, since domestic units are made up of sets of individuals whose activities are interdependent: they coordinate their activities, delegate them within the group, give services to one another and share numerous parameters affecting their daily lives. Their common dwelling, for instance, is the station of return and reunion for all members, so although the individual level is of basic significance in view of the capability constraints on time and path allocation, this level must obviously be transcended. In dealing with large and complex systems, some kind of aggregation procedure is necessary, on the one hand, as is some sort of sampling on the other, since it is too time consuming to cover all individuals and households through field-work. It is then tempting to rest content with the sampling of individuals, snatching them out of their household context and aggregating their data by statistical methods. But this is a non-structural and non-contextual approach which may be self-defeating in many ways (cf Carlstein and Thrift 1978:225-39). For one thing, much is gained by sampling entire households rather than individuals, so that the organizational and structural factors at this level are left intact. The time-geographic models facilitate a contextual approach rather than an atomistic one.

Instead of treating the daily round as an activity *system*, it is often not recognized as such, and pertinent information on it may be scattered in some fifty odd places in a monograph. To piece such data together is possible but time consuming. Frequently, however, a detailed account may be given of the daily round of one or a few sample households. Reining (1970), for instance, analyses a Haya household in Tanzania, and describes the activities of a husband and wife with accurate chronology. Translated into the sociological 'time-budget' format by the activity classification used later in this chapter (cf also Table 9:2), the outcome is a time use account, Table 9:1, which can be placed against the backcloth of Buhaya society and habitat (and compared to similar data elsewhere, e.g. in Rald and Rald 1975, Kamuzora 1980 and Cleave 1974).

A good deal of precise activity information can be extracted from similar time use accounts. In this case we can see that about 2 hours are spent by each sex on cultivation and that the woman has an impressive amount of domestic work, all in all 8½ hours. The various activities reflect and impinge upon different features of technology, social organization and habitat, and time use accounts can thus be very revealing (Table 9:1).

Any study of annual time allocation must in practice be based on daily round samples. Large sets of data (in terms of people or periods covered) have usually been treated by statistical methods, an outstanding example being the work by Szalai and his associates (1972) on industrial societies. However, *statistical* data handling and aggregation procedures do have their limitations and if too empiricist a point of departure is chosen, such an approach would tend to obfuscate essential *structural* features. The daily round is a system of interrelated and 'entangled' paths and activities, not a random cluster. Furthermore, activities are sequentially ordered in the implementation of projects, and individuals *coordinate* their activities with other individuals and entities in the habitat as shown in the previous chapters. It is only when this *activity structure* is made explicit that we can adequately assess the impact of various changes, for instance.

When the above data on the Haya are conveyed in a day-path graph (Figure 9:1), the daily activity system is made more concrete and this together with the time use account, greatly enhances the potential for analysis (in spite of the fact that the spatial dimensions are maltreated in the figure for lack of data on activity locations). The also conveys important data on who interacts with whom and when.

Similar graphs also reveal quantitative proportions among activities but are less sensitive to the way activities are classified, since the interspersal of mobile activities among stationary ones, for instance, indicates a natural discontinuity between activities, especially collective activities. Hence, sets of activities can be studied in their context without it being absolutely necessary to apply a water-tight classification scheme so that all activities become discrete, non-overlapping

TABLE 9:1

TIME USE ACCOUNT OF A HAYA HOUSEHOLD FOR A 24-HOUR PERIOD

Activities:	Total durations:	
	Woman	Man
Sleeping (SL)	8 h. 30 m.	8 h. 30 m.
Eating (ET)	1 h.	1 h. 45 m.
Personal care, rest, 'leisure' (HY, ID)	1 h.	4 h. 20 m.
Fetching fuel/firewood (FW)	0 h. 15 m.	—
Fetching water (WT)	1 h. 15 m.	—
Fetching food (FP) > (D)	0 h. 45 m.	—
Cooking (CK)	2 h. 45 m.	—
Housework (D)	3 h. 30 m.	—
Cultivation, farm work (AC)	2 h.	2 h. 05 m.
Visiting (V)	3 h.	2 h. 10 m.
Civic duties (P)	—	1 h. 45 m.
Animal husbandry (AN)	—	3 h. 25 m.
SUM TOTAL	24 h.	24 h.

(Source: Reining 1970)*

and additive. The paths always have to add up to 24 hours per day and per person, but it is not methodologically necessary that activities are classified in such a way that they also do. They only have to be consistent with the paths. (Cf Figure 5:10, for instance.)

A major problem in conventional time-budget analysis has been to classify activities so that they do not overlap, i.e. when adding up the activities of each person in a diurnal cycle, the total should be neither more nor less than 24 hours. But some activities do overlap. Child care, for example, may be combined with food preparation. One way (among others) of solving this is to account half of the total period to each activity, as did White (1976) when studying a village in Java.

Although the exact timings may not be given in the sundry ethnographic records available, verbal description combined with the approximate time coordinates of some activities may suffice for a reconstruction of daily rounds. So even when the exact times or the *cardinal* relations are not recorded, many *ordinal* relations are still there, as in the case of the Gusii of Kenya (taken from LeVine and LeVine 1966:37-40; cf Figure 9:2). Thanks to certain unequivocal relations in the configuration of paths, a time-geographic reconstruction can give a reasonable quantitative estimation of time allocation.

* We do not pretend, of course, that these data are representative in a statistical sense. There is great variation in the daily rounds of households, as shown by Rald and Rald (1975) for the Haya (Buhaya).

Activity classifications and categories

Apart from the problem of overlapping or parallel activities, there are other problems of activity classification relating to the actual choice of categories. Categories are the 'catchment-basins' of real world processes, and they can either be catch-alls or more selective in what they include.

Most local studies do not classify *movement* in space as a specific category but include this in other activities, e.g. walking to the fields is included in cultivation. This has the distinct disadvantage of making it very difficult to interrelate the activity system and the structure of the habitat-cum-settlement system. Hence changes in settlement cannot be connected with impacts on activity and vice versa. Of course, it may often not be necessary to record movements within a station such as the dwelling and its backyard, but all movements *between* stations are of great importance in view of the fact that so much time is actually spent on movement in one form or another in most societies. In pre-industrial societies, for instance, activities such as fetching water and firewood, cultivating the fields, and visiting relatives, shops or marketplaces, have a large movement component.

If we take the case of the Siane cultivators of New Guinea, the men go to the fields every third day on the average (according to Salisbury 1962), while the women travel this way every day (Figures 9:3, 5:10 and 5:11). For every 12 hours of gardening, the men spend 4 hours travelling, a maximum ratio of 3:1, while for every 5 hours of garden activities a woman spends 2 hours travelling, a ratio of 2.5:1. However, as Figure 9:3 indicates, in a six day period, a woman may spend some 12 hours walking to the gardens while the men spend only 4 hours. By delegating gardening to the women to this extent, the associated travel is also delegated to them. Even if gardens are half the distance away, travel still absorbs a good deal of daily and weekly time. To compare two otherwise similar villages and finding that in one, the women spend an additional hour doing gardening per day (the difference between 1 and ½ hours travel in each direction) would not make sense unless travel time is specified.

Activity classifications must also be sensitive to technology and *social organization*. Starting with the latter, it is important to register whether given activities are carried out singly or jointly with other individuals. Collective activities presuppose the bundling of paths and the synchronization and coordination of individual activities. There are many constraints and structural concomitants of joint action, as we will show at great length in Chapter 11. If the classification used in time-budget surveys is neutral with respect to group size and coordination, the fundamental dimension of group forma-

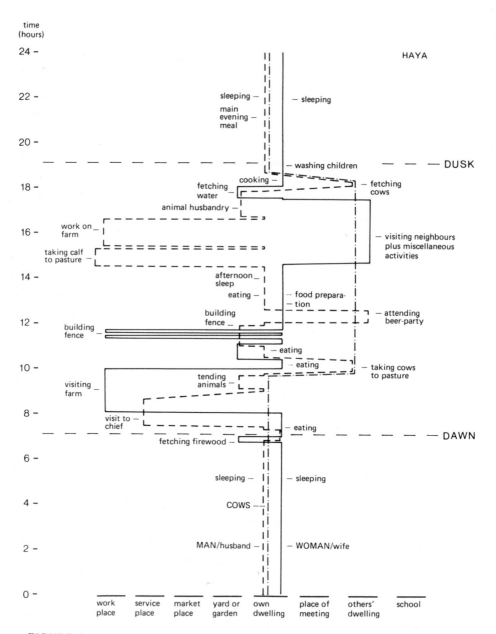

FIGURE 9:1   A daily round of a Haya (Buhaya) household showing different activities and the kind of place for those activities. Three paths are depicted, that of a man (husband), a woman (wife) and that of their cattle ('cows'). There were no data on the children, unfortunately, in the source material (Reining 1970). The places are not classified by distance in a continuous space. Instead a category space is used (for lack of data on spatial location). Still this kind of graph gives a better picture of activity sequence and daily round structure than numbers only.

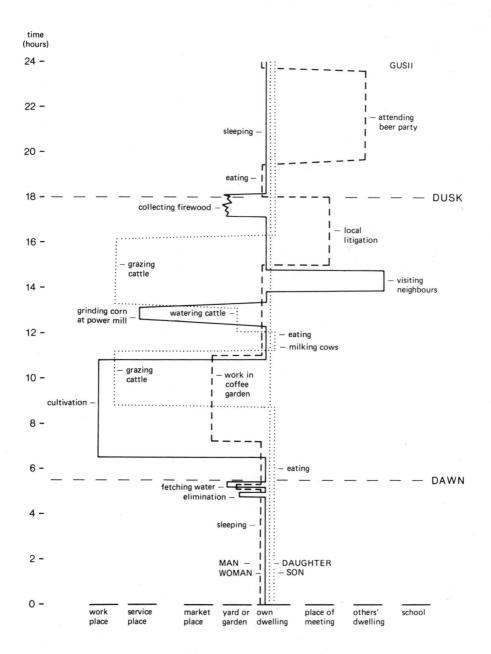

FIGURE 9:2   A reconstruction of a feasible daily round among the Gusii. Although the exact times of some activities are not given in the source material (LeVine and LeVine 1966:37-40), those times given together with certain *ordinal* and sequence relations make it possible to make a quantitative estimation of how Gusii time is spent in a typical daily round. The same category space is used in this figure as in Figure 9:1 which facilitates comparisons to similar graphs.

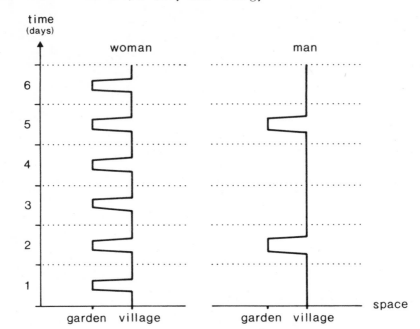

FIGURE 9:3   The frequency and duration of travel and cultivation activities affecting a Siane woman and man. The total amount of time spent on travel is a function of both frequency and duration. By not having to do gardening every day, the men save a great deal of travel time which can be diverted to other activities. (Cf Figures 6:9 and 6:12.)

tion and person-to-person interaction is simply lost and cannot be inferred from the data material.

As regards time allocation and *technology,* this brings us to the greatest catch-all category of all, viz. that of *work* and *labour.* Many scholars seem strongly to favour crude work-leisure dichotomies, and talk of average work days, etc. without specifying what they mean by work. To many economists and others, work is primarily identified as those activities for which payment is received. This may be a useful definition in economics focused on monetary flows, but it is inadequate for other approaches. Many structural changes in society and habitat can certainly be traced back to changes in the 'work' situation of certain population categories, yet it is hardly possible to determine what influences what, if 'work' is an undifferentiated category regardless of the enormous variations in the kind of ecotechnology to which a given population may be exposed (especially in urban-industrial societies but also in preindustrial ones).

In preindustrial societies, it is also common that time use for cultivation is only recorded for the main crop, whereas the time spent on cultivation of subsidiary crops is neglected (White 1976).

Moreover, there are a good many activities more or less intensively performed and more or less mixed with activities such as socializing, child care or ritual that also impinge on how 'work' should be classified, if at all, as a separate category. Johnson is right in saying that:

> Depending therefore on which measure of productive labour we wish to use, we could conclude that Machiguenga /cf below/ men spend less than two and one-half hours or more than eight hours per day in essential subsistence activities, and that women spend less than one hour or more than nine hours per day. The advantage of providing a *complete record of the activities performed* by different categories of individuals should be clear: comparative theorists are then free to define variables in accordance with their theoretical aims rather than having to accept the incommensurable figures each idiosyncratic fieldworker may choose to publish. (Johnson 1975:307)/Emphasis added/

One example of a very broad definition of 'work' is that by Rald and Rald (1975) in their study of the Haya agrarian system:

> A major viewpoint adopted here is that when we look at *adult* people's life as an entirety, all kinds of work, whether agricultural, domestic, or related to social obligations, are considered *productive work,* as opposed to *leisure.* This means that for the daylight activities even taking a meal or staying in hospital is considered work. (Rald and Rald 1975:70)/Italics added/

In this case, the traditional work-leisure dichotomy, which has been extensively criticized in the time use and leisure literature (cf de Grazia 1962 for an early study), is maintained and leisure takes on the character of a residual activity. Yet it is compounded by the marxist notion of 'productive work' as a condition for consumption and that 'consumption reproduces the conditions for production' (cf p. 52 above). Hence consumption may also be 'productive activity', since it reproduces the labour force. Both the conventional economics and the marxist concepts of productive versus unproductive activities, and of work versus leisure, are rather crude, and they are not all that meaningful distinctions at the every-day grassroots level where praxis is manifested.

Even Connell and Lipton (1973, 1977) dealing with 'labour situations' in 'developing countries' concede that what they are really proposing is a general time allocation approach rather than a narrow 'labour' utilization approach, obviously in an effort to prevent a catch-all category such as 'work' from obscuring what is really going on at the local level and with what effects.

Much of this argumentation can also be aimed at the use of *ideal type* classifications, as if cases were pure and unaffected by other than dominant activities. The activities which lend their name to hunter-gatherers, pastoralists, swidden agriculturalists or irrigation cultivators are generally not even dominant in a quantitative sense (by that token we would probably all be 'sleepers'). Again, Johnson phrases this very sensibly:

> We might follow Sahlins (1972:57) in considering only the amount of time spent in gardens as productive labour. Apparently once a group has been classified as slash-and-burn horticulturalists, Sahlins takes their garden time as an estimate of total productive labour. This definition, applied to adult married Machiguenga, finds the men productively engaged only 18.5 per cent and the women only 6.6 per cent of the time ... But I think this approach is *overly topological*. Simply because we might classify the Machiguenga as slash-and-burn horticulturalists does not mean that garden labour provides all their subsistence. (Johnson 1975:305-6) /Italics added/

When time allocation studies end up at an appropriate level of resolution and depart from activity classifications that do justice to real activity complexity (without, of course, complicating things unnecessarily), then the true implications of the daily mixture of activities rather than categorical streamlining will emerge.

Time input per areal unit versus per population unit

A common obstacle to interpreting the time and labour utilization data in the literature is the propensity of many field-workers to report their data on an *areal basis*, i.e. per hectare or acre. Otherwise excellent surveys such as that by Rald and Rald (1975) on the Haya or Cleave's survey (1974) on labour use in a variety of cases of African smallholder agriculture are difficult to use for our present purposes for this reason. From a purely agricultural viewpoint, some comparisons between farms, cultivation systems and regions may be facilitated by reducing time allocation data to spatial units, and the method portrays one interesting dimension of *land use intensity*. But from a general time allocation and structural viewpoint, data on time inputs per hectare cannot be properly related to other uses of time.

Even when time inputs are given for land use categories and when these data are supplemented with information on the distribution of land of various kinds among households, it is difficult to arrive at the distribution of time inputs among household members and among age-sex categories.

There are studies, however, in which time input per spatial unit and kind of cultivation activity have been used to assess total time inputs in local agriculture. In this case, the sampling is made on the basis of land holdings rather than directly of the population. Tax (1953) thus measured (by sampling) how much time was spent on different operations (by men, women and even children) in a few sample plots, and reduced these findings into a per acre form. He then multiplied these mean time inputs by the size and composition of all the holdings operated in the village of Panajachel in Guatemala, covering the specific activities from land preparation through the entire cultivation project. Pospisil (1963) used the same method among the Kapauku of New Guinea. While this method may be a reasonable compromise as far as sampling is concerned, the estimates of time inputs arrived at should preferably only be used to cross-check data directly collected on time allocation. And these direct time allocation data should cover all activities of the population in a given time span. (Cf Carlstein and Thrift 1978 for further points on sampling.)

The main problem of per hectare data is, of course, that they only make sense for a limited gamut of activities. It is obviously not very enlightening to report hours per hectare of child care or political participation, for instance. Even the time input per areal unit for pastoral activities is not of much validity. So for analysts interested in general time resource problems, it is frustrating to find that so many field-workers writing about rural societies are all too inclined to report the agricultural data on a per area basis *only,* rather than *per population unit and per population category.* Farm management surveys in particular are biased towards purely agricultural activities and omit most other activities, from the fetching of water to schooling. They thereby miss out the whole manner in which cultivation activities are embedded in other activities and compete with them for limited time resources, quite apart from the tendency in these studies to give time use data on a per area basis.

Reducing time use data to *per capita* figures is another way of obfuscating analysis. When time use is described for an 'average' man and woman (e.g. as in the case of Australian aborigines, cf McCarthy and McArthur 1960, Sahlins 1972:15-6), one never knows which categories of the population they are representative of. What about children, the aged and other age-sex-skill etc. categories. These details are not revealed when all categories are averaged into a composite per capita index. This destroys the validity of the data. Time use information is only of real analytical utility when the *distribution within the population* is given, by age, sex, skill and specialization, etc. as emphasized already in Chapter 8.

Analysis of time use and the distribution of activities within a local society also assumes that proper *demographic* information on age, sex, skill, specialization and other characteristics is available, i.e.

that demographic data and time allocation data are collected together and can be interrelated. Generalizations of the kind that people under 18 years of age 'engage in little work' are simply too crude. Because of great variations, not only in which activities some field-workers define as work, but also with regard to the age of entry into 'full work' or employment, figures on average per capita time expenditure on 'work' are nearly meaningless or at best only indicative. Moreover, demographic data must also be broken down for *households* of varying membership or composition. The spatial distribution of residences, work-places and other stations must also be a part of this picture, i.e. the spatial aspects of demography. Many field-workers who have lived with a local population for quite some time often fail to furnish this information (not even in appendix form), somehow assuming that such data are too mundane to fit into their advanced scheme of analysis.

To encourage comparative studies and theory building in the future, more standardized procedures for field-work, data collection, processing and presentation would be highly desirable so as to remedy the patchy and uneven information available up to now. Some of the dubious sampling procedures thus far employed should also preferably be revised.

GENERALIZING THE 'LABOUR UTILIZATION IDENTITY'

In spite of incisive criticism, standard categories of 'employment', 'unemployment' and 'underemployment' in economics have been rather carelessly applied to rural communities in Third World countries. Notions such as the eight-hour working day, with all its connotations of industrial wage labour, have been transplanted. Apart from its normative underpinnings, this concept also fails as a yardstick in the measurement of 'work', even in those cases when 'work' has been consistently defined. (For a critique of 'Western' concepts of employment, cf Myrdal 1968 and Connell and Lipton 1977.)

The evidence collected in intensive village studies, farm management surveys and similar materials demonstrates many of the inherent weaknesses of conventional employment concepts. Myrdal and his research colleagues (including Boserup and Lipton) were thoroughly disgruntled with the heaps of 'useless' employment statistics produced in Asian countries, since this material was more misleading than illuminating. Instead Myrdal (1968) and Lipton (1961, Connell and Lipton 1973, 1977) advocated a start from scratch in this field, i.e. observing and measuring the 'raw realities' of what people actually did and for how long:

We do believe that substantial progress can be made by behavioural techniques, and even by utilizing the method of questioning people, if the questions asked are adjusted to their actual life, are stretched over a short period, and are amenable to cross-checking. Even from a car or bicycle one can observe, on one's daily round in a number of places, when people go to work, and how many are doing what in the fields...

... this approach is a clear and marked advance over the conceptual system underlying the great mass of official statistics... At base, *the labour utilization approach requires behavioural studies founded on observations of the raw realities.* Such studies are all too few in number, yet this line of investigation is far more promising and fruitful than one calling for interviewees to answer questions shaped by Westernized preconceptions. (Myrdal 1968:1027)

This approach was essentially that of the village studies programme at the Institute of Development Studies in Sussex, England, in which the income resulting from labour was regarded as a function of:

a) the proportion of the population of working age *(W/P)*;
b) the proportion of such a work force participating in work *(L/W)*;
c) the average hours each person works in the period *(H/L)*; and
d) the productivity of work, i.e. output per man-hour *(Y/H)*.

Production per capita is thus a function of demographic age structure, participation rate, duration of work, and productivity of work per person-hour, as in the following expression:

$$\frac{Y}{P} = \frac{W}{P} \times \frac{L}{W} \times \frac{H}{L} \times \frac{Y}{H}$$

All these dimensions (except productivity) have already been illustrated graphically in Figures 8:3-9, but it should also be remembered that the detailed data on time allocation for 'work' and other activities must be based on the analysis of household daily rounds, as depicted in Figures 5:10, 9:1 and 9:2, for example.

As for the *productivity* component, this can be assessed in various ways for each activity category by analysing the relation between time input and some form of specified output. The productivity component is also affected by the relation between stationary and mobile activities, i.e. by the *travel time* necessary as an input in given activities, since the omission of travel time, for instance in fetching water and firewood or going to the fields, underestimates the time input, and hence overestimates the productivity of time for a given activity requiring travel, such as cultivation. The travel time component should always be specified except when distances are so short as to be negligible, say the movements within a house or its backyard or in visiting an immediate neighbour. But movements on a larger scale should be accounted for separately. (This is also of importance in time and energy budgeting, when the human energy inputs of a given activity are related to output, e.g. the calories necessary to produce a given volume of food, food being energy output in this case. Then input energy is generally calculated as the duration of the activity multiplied by the energy expended per time unit. For an excellent example of modern energy analysis at the micro level, cf Bayliss-Smith 1977.)

Another very crucial dimension to incorporate in the above labour utilization formulation is the dimension of sex. It is imperative for any analysis at depth that data are sorted out by sex so that the relative participation and division of activities of women and men can be clearly distinguished.

In the subsequent sections, the mobilization of human time for various activities (including 'work', however one chooses to define the latter), is analysed along the following essential dimensions:

1) the mobilization of human time resources by *age*, i.e. in the life cycle and with respect to relevant age categories of the population;
2) the mobilization of time by *sex*, i.e. incorporating the structure of sex roles and how they impinge on time allocation;
3) the mobilization of human time according to *skills or capabilities* and with respect to occupations;
4) the mobilization of time supply with reference to *domestic units* and their composition, as well as in terms of the *local population* of a settlement unit (e.g. a village) as a whole.

Other important dimensions of human time mobilization would be the following:

5) the structure and variations of how time supply is mobilized in the *daily round*;
6) the structure and variations of time supply mobilization in the *annual cycle*;
7) the effects of *membership composition* of households or local units on the scope for specialization and the delegation of activities among population categories;
8) the relation between time allocation and overall population size or *changes in population size* (e.g. effects of the factors of birth, death and migration/commuting on population time volume; or selective changes with respect to age, sex, skill etc.)

In the preceding chapters, some essential aspects of the structure of time allocation in the daily round were captured by means of prism analysis, and some additional features are shown in the graphs of household daily rounds (Figures 5:10, 9:1 and 9:2). However, since many of the data available are not specific enough to describe daily round structures, most of the analysis below of intensification in the allocation of human time resources is conducted at a somewhat more aggregate level. (The analysis of daily rounds will hopefully be carried out in greater detail in a forthcoming study on innovations as they affect daily rounds and activities in the annual cycle, two related time structures of activity systems.)

Many of these dimensions can be depicted together graphically which, apart from giving a useful synoptic picture, gives an idea of how the various dimensions are interrelated and circumscribe the intensity of time use with respect to different activities, such as 'work'. In much of the subsequent analysis we shall draw upon such synoptic accounts of how human time is mobilized in different local societies. The period of time chosen may be either a day, a week, a year or some other relevant period. Moreover, data furnished for a year can be reduced to an 'average' day to facilitate comparison of levels of intensity for various activities. In condensing information for a year into an average day, the information on seasonality is obviously lost, but we may get away with that in the case of a discussion of level of intensification associated with different ecotechnologies, although such an aggregate and averaged out description would be of far less use in the analysis of most other problems, for instance, the effects of innovations on the structure of society-cum-habitat. At any rate, the basic demographic dimensions of age and sex are incorporated into the graphs used (cf Chapter 8, esp. Figures 8:1-12). Age-sex categories are fundamental in delimiting the range of participation in activities, e.g. 'work', as in Figure 8:4.

Thus we may illustrate the discussion of Connell and Lipton (1977:5) in which two kinds of 'job problem' in contemporary Less Developed Countries are outlined:

> ... the labour utilization component identifies two distinct types of job problem: the 'low participation' work problem of integrated, modern 'advanced' villages with high-yielding land, great inequality, and high man/land ratios; and the 'low duration' work problem of remote, largely subsistence, overwhelmingly agriculturally 'backward' villages with low-yielding land but less inequality and lower man/land ratios. (Connell and Lipton 1977:5)

The duration of work is thus horizontally delimited (as is sleep in Figure 8:10), while participation in work is vertically delimited (as in Figure 8:4).* In the subsequent graphs by means of which we compare activity intensities in different societies, these and other dimensions of time resource mobilization are incorporated.

TIME RESOURCES AND ALTERNATIVE CONCEPTS OF INTENSIFICATION

Intensification is a multi-dimensional concept and as such, it has been employed in a more or less rigorous fashion. Some scholars, like Boserup and Wilkinson (1973) go by rather clear definitions, while Harris (1975, 1977) uses a very broad but correspondingly vague notion of intensification which amounts to virtually any form of resource mobilization. This, unfortunately, reduces the analytical value of his propositions considerably.

Any analysis of the intensity of resource utilization (and changes thereof) must, first of all, specify the resource(s) to which the intensity concept is applied. We have thus come across three forms of intensification so far:

1) Intensification in the occupation (use) of *land*, i.e. *space-time*. This refers to the increased volume of space-time occupied in a region which is bounded in space and time and can be said to have a space-time budget (cf Figures 5:1, 5:2 and 6:3). Intensification is when an increasing volume of the limited regional space-time budget is packed with parcels. A chief form of such intensification is that proposed by Boserup (1965) as increased frequency of cropping (cf Figure 5:4).

2) Intensification in the occupation and use of *human time,* i.e. the increased packing of certain activities in a population time-budget, delimited on the basis of an observation period and the size of the population. One typical form of time use intensifica-

---

*   According to Connell and Lipton, the 'low participation' job problem is typical for densely settled regions with more commercialized agriculture and market mechanisms of land allocation. This is increasingly the case in India and Pakistan today while, by and large, most parts of Africa are less commercialized and certainly have lower average population densities (man/land ratios). Connell and Lipton thus found a broad correlation between participation in agricultural work and access to land, a thesis which ties in partly with the intensification theoretic aspect that access to land is a function of how many people there are to share a given area or region. There are, of course, many additional determinants of access to land, just as there are a whole spectrum of 'job situations'. However, a good start is made when we are able to show the factors that interact to make up these different job or time mobilization profiles.

tion is that pertaining to the broad class of activity described as 'work'. Intensification with respect to work thus implies an increased volume of work activities packed into the population time-budget. A common economic term for the latter is increased *work-load*.

But there is also the concept of rural (agricultural) intensification that is conventionally used in economic theory, viz.

3) The increase in any input such as human time, water, fertilizer or capital equipment in relation to a given land area (cf p. 190-1). A special case of this definition is when more labour(-time) is added to constant land, and this is a source of confusion, since 'labour' intensification per land unit is something different from 'labour' intensification per population time unit. In this chapter when we discuss 'labour' or 'work activity' intensification we do not refer to increased labour units per hectare or the like (unless this is explicitly stated) but to increased time inputs for given activities in relation to total population time supply.*

Although all three forms of intensification may take place parallel with each other, this does not imply that they are one and the same thing, nor that they are always parallel as a society evolves historically. There are numerous instances when the input of human time per parcel of land is increased without there being an increase in the frequency of cropping, for instance. It is also important to be aware of the level on which the analysis is conducted. If a given region is more densely settled in terms of the average size population living there, more human time is generally invested in the region so that time inputs per land input is rising. In this sense, intensification of the third form may be said to have taken place (more time input/ area). But this may or may not have led to intensification in terms of more time devoted to agricultural activities as a portion of total population time.

---

* Definition 1 above is in principle inclusive enough to cover both increased temporal intensity and increased spatial density or spatial expansion (cf Figure 5:2), since both imply that an increased portion of the regional space-time budget is packed with parcels occupying space-time. The elimination of less intensively used elbow room (Figure 5:3), as when plants in a field are more regularly spaced, may be another kind of intensification subsumed under definition 1. However, intensification in the use of space along the temporal dimension is the main form of intensification in land use that was covered in Chapters 3 to 7. Definition 3, by contrast, which is the conventional economic definition, is not used analytically in the present study, since it is ambiguous in terms of which resource is used intensively in relation to which other resource.

In view of the kind of data available in local case studies, distinctions between the three forms of intensification mentioned are often impracticable to maintain; the paucity of data usually prohibits accurate calculations of work-loads associated with different systems, for instance. Insistence on high standards of assessments in this field would thus imply that the number of case studies for comparison would have to be cut down. (The latter solution was chosen in this chapter, however, since it deals explicitly with time resources.)

In cases where broad comparative analysis is aimed at, a more pragmatic definition of intensity must be adopted, as in the ambitious approach by Brookfield (with Hart 1971) who analysed forty-four Melanesian systems of livelihood, and who did not rest content with the frequency of cropping as an adequate factor for classifying the level of land use intensity:

> With Boserup ... we must regard 'intensity' as essentially the degree to which *technology is applied to land* so as to *economize in its use*, while gaining roughly *equal or greater output* per hectare. This is different from the strict sense of the term /cf p. 190/ ... but it is more operational in terms of the sort of empirical data we have at present in Melanesia. (Brookfield with Hart 1971:92) /Emphasis added./

While this definition places most stress on land use intensification, it implicitly incorporates the effects of increased work-load by the thirteen variables on cultivation *practices* (i.e. activities demanding time) used in Brookfield's composite index of intensity. The more such technological practices were applied (related to thoroughness of tillage, degree of drainage, extent of mounding, etc.), the greater the average work-load could be expected to be. Brookfield's intensity ranking thus incorporates all three definitions mentioned above.

It is still chiefly a measure of intensity rather than productivity or production volume, however, since it does not look into the general relations between input and output, but rather at the inputs necessary to reach a more or less constant level of output per land unit or per household.

The analysis of these forty-four societies reveals that those ranked as low intensive are generally of the swidden type and have low population densities, while those with high intensity rankings by and large are those with high population densities. Given that total intensity scores range from 0 to 26, some of the societies mentioned elsewhere in the present book are ranked as follows: Tsembaga 2, Chimbu 17 to 18 and Kapauku 21. Broadly speaking, this supports Boserup's model albeit with a number of qualifications.

Time use intensification

The performance of given activities has something of the quality of a rubberband about it in that an activity (resulting in a specified quantity of output) can often be carried out either at a high *tempo* or at a very slow *pace,* regardless of whether we are dealing with movement (running versus walking), preparing a meal, manufacturing a piece of equipment or telling a story. Just as the tempo or pace of performing a given activity varies between individuals it may also vary for the same individual between different occasions. There may also be cultural or regional differences of tempo; the pace of walking in a London street may on the average differ from that chosen by peasants walking to their fields in Indonesia. The question of pace and rate is certainly related to intensity, or the packing of activities in time. Methodologically in this chapter, however, we shall not regard doing activities faster as a dimension of intensification, as we have defined it elsewhere, but rather as a dimension of *productivity* of time, i.e. the relation between input and output. In the analysis of intensity of human time use pertaining to given activities, we assume that there are no differences in tempo between individuals and societies. In this sense, we are dealing with averages within the local communities examined. This kind of *ceteris paribus* assumption can certainly be relaxed to the extent that the basic data allow this.

As a matter of definition, intensification in the use of human time for a certain activity is taken to imply that an activity receives an increased allocation of time as a portion of total (local) population time supply. We are, in other words, dealing with the packing of activities in a population time-budget, a theme presented at some length in Chapter 8.

Assuming that there exists a kind of activity that can be meaningfully categorized as 'work', an increased allocation of time to this activity in the population time-budget would amount to *intensification of time use with respect to this activity, i.e. work.* It also follows that if some activity is intensified in this sense, some other activities must be 'deintensified', so to speak. They must, in other words, be correspondingly displaced.* Human time is always filled with some kind of activity, sleep or rest, if nothing else. (In this sense, it is hard to equate 'activity' with the now popular concept of 'practice', since it sounds somewhat awkward to say that someone is practising sleeping, or that sleep is a 'practice'.)

---

* This is what was referred to as an 'opportunity cost' in Chapter 8. It may also be interesting to note that many displacement effects are unintended consequences of human action/activity.

Hypotheses on the increase in work-load, i.e. labour-time intensification

On the assumption that there is a general category of activity that can be fruitfully designated as 'work' (an idea treated with a great measure of scepticism in this chapter), most theoreticians on cultural evolution and intensification hold the view that the work-load increases as we move from hunting-gathering forms of technology to shifting cultivation and on towards more land-use intensive forms of agriculture. This is the main thesis proposed by Sahlins, Harris, Boserup, Wilkinson and others.*

By and large, these scholars also support the view that the *productivity of labour-time declines* with land use intensification, so that people have to increase their work-time inputs merely in order to *maintain* their standards of living and output, not primarily to increase their output per capita. Wilkinson is explicit on this point:

> /Economic/... development is primarily the result of attempts to increase the output from the environment rather than produce a given output more efficiently. Under the impact of ecological problems the *productive workload* tends to grow throughout. More *intensive* peasant agricultural techniques often have a lower *productivity per unit of labour* than more primitive extensive methods. (Wilkinson 1973:4-5) /Italics added/

> ... The development of clothing will serve as an illustration. The easiest /least time demanding?/ way of clothing oneself is to make use of the skins of animals that are eaten. This is the most primitive method. When the supply of leather becomes inadequate to meet the population's growing needs, people are forced to develop textiles from natural fibres such as bark, flax, wool and cotton. The necessity of spinning and weaving these fibres *greatly increases the work required* to produce clothing... (Wilkinson 1973: 4) /Italics added/

> As basic resources become scarce, the growing needs which cannot be satisfied within the traditional framework provide the single most important spur to development... whichever shortage is felt first, it presents the same threat to society's livelihood. Alternative sources of subsistence have to be developed: either methods must be changed to *exploit traditional materials more intensively* or *new resources* must be found to substitute for the old. (Wilkinson 1973:54) /Italics added/

* In this volume, we cannot mobilize enough empirical material to test this hypothesis; a contribution is rather made to make it testable to begin with.

Similarly, Boserup, on whose work Wilkinson bases many of his arguments, points out that land use intensive systems operate in such a way that more labour input per capita is necessary to *maintain* a given level of output, i.e. enough to satisfy every household's basic food requirements, compared to systems with long fallow and low land use intensities.

> When the growth of population in a given area of pre-industrial subsistence farming results in lower average output per man-hour in agriculture, the reaction normally to be expected would be an increase of the average number of hours worked per year /and per person/ so as to offset the decline in returns per man-hour... (Boserup 1965:43-55)

More or less implicitly, Boserup chooses the output necessary to support an average individual or household as her yardstick. In investigating the amount of labour required to achieve this level of output, she finds that as societies get more densely settled and as fallowing is reduced (the cropping frequency raised), it takes *more work-time per average person to reach the standard level of food subsistence.* On the other hand, it takes less land to do so.

This general finding, which forms the core of her agricultural intensification hypothesis, is subject to the following little noticed reservation (a reservation undermining some of the arguments by potential critics):

> Quantitative indications of output per man-hour /i.e. productivity of labour/ in primitive agriculture are found in a few anthropological studies, farm management surveys and other sources. But the differences in output per man-hour emerging from these studies are *only very partly* to be explained by differences in fallow systems and methods; to a large extent they reflect differences in soil, climate, make of tool and quality of the human and animal labour used in cultivation. (Boserup 1965:28) /Itals. mine/

Given this reservation, the implication would be, of course, that if people want to alter the nature or culture given preconditions of soil, moisture and topography, e.g. by irrigation or terracing or other techniques, this will involve time demanding activities, both in terms of investment and maintenance. This is something consistently emphasized by Boserup. Hence, the work-load per capita tends to increase. Intensification with respect to labour-time is thus one way of compensating for reductions in the average productivity of labour in preindustrial food production systems.

It is also interesting to note that Boserup expressed doubts at an early stage about the idea that people in primitive societies had to work very hard to produce their subsistence rations of food:

The reason for this failure /of economists/ to see the possibilities of lengthening the working day in primitive societies is probably to be sought in some vague assumptions to the effect that members of small tribes of very primitive agriculturalists must be working very hard in order to produce and collect enough food for subsistence, or that they are so weakened by insufficient nutrition and malaria that they are unable to work more than they actually do ... However, most studies by anthropologists and others who have been living for long periods among primitive peoples paint a picture of their daily life which makes it difficult to believe that they *fully utilize their capacity for work*. The impression left by such studies is that members of sparse tribes of primitive agriculturalists usually work much shorter and less regular hours than do members of densely settled peasant communities. (Boserup 1965:45) /Emphasis added/

Harris is of a similar opinion:

The development of farming resulted in *an increased work load per capita*. There is a good reason for this. Agriculture is a system of food production that can absorb much more labor per unit of land than hunting and collecting. Hunters-collectors are essentially dependent on the natural rate of animal and plant reproduction; they can do very little to raise output per unit of land (although they can easily decrease it). With agriculture, on the other hand, people control the rate of plant reproduction. This means that production can be intensified without immediate adverse consequences, especially if techniques are available for combating soil exhaustion. (Harris 1977:11) /My italics/

One implication is that as long as population density is low and the rate at which natural animal and vegetative resources are tapped is correspondingly low, hunters and gatherers can get both a high quality diet and enough to eat. They can enjoy plenty of leisure and are the 'original affluent society' as Sahlins (1968, 1972) put it.

Wilkinson (1973), like Harris, generalizes the trend of increasing work-load to cover also industrial societies:

As the workload increases, societies bring additional sources of power to their aid, first animals, then wind and water power and finally mechanical sources of power. During industrialization the introduction of an efficient division of labour and the development of laboursaving machinery becomes a necessity as the workload reaches crisis proportions. (Wilkinson 1973:5)

We will leave the examination of industrial societies to Volume 2 of this book, but let us begin here by further qualifying the hypothesis.

Mapping the increase in work-load and activity intensification

In general outline, the increase in 'work-load' can be mapped in the manner presented in Figures 9:4a-c. These figures assume a population which in terms of age distribution exhibits neither extreme growth nor a stationary growth rate. To illustrate the main drift of the argument, we have a core of the working force consisting of men and women of the ages 20 to 50 years. They are assumed to allocate a certain volume of time to a given activity A, in this case 'work', as in Figure 9:4a.

The intensification of activity A would, assuming that the productivity of time is constant, entail expanding the time volume allocated to activity A in this population time-budget.* Additional time for A can be mobilized along two lines, either by extending the hours of those already engaged, as shown by segment C in Figure 9:4b or else by supplying more time for 'work', or some other activity, by spreading the 'work-load' or activity more evenly among age categories, as shown for the female part of the population in the same figure (expansion in terms of segment B). Of course, more time for some activity can be supplied by a combination of spread among age-sex categories and extending the hours.

The general hypothesis of increased work-load with cultural evolution and land use intensification thus entails a mobilization of human time in both ways, both by a spread among age-sex categories and by an extension of the hours, so that the conditions shown in Figure 9:4a, that supposedly pertain to hunting-gathering societies, would eventually end up like something in Figure 9:4c via measures of the kind shown in Figure 9:4b. Human time allocated to 'work' is thus intensified *in relation to the total time-budget* of the population. It would thus follow that, with these shifts, certain other activities must correspondingly have been reduced in volume, i.e. displaced or allocated away. In terms of socio-cultural output, this need not necessarily have been reduced as time inputs were, because that would also depend on the *productivity* of time. If productivity is increased to the same extent as a given activity is reduced in volume, output would have remained constant, while productivity would have increased. (For convenience of exposition, we have assumed so far in our discussion of intensification that productivity of human time for given activities remains constant, an assumption which is unrealistic and which will be relaxed later on.)

---

*   There is no reason why intensification should apply only to one single kind of activity rather than several kinds. Of course, other activities would be displaced on a corresponding scale.

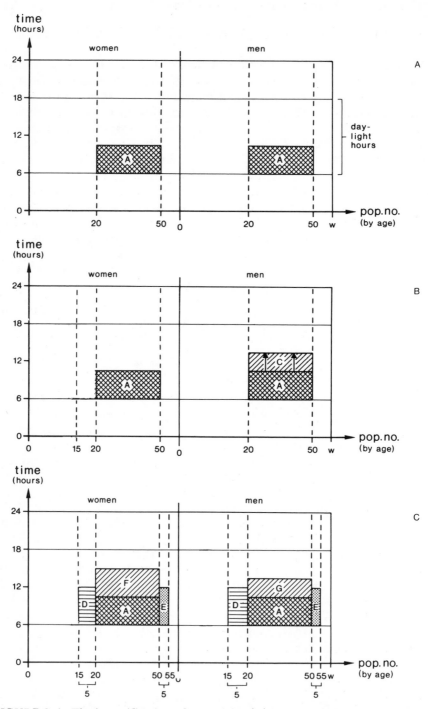

FIGURE 9:4 The intensification of an activity (A) in a population time-budget. Activity A is extended by segments affecting different age-sex categories (cf text).

The case of the Lele and Bushong

Methodologically, the set of dimensions of (labour-)time mobilization shown in Figure 9:4 have bearing on the much discussed case of the Lele and Bushong in Central Africa as presented to anthropology by Douglas (1962, 1963). For the moment we may ignore the substantive issue that both these peoples also practised cultivation rather than pure foraging. A main point Douglas wished to make was that it was not the environment (poor soils and the like) that determined level of production but that this was attributable to socio-cultural factors, particularly the system of rewards for labour which in the case of the Lele was anything but conducive to a broad scale mobilization of labour-time in food and goods production. Sahlins took up this case mentioning that:

> One of the main conclusions of Mary Douglas's brilliant comparison of Lele and Bushong economies is that in some societies people work for a much greater part of their *lifetime* than in others. 'Everything the Lele have or do,' Douglas wrote, 'the Bushong have more and can do better. They produce more, live better as well as *populating their region more densely* than the Lele' (1962, p. 211). They produce more largely because they work more, as demonstrated along one dimension by the remarkable diagram Douglas presents of the *male working life-span* in the two societies. (Sahlins 1972:52) /Emphasis added/*

Douglas' graph showed that the Lele men started active work around the age of 30 years and retired at 55, while the Bushong men not only started work at 20 but also retired much later, around the age of 65 (Figure 9:5). Sahlins concludes that,

> Beginning before the age of 20 and finishing after 60, a Bushong man is productively occupied almost *twice as long as a Lele*, the latter retiring comparatively early from a career that began well after physical maturity (Sahlins 1972:52).

From Figure 9:5 then, one might easily infer that the work-time volume of all Lele men would be only half that of the Bushon male population, and further that the withdrawal of the young men's time would be as serious a loss as that of the old men. This, however, does not follow, because if one takes the men between 55 and 65, i.e. a 10 year cohort, they constitute only about 9 per cent of the male

---

*   Curiously, Douglas noted that the people who worked the hardest were also the people who populated their region most densely, quite in accord with intensification theory as expressed by Boserup, Wilkinson and Harris, while Sahlins interpretation would go via the political system and its concomitant population density.

population between 20 and 65, while the men between 20 and 30 (another 10 year cohort) make up as much as 33 per cent of the corresponding population. The men between 30 and 55 would add up to, not 50 per cent as it would seem from Figure 9:5, but 58 per cent of the population between 20 and 65 years of age (assuming the mentioned population age structure). So the work span in the life cycle is not the sole determinant of the potential 'labour' force. This may not invalidate other arguments put forth by Douglas or Sahlins, but it illustrates the importance of assessing time allocation in relation to demographic structure (and the need to furnish data about the latter).

Another point emerging from this analysis is that more is to be gained by mobilizing the younger age categories than the elder ones, following from these age-pyramid effects. Alternatively, less is lost to society by letting the elders withdraw from active production, while it is costly to let the youth get away. This is even more obvious in the contemporary Less Developed Countries, where population growth is rapid and where the age-pyramid would have a very broad base (cf Figure 8:3), i.e. where as much as half of the population would be below 15 years of age. (Conversely, in a country like contemporary Sweden, with a stagnant population and a larger proportion of the population being elderly, more of time for work is lost by setting the retirement age early. Of course, many conditions are so different in the latter case that comparisons of consequences of age structure for society must be made with utmost caution.)

However, as for the time allocation studies made in anthropology so far, yet another very important methodological issue is clarified in Figure 9:4, viz. that unless the *range of participation* in a given activity (for instance 'work') is specified with respect to sex and age, it is not very enlightening to know that 'the women do only two hours of agriculture a day'. Who are these women? Are they between 15 and 60 or 25 and 50? The overall work-load or intensity of a particular activity cannot be specified in terms such as 'the adult women' or 'the average man'. If the definition of the 'average' woman (in Figure 9:4b) is extended from the age bracket 20 to 50 years to that of 15 to 50 years (+ 5 years), the total time supplied to the activity in question may increase by 22 per cent or so. This difference is often greater than the hypothesized increase in workload with the transformation of a society from one of hunting-and-gathering to one with shifting cultivation, according to the evolutionary theory proposed by Harris (1977), Sahlins (1972), Wilkinson (1973) and Boserup (1965). In other words, with the kind of methodological precision that these issues have been handled so far, evidence of the kind presented by Sahlins (1972) is in many respects discouragingly inconclusive, in spite of the fact that Sahlins' compara-

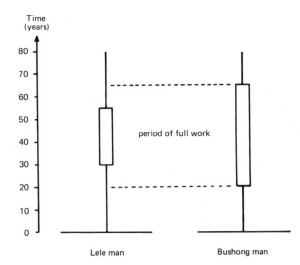

FIGURE 9:5   The span of active work in the life-cycle for a Lele and a Bushong man, as envisaged by Douglas (1962). The Lele man enters into full agricultural responsibility at 30 and retires at 55, while the Bushong man starts working at 20 years of age and retires at around 65. (Douglas graph did not show the situation for the women where one may suspect the difference was less.) *

tive analysis is one of the best to date in anthropology. The same is the case with the similarly ambitious comparative study by Minge-Klevana (1980). The data they have had to rely on are simply not up to the standard required by the hypotheses that are being tested, except perhaps in a handful of cases. In the present study of time resources and the capacity for interaction, we have tried to confine our analysis to the limited set of studies for which time allocation data are of reasonable quality in view of the problems under investigation.

*   This graph used by Douglas to describe the Lele and Bushong spans of working life for males is a simplified version of Linton's (1942) more complex graphical description of life cycles and social structure among the Cheyenne and the Tanala. Along the vertical dimension, Linton placed age, while the horizontal dimension consisted of an ordinal ranking of prestige associated with different status positions individuals could attain in these respective societies. Each status position was connected by possible career paths for men and women, thereby illustrating both possible life strategies in each society and the wider social structure. Although time-space graphs have been used in this book to depict ecological relations mainly, it is possible to define other kinds of positions than stations located in geographic space. One may, for instance, maintain the time dimension inherent in the life cycle but use 'category' spaces related to social institutions along the horizontal axes. These institutional positions can later be correlated to spatial positions, just as shifts in marital status may be related to changes in residence, to take but one example.

HUNTER-GATHERERS

The mobilization and allocation of time among hunters-gatherers-fishermen (food foragers) has been subject to far less ethnographic inquiry than that among agriculturalists. One reason for this is that few *pure* food foragers remain in this century. It may not be entirely too late to collect additional ethnographic information on what Lee (1979)* calls the *foraging mode of production*, but foragers are rapidly being influenced by new ecotechnologies and social orders which make them less genuine as an ideal type in the evolutionary schema.

Not only is the collecting of information on foraging modes of production and ways of life hampered by the fact that most societies employing this ecotechnology are extinct, but when field-workers are living among hunter-gatherers, participant observation is impeded by the indivisibility constraint of the observer.† Foragers scatter in different directions at various times of the day which makes quantitative assessments of the activities of an entire local population nearly impossible, even if the field-worker had the feeling for time-and-motion study that could furnish data for time-geographic interpretation in depth. Lee, for example, living among the Bushmen (San) would have liked to carry out time-and-motion studies of activities, but had to be less ambitious in his study of 'work':

> The beauty of the study of *work* is that work can be precisely *quantified* and can be tied into a whole gamut of social and economic variables. Underlying the network of social relations anthropologists are so fond of studying is a network of *energy relations* to which we pay little or no attention. Yet the *basic units of social behaviour and interaction* have never been satisfactorily defined and isolated, although the basic units of energy relations are relatively easy to define and measure. The advantage with the *study of work* for anthropology is that it *anchors the ephemera of social life* on the foundations of the natural sciences. (Lee 1979: 250-51) /Emphases added/

Lee also takes up the relationship between time, energy and motion:

---

*   Unfortunately, Lee's major study of the !Kung Busmen (1979) was not available when Chapters 1 to 7 were written and printed, but in this chapter we are using Lee's recently published data on time allocation to various activities. In his recent publication Lee also refers to the Bushmen more properly as the San, so in this chapter we use both names for this ethnic group. Lee's concept of 'the foraging mode of production' is also from his 1979 publication.

†   Anthropology and other social sciences could do with field-workers having the qualities of individual A in Figure 8:22, p. 326.

> Units of *time* and *energy* are the ones most commonly used in the measurement of work... Precise measurement of energy expenditure expressed in calories /or Watts for that matter/ requires the use of a physiological monitoring apparatus, and the minute-by-minute sampling frame of a *time and motion* study. In neither area did my own methodology approach this standard ... (Lee 1979:253) /Italics added/

Of course, it is not only 'work' which is quantifiable but *all activities* can be measured in terms of time, as well as energy. Time relations between activities and time resources for activities are just as important and structurally significant in societies which operate at a lower tempo or pace as in urban-industrial societies, where people are more perceptive of human time as a resource for activity. Moreover, many energy studies on human work have to go via time-budget analysis (for a recent survey, cf Piementel and Piementel 1979). Since the energy factor has been made subject to theoretical analysis to a much greater extent than human time has been (cf the study by White 1959, for instance), energy based analysis has been more established in anthropology than time allocation theory, but even when activity analysis is placed in a temporal framework, this scheme will not really be effective unless there is an explicit spatial framework as well. Otherwise it is unfeasible to deal with motion, transport, settlement and habitat. Hence the conceptualization of individuals as paths in time-space offers an instrument for integrating human time, energy and space into one activity systems theory.

!Kung Bushman/San time allocation

Although the data on time and work for this society is better documented than for any other foraging society, these data are not detailed enough to permit the construction of a graph similar to that for subsequent societies analysed, cheifly because of lacking precision with respect to age-sex categories. On a per week basis, the ('average') man spends 21.6 hours on subsistence work, 7.5 hours on tool making and 15.4 hours on housework, making a weekly total of 44.5 hours per week or *6 hours and 20 minutes per day*. Women spend only 12.6 hours on subsistence work and 5.1 hours on tool making and fixing, which for both categories is less than the men, but they spend more time on housework instead, i.e. 22.4 hours, giving them a total work week of 40.1 hours or *5 hours and 45 minutes per day* on the average. Men thus do a substantial portion of housework as well, since they collect firewood and do most butchering and cooking of meat (2.2 hours of housework for men and 3.2 hours for women).

As for manufacturing and maintenance of equipment, these activities amount to some 65 minutes per day for the man and about 45 minutes for the woman. All these data were collected for the Dobe camp during midwinter dry season, i.e. the season when people are clustered around the permanent water holes (cf Chapter 3). The conclusion Lee draws from his data is that the !Kung San 'appeared to enjoy more leisure time than the members of many agricultural and industrial societies' (Lee 1979:259).

The problem with the time allocation data presented for the San is that although we get a rather accurate picture of the distribution of work between the sexes, we cannot pinpoint the work put in by specific age categories. But Lee does furnish some clarifications. Although there is a general tendency for work effort to decline with age (i.e. after 40 years of age), the hardest workers among the men are found in the 40 to 59 group, while for the women they are in the 20 to 39 age bracket. The elderly above 60 are very moderately active in subsistence (food quest) tasks. As we pointed out in the previous section on Lele and Bushong, however, it is the contribution of the youth which is more decisive in terms of general level of intensification. Lee notes that 'only two people under 15 years of age did any significant subsistence work'. The rest of the children were 'footloose and fancy-free'. There were no young people between the ages of 15 and 19 in the (rather small) sample, however, so it is difficult to determine the age of entry into active work. Adolescent girls do not start working properly until they marry, so this factor seems to be more decisive than actual age.

Be this as it may, the basic conclusion is that the San can subsist very well with only a relatively modest input of time into work such as foraging, manufacturing/repairs and housework (Lee 1979, Ch. 9). Given that the hunting-gathering Bushmen (San) spend only 12 to 14 per cent of their total time supply on 'work', how representative are they of other foragers? Judging by other more or less quantified descriptions amassed in the literature by Lee and Devore (eds. 1968) and Sahlins (1972), it seems plausible that, on the whole, contemporary foragers (i.e. the few remaining into the 20th century) living in marginal tropical or sub-tropical habitats have had a rather low work load. On the other hand, one should be cautious in generalising this state of affairs to more advanced societies based on hunting-gathering-fishing, i.e. to societies of the kind represented by the Northwest Coast Indians, many of which were not located in ecologically marginal environments and which had complex social structures. The only hypothesis which can be refuted with a good safety margin is the idea that hunter-gatherers had to work extremely hard to survive and that the advent of agriculture brought relief to this condition.

Some concluding remarks on food foragers

While food foragers are of principal and theoretical significance, they form a nearly negligible minority in today's world compared to agriculturalists. We therefore choose to leave most unsaid about them and place our emphasis on cultivators. This is even more justified in view of the weak data base available for foragers. It would take a book in itself to reinterpret the essentially qualitative data on hunters and gatherers so as to get a better picture both of time allocation mechanisms and the approximate amounts of time spent on different activities by various age-sex categories of the population.

The absence of pure foraging societies in our contemporary age should not obscure the fact that hunting, gathering and fishing *activities* are by no means extinct. Far from it, but these activities are mixed with what are from the evolutionary viewpoint more recent ecotechnologies such as agriculture and pastoralism.† It would seem, however, that even evolutionary anthropology would benefit from the analysis of mixed types rather than ideal types. The Machiguenga, for example, know and practise cultivation but they still have a large foraging sector (cf the next section). It also seems more than likely that through a general theory of the living and activity possibilities associated with different ecotechnologies and activity systems, we may be able to infer a good deal about pure types although our observations may have been based on mixed types. This is one rationale behind the construction of time-geographic theory as presented in this book. Through a process-cum-structural approach in which due attention is given to living and activity possibility boundaries and the aggregation of (positive and negative) constraints, the possible types of society-habitat might be better explored by means of data that otherwise leave much to be desired. *

† When it comes to pastoralism, there is a similar shortage of time use data. This is not by far so serious as in the case of hunter-gatherers, since pastoralism is still a very widespread ecotechnology and more fieldwork is possible in the future.

* Such an approach also helps us to get away from ideal type classifications. There may be a conflict between storage and mobility in many hunting-gathering societies, for instance, but in many cases this conflict could be resolved in a variety of ways depending on environment, transport technology, division of activities between individuals, and so on. (For a recent critique of some current ideas on the role of storage and the temporal immediacy of consumption in foraging societies, cf Ingold 1980.)

SHIFTING CULTIVATORS

We will now examine some shifting cultivation societies for which time allocation has been satisfactorily documented to serve our present purposes. Needless to say, while there are now a good many monographs on societies with shifting cultivation, all too few of them contain time use information other than incidental references to certain sub-sets of activities. Two exceptions to this rule are the studies of the Machiguenga (Johnson 1975) and the Ushi (Kay 1964). Apart from practising agriculture, both these societies have a sizable gathering-hunting-fishing sector as well. It would have been desirable to compare a wider range of societies with shifting cultivation ecotechnology, but even with two cases, there is more variation in time allocation patterns than we may explain on the basis of existing intensification theoretic statements. The whole story is more complex than a simple aggregate increase in 'working hours' as a technology becomes more land use intensive.

The Machiguenga of Peru

An outline of Machiguenga time use is presented in Figure 9:6, the figure being a graphic presentation of Johnson's data (1975). The classification of activities and key to abbreviations used for activities in this chapter is given in Table 9:2, p. 367.

The Machiguenga studied by Johnson and his colleagues lived in the community of Shimaa, along a tributary of the Upper Urubamba River in the Department of Cuzco in Peru. These people,

> ... typically live in small groups, either as single families, or as clusters of closely related families, varying from less than ten to as many as 30 individuals in a settlement. They derive most of their food from slash-and-burn gardens, cultivated largely by the men, supplemented by smaller quantities of food from the tropical forests and rivers, in the form of fish, grubs, wild fruits, and occasional large game such as monkeys or peccary. Small family groups are almost completely self-sufficient, except for occasional needs for trade goods such as aluminum pots and machetes. Machiguenga hunt with bow and arrow, manufacture their own cotton fabrics, and in the past have preferred to do without trade goods rather than become closely bound with traders or missionaries. (Johnson 1975:302)

* Although pastoralists were presented subsequent to foragers in the previous chapters, the time use material that the present author has had time to retrieve on pastoralists is of such uneven quality and coverage, that there will be no section comparing pastoralism to other ecotechnologies in this chapter.

Johnson's data are excellent in terms of time allocation for the relevant age-sex categories, but a minor flaw in his tables is that the exact number of individuals in each category was not stated. However, he did specify the number of observations for each category, and since his sampling was random, the number of observations should be roughly proportional to the number of individuals (the assumption underlying Figure 9:7). Moreover, we know that there were 15 female and 20 male adults. Still, it is important that demographic structure is properly stated together with time use data.

A quick glance at Figure 9:7 reveals that the balance of activities between the sexes is fairly even, and that the main male activity is agriculture (AC) closely followed by hunting, gathering and fishing (HGF), i.e. 2½ and 2 hours respectively. The main adult female activity is domestic work (D), such as cooking (CK, 2 hours and 24 minutes) and child care (CC, about 1 hour), but women also spend a fair amount of time on manufacturing (MF, 2 hours). Although Johnson's data are good in terms of their age specification (even toddlers and small children between 2 and 5 are analysed separately), there is no category for the elderly. At any rate, the adult women (14 years and upward) each spend less than an hour per day on actual food production (AC, FG), while men spend only some 12 minutes per average day on domestic activities such as child care and cooking. The 'work-load' for food getting and food processing activities thus comes to about 4 hours per day for males and females above 14. Children above 6 do contribute to these activities, however, again with a marked division of activities between the sexes, boys engaging little in domestic activities but contributing more to agriculture and hunting-gathering. If one would like to define 'work' as including manufacturing and child care, the total figure for men would be 6 hours and for women 7½ hours. Excluding child care, it would be about 6 hours in either case. (One may, of course, go on elaborating like this and define 'work' to suit any objective.)

In terms of the aggregate intensity of time use with respect to food production and food preparation, the Machiguenga men spend nearly 4½ hours on agriculture (AC) and foraging (HGF) plus about 15 to 20 minutes on domestic activities (cooking, CK, and child care, CC). The women spend about 1½ hours on food production (AC, FG) and together with domestic activities the 'work-load' thus comes to a bit less than 5½ hours. This applies to an age range of 14 years and upward. In the Bushman case, food production and domestic activities come to about 5 hours for the women and 5 hours and 20 minutes for the men. The age category for the Bushman population discussed here was somewhat less broadly defined, but in view of the margins of error in data collection, there is no significant difference in work-load between Bushman foragers and Machiguenga foragers-cum-cultivators. Consequently, it is difficult to tell whether or not the discrepancy in work-load in these two cases can be ascribed to differences in ecotechnology.

Turning to the allocation of time to various activities in terms of age-sex differences, it is interesting to note that the balance of 'work' (AC, HGF and D:CK, CC) is more even between the sexes among the Machiguenga and the Bushman populations than between the Ushi men and women (described in the next case). Hence this varies a great deal regardless of ecotechnology. There is, in other words, little room for technological determinism on this point of sex differences.

The data by Johnson are very good in that rather narrow age categories are registered at the same time as the activity classification has wide coverage. He thus furnishes information on time spent on activities such as schooling (SC), visiting (V), hygiene (HY), manufacturing (MF) and idleness or rest (ID) as well as on miscellaneous activities (X). For married adults, even sub-categories of activities are given (Johnson 1965:308). In terms of age categories, 'non-work' activities are usually neglected in many of the work-biased studies available. This automatically makes age groups below 'working age' unaccounted for. Johnson, on the other hand, also describes how toddlers and small children spend their time.

The adult sex role pattern starts coming through from 6 years and upwards, and by the age of 14, it is more or less fully developed. The boys of 6 to 13 spend more time visiting and in school than do girls of corresponding age. Unfortunately, Johnson did not specify any category of elderly people, so we must assume that time use is much the same for them as the other adults. This is by no means physically impossible. Among the Ushi (described next), the elders work even more than the young people. This may possibly reflect performance at a slower pace, since the pace is rarely measured as such in these kind of studies, only duration. Machiguenga women help with agriculture by doing a good deal of harvesting, which is not strange considering that this is often done under time pressure or is a prior stage to food preparation.

The Machiguenga do practise agriculture as their slightly dominant form of food production, but nearly as much time is spent on hunting-gathering-fishing. However, the yield from agriculture in food energy received per hour of time input is greater than from foraging. In manufacturing, the women produce cloth while the men do most woodworking.

When interpreting Figure 9:6, it should be remembered that the time for each kind of activity has been averaged out within each age-sex category. No difference is made between children attending or not attending schools, nor is there any dimension of specialization.

TABLE 9:2

ACTIVITY CODE AND CLASSIFICATION

AC	Agricultural or cultivation activities (in general)
ACI	Agriculture, irrigation of fields
AN	Animal husbandry (tending livestock, guarding, feeding, etc.)
CC	Child care
CK	Cooking or other food preparation
D	Various domestic activities (unspecified)
ET	Eating, having meals
FW	Fetching firewood, preparing firewood or fuel of some kind
F	Fishing
G	Gathering, collecting (esp. wild foods)
H	Hunting
HB	Housebuilding, erection of shelters
HY	Hygienic activities, washing oneself
I	Illness; visiting physician or hospital
ID	Idleness, unspecified leisure, rest
L	Litigation, court visit, participation in legal procedures
M	Migration (away from own locality for longer period)
MF	Manufacturing, construction, handicrafts
P	Political activity, councils, committees, decision-making
R	Ritual, church, prayer, ceremonial
SC	School, formal educational activity
SL	Sleeping
TD	Trading, marketing, shopping
TS	Traversing space: travel, movement, transport (when specified separately from other activities)
W	Wage work, miscellaneous work giving cash payments
WT	Fetching water, pumping domestic water (not irrigation)
V	Visiting (e.g. friends, relatives, neighbours), socializing
X	Miscellaneous unspecified activities (residual category)

(Domestic activities, D, when unspecified include CC, CK, FW & WT.)

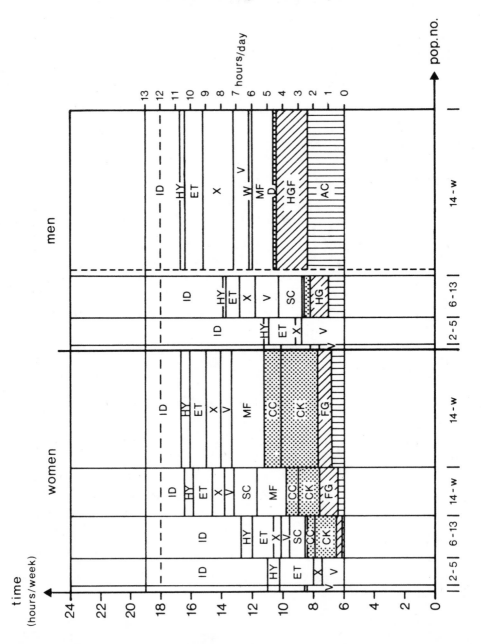

FIGURE 9:6

MACHIGUENGA, PERU

(The blank category among the men above 14 years of age covers the unmarried men who were too few for random sampling of activities. Time scale: 1 mm = 15 min. Source: Johnson 1975.)

FIGURE 9:7

USHI, ZAMBIA

(The survey did not cover females 1-15 years of age or males of age 0-20. Time scale: 1 mm = 15 min. Source: Kay 1964.)

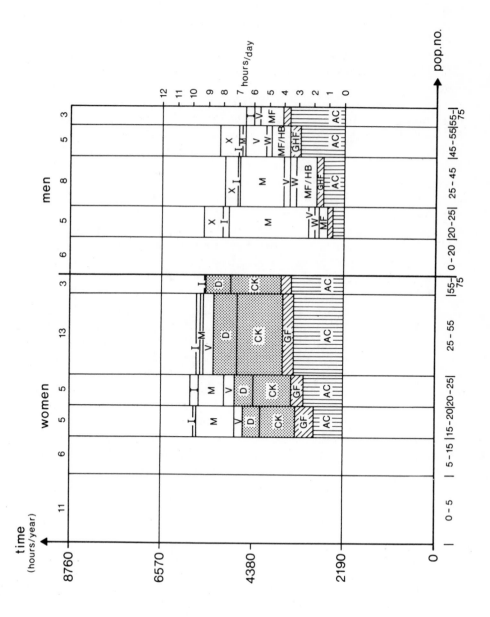

The Ushi of Zambia

An interesting example of a thorough study of time allocation, land use, ecology and socio-economic organization of a shifting cultivation society is that of an Ushi village (Chief Kalaba's village) by Kay (1964). The village population fluctuating around 70 individuals was observed for an entire year, and seasonal variations in time use were analysed in considerable detail. Daily records were kept by local investigators of what the 47 or so adults were doing on the days they were in the village. Time use was given for whole hours and although the hour by hour mesh in which data were collected is not very fine, it is nevertheless adequate when data are registered for a whole year and when this aggregate record is reduced to an average day, as in Figure 9:7.

The Ushi are shifting cultivators employing the well-known Central African *chitemene* technique. They have no cash cropping (at least they had not when the village was studied in the early 60's), and their staple crop is finger millet, although cassava, a tuber, is a recent important innovation (cf. p. 199-206 above on pseudo-intensification). There is no shortage of land or trees to cut in the general region and 'the extent of land held is largely controlled by the availability of labour and, in the absence of markets, by the subsistence requirements of the household' (Kay 1964:33). The Ushi thus conform neatly to the original assumptions made by Chayanov (cf Figure 6:13). As for domestic livestock, animals play a very minor role in the overall economy, so pastoral activities do not even show up in the graph on Ushi aggregate time use (Figure 9:7).

One curious feature of the Ushi village in question is that the distances to plots were rather long. Only 50 per cent of the gardens were less than 5 miles (8 km) distance from the village, while 80 per cent were less than 7 miles (10 km) distant. This means that a considerable amount of time must have been absorbed in travelling to the fields even though the frequency of travel may have been adjusted so as to save time (cf p. 229-32). This travel time seems to have been registered as a part of the cultivation activities proper.

Turning then to the aggregate pattern of time mobilization for different activities, we can note, first of all, that gathering, hunting and fishing (GHF) is a relatively small sector of activity compared to agriculture (AC) among the Ushi. We can also note that the aggregate pattern in terms of the amount of time devoted to 'work', defining work as foraging (GHF), agriculture (AC) and domestic work (D, CK), is not very instructive unless we start observing the distribution among age-sex categories. (The activities mentioned are shaded in Figure 9:7).

The young men and women did considerably less than their elders, and there was a marked difference between the sexes; the average working year of the young women was 944 hours, and that of the young men was only 427 hours. With few exceptions the young men were relatively indolent in respect of agricultural work, and the young women relatively industrious. But the old men, well established in village life, were certainly not lazy, and indeed one man did far more work than any woman. It should, however, be remembered that these figures measure time only, and not efficiency — for which no data are available. (Kay 1964:49)

In terms of sexual allocation of time, it is evident that the load of food producing and domestic activities puts heavy pressure on the women, and while it is the women who certainly do most cultivation among the Ushi, the men do most agricultural work among the Machiguenga. Furthermore, there is a marked increase with age in the volume of time allocated to foraging and agriculture among the Ushi. This is clearly brought out, since the age category 55 to 75 is separate in the Ushi data. It can also be seen that a good deal of the male time is spent on migrations (M) elsewhere, something which is also reflected in the skewed ratio between men and women (whereas the Machiguenga population is more balanced in its sex composition). The men, on the whole, make a very meagre contribution to the subsistence activities (GHF, AC and D, CK) among the Ushi, while these activities are volumewise much more evenly distributed among the sexes in Machiguenga society. It is only, typically enough, at certain peak seasons of the year when the division of these activities is more even between the Ushi men and women:

During the urgent, wet season of cultivating, sowing and planting — which comprised one-third of the total agricultural work — there was no division of duties between the sexes, and men and women of each household frequently worked together in the same gardens... The little weeding done was mostly carried out by the women, but the men did almost a quarter of it... During the dry /slack/ season, when the pace of work was much slower and the various operations were more distinct, a strong division of labour appeared. (Kay 1964:47)

If one includes cooking and other domestic activities, the workload carried by the Ushi women is relatively heavy, which in part reflects household and cooking technology, but also the uneven allocation of 'work' between the sexes. However, both among the Machiguenga and the Ushi, rather little time is spent on food production activities, although the volume of agricultural activities is greater in the Ushi case.

SHORT FALLOW CULTIVATORS

In this section we will take up agricultural societies with short fallow
systems at varying levels of land use intensity and regional population densities. Many characteristics of these types of systems have
been discussed in Chapter 6, so we will not bother to repeat them.
However, the common denominator for the populations discussed
here is that their agricultural systems are rainfed. (The additional
activities associated with irrigation are left for the next section.)
Most cases discussed here are subject to a rainfall regime with either
bimodal rainfall or else rather evenly distributed rainfall throughout
the year. This implies that irrigation is not necessary in order for the
cultivators to reduce *seasonal* fallow and thereby practise either
multi-cropping or the nearly continuous kind of tuber cultivation
found in New Guinea. The distribution of rainfall in the annual cycle
as a factor of land use intensification has been discussed *inter alia*
by Brookfield (with Hart 1971) and Turner, Hanham and Portararo
(1977).

The Raiapu Enga of New Guinea

The Enga of New Guinea are one of the most densely settled populations in the Central Highlands, and this region is, in turn, much more
densely populated than the lowland region where shifting cultivation
is generally the main form of agriculture. In the highlands, however,
a wide range of ecotechnologies of food production can be found at
different levels of land use intensity. One technique is often employed
parallel to another as among the Kapauku (Pospisil 1963) and the
Enga, of which we will discuss the Raiapu Enga section described by
Waddell (1972, 1973).

The Raiapu Enga practise both continuous intensive agriculture
with mounding and extensive shifting cultivation, each on suitable
land of its own (cf p. 222 above). In Figure 9:8 which summarizes
Raiapu Enga time allocation, these different forms of agricultural
time use are not shown separately. However, it is the women who do
the work of sweet potato cultivation in the continuously used *open
fields*, while the men do most of the work in the seasonally used
swidden gardens, referred to by Waddell as *mixed gardens*. The open
fields constitute intensive monoculture and are located in the terrace
sections of the mountain slopes. Sweet potatoes are the only crops
grown there, and they contribute two thirds of the Raiapu Enga food
intake. These potatoes are also the staple food supply for the pigs.
The mixed gardens contain subsidiary crops and are extensively cultivated swidden clearings which are located on the dissected gorge

slopes of the mountains. These intercropped gardens are cultivated on a seasonal basis only. It takes 6 to 8 weeks for the men to clear such plots which are part of a 10 to 15 year swidden cycle with a single planting in each cycle. They are worked fairly irregularly by the men. Once planting is completed, little maintenance is required.

The bulk of the agricultural work which is on a regular and more continuous basis of about 4 hours per day is thus left to the women, while the men spend on the average less than 2 hours a day on agricultural activities. However, in spite of the fact a substantial pig population gains its chief diet from human work in open fields, we find that male and female time put into food production among the Raiapu Enga is not much, if at all, greater than among shifting cultivators like the Ushi of Zambia mentioned in the preceding section.

The Enga food production system is quite an efficient one, not least because of its dispersed system of residence and the practice of what is a kind of infield-outfield system, the infields being the open fields which are fenced in so that the pigs cannot enter them and spoil the crop. The Enga residences are located along this fence and the pigs are allowed to forage outside in what is the fallow land of former mixed gardens. The mixed gardens in use in a given year are also fenced in and (as mentioned on page 222) fencing is a time consuming investment necessary to keep pigs and gardens unmixed. Once built, however, fences save much time in pig husbandry, since pigs come back to their houses in the evenings where they are fed sweet potatoes by the women. Pigs thus take care of themselves and are not very time demanding which is reflected in the population time-budget (Figure 9:8). Had the pig houses (and women's houses) been located differently in relation to the open fields, the parameters in the daily round would have been different and much more time would have been spent commuting to the gardens by the women. They would also have had farther distances to carry sweet potatoes to the pigs. In the system of settlement actually used, women have on the average a 7 minute walk to the open fields although the men spend some 20 to 30 minutes as they intermittently go to their mixed (swidden) gardens, where crops used mainly for ritual occasions are cultivated (cf Figure 6:9, 6:12 and 9:3 relating to distance, frequency of movement and time spent on travel). Consequently, by having dispersed residence but also a short distance between stations most frequently used (open fields and dwellings), much time and energy for travelling is saved. Time spent on cultivation and pig raising can thus be kept quite low considering the high output from and level of intensity of the Raiapu Enga food production system. (For a discussion of settlement pattern, land use and mobility, cf Waddell 1972:176-82.)

The allocation of activities between the sexes exhibits a pattern where most of the agricultural work is carried out by the women, as was the case also with the Ushi (and many other peoples in sub-Saharan Africa). The women in the ages 20 and upward put in some 4½ hours per day on the average, i.e. for every day of the year. A minor portion of this time is spent on cash cropping (AC/W). This leaves less time for women to do domestic chores, and according to Waddell, cooking by women takes less than half an hour per day! This is possible only because it is the men who collect most of the firewood (FW). Such wood is largely a by-product of their fencing activities at swidden (mixed) gardens. The fences around the permanent gardens need not be constructed so frequently, so time is thus saved for the men, just as the introduction of steel axes a few decades earlier had this effect (cf Salisbury 1962 for a similar case). Fencing is also minimized by the contiguity of permanent fields (the infields). Fencing activities may alternatively be regarded as part of the pig husbandry activities, since it is the pigs which necessitate this activity. Fencing is, however, subsumed under agriculture in Figure 9:8.

The Raiapu Enga case is the only one described in this book where the men do more domestic activities than the women (Figure 9:8). They do about as much cooking as the women but are more involved in preparatory tasks such as firewood collection. However, Waddell's data are perhaps underestimations of the domestic work sector:

> The data were gathered through a combination of observation and daily interviewing in which all activities were considered except for resting, eating and sleeping, and a few others carried out on a rather casual basis (often after nightfall in and around the house). ... Of the last the most important ... are the routine preparation of food, making string bags, caring for young children, informal visiting ... However, some attempt has been made to estimate the labour requirements of several of them ... (Waddell 1972:82)

Child care may be one such underestimated activity. Since the time use of children below 15 is not accounted for, we cannot assess the extent to which children are cared for by older siblings, for instance. In many societies, child care is rather incidental to other activities anyway, so we may suspect that the domestic chores of the women are underestimated. On the other hand, the high level of male time put into domestic chores may be related to the fact that most of their other activities are *located near the dwellings* anyway, in a manner resembling that of the Siane (cf Chapter 5 and Figure 9:3).

Both men and women spend about an hour each on rituals, ceremonials and related activities. We can also distinguish two separate categories of males that are not based on age or sex alone. These comprise the divorsed or unmarried men. Males of category *b* who are unattached to any other household have to spend more time doing 'female' agricultural work. These men have a work-load comparable to that of the adult women. The unmarried or divorsed men who are attached to some household enjoy many of the services rendered by the females belonging to it (category c in Figure 9:8), and their time use is very similar to that of the married adult men.

Sweet potato cultivation in open fields is, like pig husbandry, an activity that must be carried out with utmost continuity:

> ... even a single night away from the locality can have a disruptive effect on the agricultural routine, due to the clearly defined division of labour among the sexes and the substantial dependence of the pigs on cultivated foods. In fact, on account of the pigs, it is usual for only one adult member of a household to be absent at a time, and on no occasion were all the members known to be away ... (Waddell 1972:84)

A nearly aseasonal continuity of work at a fairly constant level of intensity is typical of this pig-tuber ecotechnology. The pigs, whose significance is great both in the sphere of livelihood and politics, impose both time demands and coupling constraints by relying mainly on cultivated foods. They thereby contribute to the pressure on land together with their human custodians. However, the pig-tuber complex is still a very successful ecotechnology in terms of keeping the overall work-load relatively low.

The village of Char Gopalpur in Bangladesh

The village of Char Gopalpur described by Cain (1977) has a popula-tion of 2043 individuals living in 303 households and cultivating a land area of 312 ha. It comes close to the average for Bangladesh in terms of population density, ecology and economy. The land of Char Gopalpur is not irrigated although it is fertilized by the seasonal in-undation of the Old Brahmaputra river. The land is cropped with rice at least once per year, but a good portion of the area is double-cropped. This is feasible in a humid tropical region such as this in spite of the seasonal differences between the hot and wet summer and the four cooler and drier winter months. The second crop is usually another variety of rice or jute or else winter vegetables. Given this climate and the emphasis on cereals, there are marked seasonal variations in time demand for agricultural activities. So the cultivation system found here is much less even and continuous than was the case with the Enga. Furthermore, Char Gopalpur is very different in terms of mean population density (660 inh. per km^2) and general land availability but also with respect to the age structure

FIGURE 9:8

RAIAPU ENGA, NEW GUINEA

(Male categories are as follows: a: unmarried; b: divorsed, widdowed, and men unmarried but not attached to another household; c: same as category b but attached; d: married. Time scale: 1 mm = 15 min. Source: Waddell 1972.)

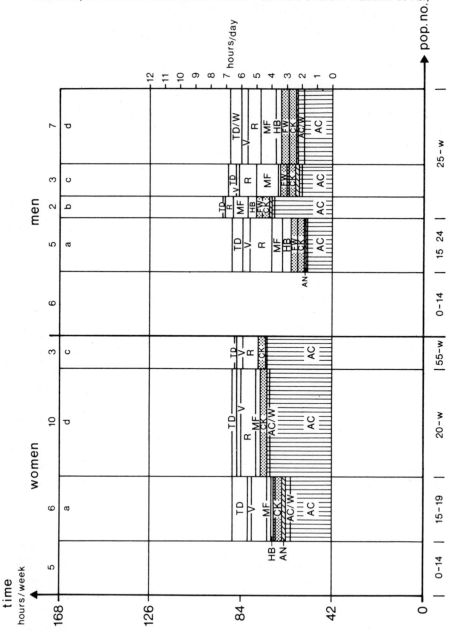

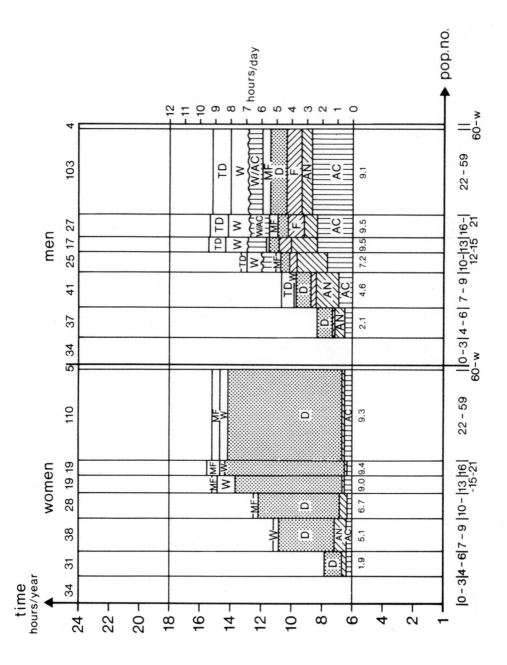

FIGURE 9:9

CHAR GOPALPUR, BANGLADESH

(Time scale: 1 mm = 15 min. Source: Cain 1977.)

of the population. Like elsewhere in present-day Bangladesh, the population is growing at a fast rate and almost 50 per cent of the villagers are below 15 years of age. This age structure is clearly mirrored in the population time-budget (Figure 9:9).

The time use data were gathered by Cain to illustrate the livelihood activities of children, and he therefore differentiated among the age categories in an exemplary manner, putting all superficial discussions about how much the 'average' man or woman works to shame. For the moment we shall confine our analysis to the intensities of time use for different activities (in line with previous cases), but we will return to the Char Gopalpur case in a discourse on time resources of children and the reproductive strategies of households. It should also be mentioned that this case and the two subsequent ones (Java and Guatemala) cover preindustrial societies, although each of them have a considerable sector of activity which is monetized, unlike preceding cases.

Looking at the aggregate time-budget for the village, the amount of time spent on domestic activities by the female population is striking. This is in part due to the sex-role specialisation implicit in the Muslim institution of *purdah*, i.e. the seclusion of women, the effect being that female activities should be located in or close to the homestead. To a great extent this precludes their participation in agricultural work in the fields. The domestic work load carried by women also bears a strong imprint of the state of household technology as well as the choice of staple food — rice — with its associated time and energy demanding activities of storage and preparation. Not only must the unhulled rice be parboiled, dried, husked and cleaned prior to consumption, but in the prevailing moist climate simple storage requires repeated winnowing and drying at suitable intermissions in the summer rains. Additional domestic tasks include the gathering of fuel and water. Although boys also perform the latter two activities, food preparation, washing dishes, sweeping, repairing floors and washing clothes are exclusively female tasks.* Women are, however, allowed to do some jobs in agriculture, chiefly the handling of less profitable crops such as potatoes and chillies cultivated in nearby gardens. Women can further get wage work in the processing of crops, husking and drying, also typically located within the village. In many parts of the world, sex roles such as these may be found to be relaxed under the influence of demographic chance as in households where parents did not succeed in getting any sons. Families in Char Gopalpur would, however, rather hire male labour than reallocate activities between the sexes.

* Cain (1977) describes the daily round of a widdower, however, who performs so much time consuming domestic work that he is unable to tend his agricultural land which is consequently rented by another household.

While women do most domestic activities, it is striking how nearly all agricultural activities and associated animal husbandry are done by the men. This is typical of the mixed farming systems prevalent in Bangladesh. The boys and the young men allocate relatively more time to animal husbandry than the adult men (16 and over) do. While girls and women may be engaged in animal care as well, it is evident that fishing is a male task, and, given a country pervaded by rivers, fishing is a sizeable sector of activity. Fish and rice is as characteristic a diet of this area as is that of pigs and tubers in New Guinea. Men also do a fair amount of wage-work per average day, mainly in agriculture. By and large, the overall picture is one of a much greater work-load both on the men, women and children than in previously discussed cases. However, with respect to intensification theory, it is interesting to note that this high level of intensity of time use is not so much a function of increased amounts of time for agriculture and animal husbandry (i.e. food production) as of the extraordinary volume of time put into domestic activities. The contrast to the Raiapu Enga with respect to domestic activities is enormous, even though one may suspect that Waddell (1972) might somehow have underestimated the volume of these activities. *

IRRIGATION AGRICULTURALISTS

The two populations examined in this section utilize land extremely intensively. Both societies cultivate two or more main crops per year which is made feasible only by irrigation. In Kali Loro, Java, nearly 40 per cent of the cultivated area is under irrigation; in Panajachel, Guatemala, it is 90 per cent in what is a hyper-intensive and highly specialized garden agriculture. Not surprisingly, we find in Panajachel that a substantial portion of the work in agriculture is associated with irrigation, although it is a very local kind of system found in a river delta immediately above a lake. We thus cannot say much about any hydraulically inspired bureaucracy à la Wittfogel in either case. In many respects, there are greater similarities in terms of general ecology, economy and social structure between Char Gopalpur and Kali Loro than between the latter and Panajachel. All three societies, however, exhibit common characteristics such as very high local population densities, a high land use intensity, low productivity of time, an uneven distribution of land, and a high degree of commercialization. In most aspects, Panajachel outranks the other two cases.

* It should be added that neither Raiapu Enga nor Char Gopalpur conform perfectly to the category of 'short fallow cultivation', since, like the Wakara mentioned on p. 224-5, they are mixed cases. Hence the Wakara, although they have a bit of irrigation is hardly an irrigation based society and their work-load can be less explained in terms of irrigation than fertility maintenance activity.

The village of Kali Loro in Java

Kali Loro village described by White (1976) is located in Central Java. It has a crude population density of about 730 inhabitants per square kilometer which is slightly above the average for the area. The ratio of irrigated land *(sawah)* to total agricultural land is 34 per cent while it is somewhat more, 40 per cent, for Central Java. Land distribution by ownership also displays the same tendencies found elsewhere, and 70 per cent of the households do not have enough *sawah* to meet their annual rice requirements. People are thus forced to seek income from activities where their productivity of time is lower than in agriculture. Two crops of rice are grown on most of the irrigated land in Kali Loro, and there is a marked seasonality, where slack periods occur between each planting and harvesting as well as between the cycles. Five months are relatively busy and seven are rather slack. High-yielding varieties of rice have recently been introduced, but the benefits have been greater for those with more than average size holdings of land.

White collected his time use data by day-after interviews evenly spread throughout the year making 60 visits per household. The data he presents have been translated into Figure 9:10, and they cover a sample of 20 households and 104 individuals (6,240 person-days). Since White was interested in the role of children, his data are excellently broken down into age categories, although there is no specific category for the aged. The population age-structure is active so nearly half the population is below 15 years, reflecting a population growth rate of about 2 per cent. The 3.5 per cent of the households who own more than one hectare of irrigated land and between them share 40 per cent of the total irrigated land are not represented.

Looking at the overall work-load and activity structure in Figure 9:10, what is perhaps most striking is the contribution of the women, who on top of some 6½ hours of agriculture and domestic work spend an additional 4 hours or so on manufacturing and trade/wage labour, adding up to 11 hours which is a high figure indeed. The men work more than 8½ hours out of which 4 are spent on agriculture and associated animal husbandry. It is the domestic work which gives the women such a high figure by comparison, although the contrast to the Char Gopalpur case in Bangladesh is noteworthy. Since the sex role structure is much more flexible in Java, there is a greater engagement of women in trade and wage-work as well as in home manufacturing.

The main activity done by women is food preparation which includes the drying, handpounding and cleaning of rice, the cleaning of other crops, and cooking. It can be seen that rice culture is as time demanding in Kali Loro as it was in Char Gopalpur. While women nowadays may still pound their own rice by hand, hand-pounding for wages has disappeared with the introduction of hulling machines. Grinding mills for rice or corn have been a tremendous time-saving innovation for the women in many rural parts of the world (cf Carlstein 1978b). The time saved for the women has been diverted to trading and manufacturing instead.

Women in Kali Loro spend surprisingly little time in agriculture and animal husbandry (about 1½ hours) in relation to food preparation. It should be something of a memento to theoreticians on intensification when food preparation takes twice the time of food production, albeit this refers only to the female population. Even if food production and preparation had been equally shared between the sexes, the production of food still takes only a bit more than 2½ hours per day while food preparation consumes only slightly less. Increases in work-load with land use intensification is not all about food production.

The active roles of children are also conspicuous in Figure 9:10. Girls quickly fall into the domestic sex role pattern. While small boys help taking care of their younger siblings, older boys all engage in the care of animals, an activity reaching its peak at the ages 12 to 14 years. The other main domestic task for the boys is the collection of firewood. The contributions of children to household production is considerable.

Placing Kali Loro village in the broader context of the 'agricultural involution' (Geertz 1963) which has taken place in Java, some of the traditional institutions for distributing work and land resources have been rapidly deteriorating in recent years, chiefly as a function of increasing intensification in land use, just as there has been increasing class differentiation based on land ownership. The pressure of people on land is thus getting thoroughly compounded by the 'pressure of people on people'.

Some studies of the employment situation in Java have indicated that lacking employment is expressed both in terms of low participation rates and short duration of work. Studies pointing to the short duration of work in the average day have often focused only on rice cultivation, forgetting many of the other activities people perform. The great mobilization of women and children, however, is hardly a good indicator of a low participation rate. White's point is rather that, judging from the data for Kali Loro, work days are long and the participation rate is high. "Like 'general' idleness, seasonal idleness is a luxury which most of the population of Kali Loro cannot afford", White (1976:281) points out, and the problem is not that there is not enough to do but rather that the *productivity of time in many activities is so low*. This has to be compensated for by higher intensity. The productivity of time is low in part because of the general

FIGURE 9:10

KALI LORO, JAVA

(It was not possible to determine the exact size of age category 0-5. The annual
number of hours per activity are given in the figure. Time scale: 1 mm = 15 min.
Source: White 1976.)

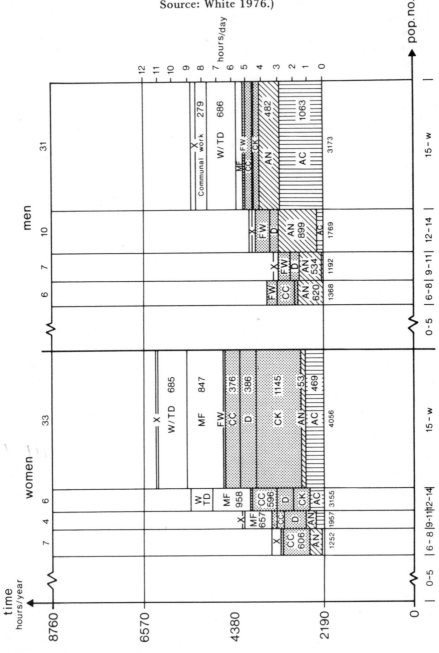

FIGURE 9:11

PANAJACHEL, GUATEMALA

(The material covers only the Indian population. Some time use differences by sex among children could not be determined, since data were sometimes aggregated for both sexes. Time scale: 1 mm = 15 min. Source: Tax 1953.)

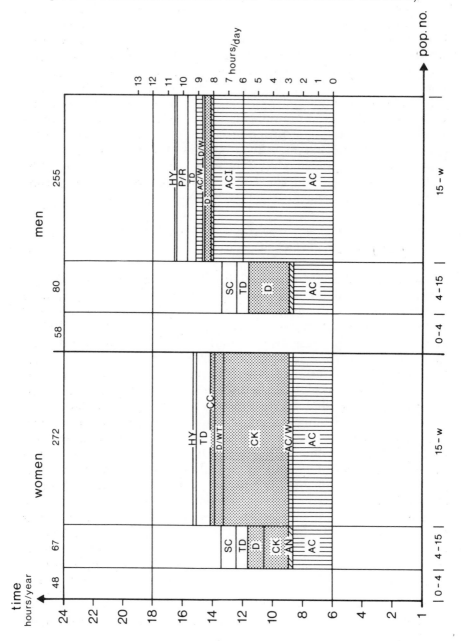

land shortage aggravated by unequal land distribution among holders. The productivity of time is the lowest for the sharecroppers with little or no land.

Yet it is in agriculture that we find of the highest returns to labour-time. In manufacturing, trade, preparation of food for sale, etc., the productivity of time is lower (White 1976: 279, Table 2). Interestingly, this productivity can be increased a good deal if a household has access to a bicycle, since this vehicle cuts down time spent on travel. This fact confirms that in future time allocation surveys, it is important to take special notice of the travel time associated with other activities. This is especially the case in intensification theoretic studies, given that there is still, by and large, not such great differences in time spent on food production between technologies operating at different land use intensities. These differences are to a fair extent influenced by travel time also, and travel time reflects a settlement system which is not structured only by technology. (Cf the earlier discussions in Chapters 3 to 6 on prisms, travel time and settlement structure.)

The village of Panajachel, Guatemala

Panajachel, a village of about 1050 inhabitants, was the object of a classic study in economic anthropology by Tax (1953). The study is meticulously quantified and has been widely used as a source of further explorations. When Tax did his field-work in the late 1930's, Panajachel was operating on a preindustrial technology. Yet it was a thoroughly monetized society, a fact lending the name to the study: 'Penny Capitalism'. Panajachel is also a typical Central American class society in which the Spanish descendants, the Ladinos, form the upper class and the Indians the intensively working lower class.

The Panajachel Indians practise a bit of dry farming on hill land, but this is of marginal importance compared to their intensively irrigated land in the delta of a river. The Indians own 37 per cent of this delta land while constituting more than two-thirds of the resident population (785 inhabitants out of 1050). The resident Ladinos allocate only 13 per cent of their land to intensive truck vegetable gardening in contrast to the resident Indians who allot as much as 60 per cent of their delta land to truck vegetables sold in nearby market places. These truck crops are all irrigated. Most remaining Indian delta land (30 per cent) is used for the cash cropping of coffee.

As far as time input in agriculture is concerned, Tax notes that:

/One must/ take into consideration not only land resources and technology, but the use of time in families of differing land resources. It would appear, for example, that the Indians would make fuller use of their land if they grew vegetables where they grow coffee; but the fact is that *they would not have time* to put all their lands to the *intensive cultivation* they employ on vegetables. Similarly, while a family with very little land can most profitably put it all in vegetables, one with a great deal of land would find itself limited by the impossibility or inconvenience of hiring the necessary labour. (Tax 1953:128) /Italics added/

Coffee cultivation is perhaps the most important innovation which had recently affected Panajachel. It is not very time or water demanding and yet gives a fair return in cash. Per land unit it does not yield as much as vegetables which is less of a problem for the Ladinos who have plenty of land compared to the Indians, who have far less land per household, but even among the Indians, land is not very evenly distributed. Since much of the work on Ladino land is by Indians hired for the purpose, the carrying capacity of the Indian population time-budget would easily have been exceeded, if time-intensive vegetables had been the dominant crop also on Ladino land. Coffee which is the main Ladino crop is not, however, so time demanding.

The population time-budget of the Panjachel Indians is described in Figure 9:11. The women do all but a fraction of the domestic work, and they get a good deal of help from the children, who are generally held to be too precious for the parents to be 'wasted' on schooling. The conspicuous feature in the Panajachel case is *the total volume of work carried out in agriculture,* some 8 hours per average day the year around by the men and as much as 3 hours by women and children. This is indeed a remarkable level of intensity. Out of this an average of 2 hours per day is spent by the men on irrigation. Outside agriculture but connected with it, the time allocated to trade, marketing and shopping (TD) is about an hour for the women and a bit less for the men. Given, however, that Panajachel is so thoroughly commercialized, trading is a crucial activity and Panajachel is part of a region with a far going economic specialization between villages, its special products being onions and garlic.

From the perspective of intensification theory, the high level of time use intensity cannot be explained in terms of ecotechnology alone, notwithstanding the extra burden of irrigation. It rather reflects the distribution of land as a function of class structure as well as the system of economic exchange. The Panajachel Indians have to work hard and long hours, and the intensity of their time use is both a compensation for the low productivity associated with lack of land and for the terms of trade they meet on a highly competitive market.

TIME USE INTENSITY AT THE HOUSEHOLD LEVEL

On the whole, the kind of time-budget and time use intensity graphs presented for local populations in Figures 9:6-11 give a good synoptic picture of how time resources are mobilized for different kinds of activities by different population categories. These graphs are dense in information and permit comparisons between different societies, although the picture provided is an aggregate one which leaves out a number of important features such as the degree of specialization. Nevertheless, these data are far better than those generally available for contemporary rural societies in view of the fact that we have amazingly little information of time resources and activities in Third World rural populations, including essential aspects such as their employment situation (cf Connell and Lipton 1977).

Just as there are differences in time use and intensity levels between local societies, we also find such differences between the households constituting these societies. The latter variations were obviously evened out in Figures 9:6-11, so it is important to go one step further and have a closer look at the domestic unit level.

Chayanov's theory revisited

The reason for devoting space to Chayanov's theory here is not so much that it delivers the best explanation of household 'labour' intensities. It is done more to counter what has become a piece of conventional wisdom in this particular field of study. In a way, the theory of Chayanov was an early and primitive attempt at time-budget analysis, but since the collection of time use data has always been one of the most time demanding forms of inquiry, Chayanov like many others after him felt obliged to resort to proxies. He thus designed his production functions accordingly:

> Unfortunately, it is very *difficult* /cf Fig. 1:3 above/ to record labour processes objectively, and we have almost no statistical materials of this sort at our disposal. Therefore, to measure labour intensity, we have to make use *not of a direct record of its expenditure in working days* but of *the results* of this expenditure, recording the worker's annual earnings and, quite conventionally, assuming that each unit of value is obtained by approximately equal labour efforts. (Chayanov 1966:76)

It follows that although Chayanov wants to measure the *volume of economic activity*, he often is forced to do so in terms of *output* rather than *input*, although, by definition, *intensity* is a relation between inputs (cf p. 190). It is the physical or financial result of activity which he measures rather than the time input.

Moreover, it should be noted that Chayanov's theory does not encompass the entire farming system or the many environmental factors or material conditions that affect the level of labour *product-ivity*. Factors such as soil fertility, location of farm in relation to markets, forms of exchange or kind of technology are not included in what is essentially a theory of an economic unit, the farming household in its homestead or the 'labour-farm' (cf Chapter 6, p. 235-38 above). Just as von Thünen makes use of an ideal type, 'The Isolated State', in his explanation of land use intensity zones, one may say that Chayanov deals with the 'Isolated Labour Farm', where it is assumed that households produce only for their own use and exchange neither labour, land or other factors of production, nor produce.

On this 'isolated' farm, the *supply* of labour is a function of the size and age composition of the farm household, whereas the *demand* for labour is determined 'entirely' by the minimum consumption requirements, which are similarly a function of household size and age composition (cf Chayanov 1966:53). The degree of labour intensity (or of 'self-exploitation') is thus governed by the relation between supply and demand, both being a function of the same variables.

Just as working capacity can be weighed by age (degree of able-bodiedness), so can consumption requirements, and for each house-hold a ratio of *consumers to workers* can be established (cf also Sahlins 1972). This (dependency) ratio will vary with the stage in its developmental cycle that a family finds itself. When all consumers are also able-bodied workers (e.g. one adult man and woman), the ratio has the minimum value of 1, only to rise as more dependents (children) enter the family. It eventually reaches a maximum, but as more children become able-bodied, the ratio begins to fall. Looking at one of his many statistical tables, Chayanov noted:

> ... other things being equal, the peasant worker, stimulated by the demands of his family, develops greater energy as the pressure of these demands becomes stronger. The measure of self-exploitation /intensity/ depends to the highest degree on how heavily the worker is burdened by the consumer demands of his family... The volume of the family's activity depends entirely on the number of consumers and not at all on the number of workers. (Chayanov 1966:78)

Hence, Chayanov's general finding is that the *intensity of production per worker* increases with the relative *number of dependants*.

In order to demonstrate his thesis of underproduction and the lack of pressure on resources in primitive societies, Sahlins finds Chayanov's rule magnified several theoretical powers when reformulated as follows:

> Intensity of labour in a system of domestic production for use varies inversely with the relative working capacity of the producing unit. (Sahlins 1972:98)

In other words, if those households who have many dependents can work more per capita than those with few dependants, the latter units must obviously be working below their potential capacity. They underproduce. Of course, it is only when a local population time-budget is disaggregated to the household level that this dimension of underproduction and differential 'time pressure' is revealed. In absolute terms it is very difficult at the local society level to establish what is underproduction or not, if one compares Figures 9:6-11, since in none of these cases do people 'work' all the time they do not sleep. No society described seems to be 'fully employed', as it were, unless we introduce some external more or less arbitrary yardstick, such as an 8-hour working day. So rather than formulating the problem as one of underproduction, it ought to be more fruitfully formulated as one of the *structural consequences* of alternative allocations of time (or land) among a variety of activities, including effects on levels of production. And in this context, a more comprehensive perspective would require that we go far beyond what is substantively defined as 'economic' activities, since they cover only a portion of the daily round anyway.

If we further consider that there are intensity variations in time use over the annual cycle of activity (an admittedly neglected aspect in this book), we come upon another problem of the synchronized economizing of resources, which bears heavily on the underproduction thesis. Chayanov noted, for instance, that,

> ... /unlike in industry/ ... a great part of the agricultural process is exclusively seasonal in nature... /due to the climate/ ... Because of this, the labour intensity curve in agriculture always shows extremely uneven development. Sowing, mowing, harvesting, and some work on specialized crops sometimes demand exceptional accumulation of mass labour in insignificant time periods, while in other, sometimes very lengthy, periods of the farm year, agriculture finds no *objects on which to use its labour.* (Chayanov 1966:74-5)* /Emphasis added/

* The need for additional inputs with which to combine human time is one determinant of how intensively human time can be used. This problem is at the core of what we have called synchronized economizing (cf. p. 232-5 and 268-71).

One conclusion emerging from these observations is that in order to explain time use intensities or define 'underproduction', it must be considered that certain short seasons have bottleneck effects on the level of output. It is not simply the average level of time input that determines the volume of agricultural output but rather the time that can be mobilized by domestic units at crucial short periods of time. (This is particularly relevant when intensity is measured by proxy in output rather than input terms as do both Chayanov and Sahlins.) If one takes a cultivation project as a whole from the stage of land preparation to harvesting and perhaps also threshing, it would not necessarily help much to mobilize more labour-time on operations such as sowing or weeding if there are not enough people for timely ploughing immediately after the first rains or to harvest the crops before they may get damaged. Given that there are short peak seasons circumscribed by uncontrollable factors such as the weather, the average level of time mobilization may be a rather poor index of 'underproduction' when seen in this logistic perspective.

Next one must then ask if households of different composition *are differentially affected by peak season time demands* in the primitive and peasant societies. Within the same locale one might reasonably suspect that everyone is rather equally affected by the seasonal variabilities of climate and weather. But this need not be so for several reasons. If the level of *technology and equipment* varies, the productivity of time may vary also, which in turn means that some households can cut peak time demands more effectively than others. In pre-industrial society, the number of draught animals at the disposal of a given household may be a very crucial determinant of how a household can cope with peak time demands. Differential access to irrigation water may be another such determinant, let alone differences in degree of mechanization that affect peasant farms in today's Third World countries, say pumps for watering. Chayanov's theory only works, of course, when *technology is homogenous* for all households and when there are *few indivisibilities* to reckon with. This situation is perhaps most likely to be found only in the most primitive isolated shifting cultivation societies practising horticulture with hoes and axes as basic implements and where land access is no problem, either in terms of distance or in terms of appropriation. In sum, the *productivity of time* may vary between households for several reasons that have little direct connection with household size and composition, yet productivity differences affect how intensively households mobilize their time for various activities.

Household intensification versus expansion

In the earlier chapters we discussed two principal forms of land (space-time) mobilization, one being *expansion* into previously un-occupied land, the other *intensification* in the use of already occupied land (Figure 6:3). Household time resources can similarly be mobil-ized either by its members supplying more of their time to given activities (intensification) or by boosting time resources through the inclusion of more members (expansion).

The gist of Sahlins' Domestic Mode of Production theory was not the simple point that the general level of output in primitive, tribal societies was low compared to technologically more advanced socie-ties but rather that output was low in relation to *existing possibilities* (Sahlins 1972:41). Human time and land resources were thus held to be *underused* and the great 'challenge of intensification' (cf p. 250 above) was 'getting people to work more, or more people to work'. However, the more-people-to-work aspect is never really followed up.

While both Chayanov and Sahlins aim to explain the *level of intensity* at which labour-time is mobilized in autonomous household units, the reasons why households *expand or change their composi-tion* are not assumed to require much explanation. Chayanov takes household change simply as a function of the *biological* progression of individual life cycles.

> Nevertheless, however varied the everyday features of the family, its basis remains *the purely biological concept* of the *married* couple, living together with their descendants and the aged representatives of the older generation. This *biological nature* of the family determines to a great extent the limits of its size and, chiefly, the laws of its composition; although, of course, daily circumstances introduce numerous complications. (Chayanov 1966:54) /Emphasis added/

Although Chayanov takes the developmental cycle of the household-family as his point of departure, he is not really interested in either population dynamics or in economic dynamics but he rests content with a cross-sectional analysis of variables, thereby reducing house-hold complexities to a simple consumer/worker index (dependency ratio).

That Sahlins tacitly accepts household dynamics to be a straight-forward extrapolation of biology as Chayanov suggests is strange, since to any anthropologist a 'purely biological' concept of a 'married' couple is a contradiction in terms. The reason why Sahlins ignores this problem can be found in his *ideal type* conceptualization of the DMP which goes too far in its *ceteris paribus* assumptions. Just as the household of urban-industrial societies is wrongly claimed to have become demoted to a 'mere consumption status' (Sahlins 1972:76), that in primitive society is demoted to a mere *force of production* in the DMP. The domestic unit is assumed to have no real relations of production *within it,* nor much of other internal relations. It is conceived of as homogenous, egalitarian, autonomous and self-centered. Its level of labour intensity — discounting that induced by surplus production — is dictated by the time it must mobilize to reproduce itself from day to day, each member working according to ability, whereupon there is pooling and redistribution of produce according to the crude consumption needs of individuals weighed by age and sex and at best fortified by an internalized social norm of minimum requirements. Producing basically for use, labour intensity is totally determined by *external* or *exogenous* biological (or 'demographic') factors. When producing for surplus above use, labour intensity is determined by *external* relations of production. To both Chayanov and Sahlins, household expansion (through the addition of juveniles) is not an alternative or supplementary strategy to intensification, but the independent cause thereof.

Demographic structures inherent in the mode of production

While for Sahlins, exogenous demographic variables lend structure to the mode of production, Godelier reverses this perspective by examining how the mode of production determines 'demographic structures'.

> /Demographic/ structures are not a *primum movens* but rather the combined result of the action of several 'deeper' structural levels, of a hierarchy of causes, the most important of which is again the mode of production; that is to say the productive forces and the nature of social relations of production which make up the infra-structure of the society. Having noted this, the significance of the fact that demography is the 'synthetic' result of the action of several structural levels, a combination of causes of varying importance, must be analysed further... (Godelier 1975:4)

It follows, for instance, that population growth cannot be taken as a cause in itself, according to Godelier. Nicely enough, he illustrates his reasoning with the case of the Bushmen, whose mode of production formed a very starting point in this book. He takes the by now famous example of how spatial mobility requirements involved in production activities affect the timing and spacing of children. A woman of child-bearing age covers an average of 2,400 km per year as she gathers food for the camp and visits other camps. She carries heavy loads of water, food, firewood *and* a small child. A child is weaned from breast-feeding at around four years and in the course of this period, it is carried some 7,800 km. Godelier concludes,

> Given that mobility is one of the necessary constraints on her
> economic activity ... the effort expended by a woman in carrying
> a young child must be maintained within limits compatible with
> the regular and efficient accomplishment of her economic activi-
> ties. Her work /load/ depends on the interval between births...
> Theoretically, an interval of births of at least three years appears
> to be a demographic constraint imposed by the Bushmen's mode
> of production, and this is verified. (Godelier 1975:19)

What better example of how 'carrying' capacity limits population
growth can we find!   Bushman adults are well aware of the time and
energy costs of children, and sometimes even practice infanticide (cf
p. 91-2), for instance to children born defective or to one of two
twins, although unintended biological factors such as the suppression
of ovulation in women due to prolonged breast-feeding also intervene
and affect the 'spacing' of children in time.

Godelier's point that demographic structures are not exogenous
to society but very much *endogenous* to it has considerable merits,
of course. Demographic events such as births, deaths, migrations,
fusion-fission, household compositional changes and general household developments over time and life-cycles are not simply biological
or demographic 'accidents' affecting society. To Godelier, such
'accidents' are structurally caused, which is not to deny that there is
a strong element of biological programming of the pace of given
life-cycle developments, once certain elementary events have been
socially generated. Hence, it is not very fruitful to take these events
out of their social contexts or remove them from the material conditions generating and structuring them.*

---

*   To Godelier, however, employing a marxist-structuralist framework, it is not
so much *individuals* that act as *structures* that act on other structures. There
are reasons to be somewhat sceptical to the essentially functionalist assumptions
that sneak into Godelier's analysis of demographic structures in relation to other
structures, since he goes too far in his insistence on functional necessities and
he further tends to impute a rationality at the system level that cannot be sub-
stantiated. In the villages of India, for instance, there may be very little of a
collective consciousness with respect to overall population growth. The latter
is, at the aggregate level, merely an epiphenomenon of a number of acts and
decisions occurring at the household level which do not slavishly follow from
mode of production structures, or from these kinds of structures only.
    Although actions/activities are structured by preexisting conditions, they also
reproduce and generate the subsequent structural conditions which may be
partly different. Structural changes as well as reproduction of structures (structuring properties of systems) are better dealt with, it seems, when there is a
two-way relation between individual action and structure, as emphasized in the
theory of structuration proposed by Giddens (1979).
    Like Godelier, Sahlins is caught in a structuralist mode of production model

Population dynamics and the household level

The reason why Godelier objects to theories regarding population dynamics as a *prime mover* is just that population growth, to take a determinate form of demographic development, has been looked upon as an *exogenous* cause or independent variable. This has usually been coupled to an analysis at the *aggregate level* of the total regional or national population. Aggregate population growth has thus been assessed in terms of its impact on 'society'. The problems of such an approach have been amply demonstrated in much of the analysis of population growth in today's Third World countries. Given this high level of aggregation chosen, the nation, the tendency towards crude man-land and production-consumption formulations, and the frequent focus on population projections, the general key-note in these analyses has been that *population growth causes poverty*. Two common corollaries have been, firstly that people in these countries are acting irrationally by producing so many children against their own good, and secondly, that family planning or population control should have top priority in combating poverty. A vital ingredient in the latter strategy by the state is to give people better education so that they come to know what they are doing and can safeguard their own basic interests.

Some of the otherwise very useful studies on intensification and involution (Boserup 1965, Geertz 1973, Wilkinson 1973, Harris 1977) assume population growth to be a prime mover and concentrate on establishing the effects thereof. Given the outmoded formulations by Malthus, Boserup was certainly justified in reversing his perspective and positing population growth as the independent variable of land use and food production intensity. Still the problem remained that demographic variables were somehow exogenized. Even though Boserup demonstrates that population growth leads to resource mobilization (land and human time) as a positive adaptation, this growth is chiefly explained as a function of reduced mortality. The fertility dimension, however, is neglected.

where household activities are structured by kinship and political relations of production (although the demographic variables affecting household time use and activity intensity are exogenized in this case). He does not, however, stress how these relations are reproduced through the strategic action of individuals or how the population is similarly reproduced through actions/activities at the household level. It is only when he discusses how 'big-men' act as intensifiers and 'move byond the narrow base of households' that he touches upon the action perspective. The merit with Sahlins analysis in the present context of population dynamics is that he struck upon a relevant level of analysis for action, that of the household, although the latter is treated as a structure only.

Strangely enough, many analysts who have been very preoccupied with treating population growth as a cause, have shown far less interest in examining the causes of population growth, and this cannot be done unless one steps down from the aggregate level to the one where the process of population generation takes place, the household and individual level:

> From the perspective of the village, continued rapid population growth is likely to intensify the pressure on resources and erode overall economic welfare. In Bangladesh, however, the village is not the locus of decisions governing human reproduction. Such decisions are made by households and individuals in response to the socioeconomic circumstances they face; and from the micro-perspective of the household, the economic welfare implications of continued high fertility (and large numbers of surviving children) are not nearly as clearcut as they appear to be from the macro perspective. (Cain 1977:202)

Recently, scholars looking at population *in society* rather than somehow above it or next to it, have questioned the formulation that population growth can be fruitfully regarded as a cause of poverty.

> Thus far we have been concerned with the general effects of population pressure and some other factors on rural economy in Java ... it may be interesting to ... consider the possible effects of rural economy on population dynamics ... (White 1976:284)

This is in line with Mamdani's examination of why family planning programmes fail in northern India:

> ... our purpose here is to understand the *material conditions of the farmer's existence*, and to specify what influence these have on the size of *his* family. (Mamdani 1972:69) /Italics added/

The arguments by Cain, White and Mamdani are compatible with Godelier's stance that structure affects action, but any examination of household expansion strategies has to pay respect to the dimension that individuals are not structural prisoners but also act rationally, and that structures are partly the outcome of rational action. White challenges the thesis of irrationality often encountered:

> ... population growth in Java has not been the result of villagers' apathy or irrationality or of the non-availability of birth-control methods, but rather a *response to an economic system* imposing *conditions of production on the household economy* such that the economic advantages to parents of relatively high fertility outweigh the economic costs. (White 1976:286) /Italics added/

Time demands on, and for, household members

Moving down to the level of domestic units, their projects, strategies and *daily rounds of activity,* the perspectives of action and structure converge. Actions-activities are to a considerable extent structured by prior actions, both of the acting individuals themselves and of others around them, just as the structural properties of the wider system facilitate and constrain the multitude of actions by particular individuals. If the structural parameters in given situations are such that household members are exposed to relatively heavy time demands, i.e. to 'time pressure', will the members in charge of household affairs seek to solve this problem by expansion of domestic time supply, i.e. by seeking to add more members? Recent research seems to indicate that this may to a considerable extent be the case in Third World peasant societies (Mamdani 1972, White 1976, Cain 1977).

'Time pressure' may not necessarily be felt because all waking time has to be spent on work, but rather because the *productivity of time* is so low that too many tasks are felt to be arduous and time consuming, or because the household members who decide things want to expand farm output without having to expose themselves to excessively high levels of labour intensity. The latter reason is not uncommon when peasant societies become commercialized so that all output above the needs of the household can be exchanged on the market for other goods, including luxuries that bring comfort or prestige.

There are several important reasons why the productivity of time may be low in preindustrial (primitive and peasant) societies. One reason is the *low level of technology,* both in food production and in domestic activities. The low level of technology in itself need not lead to excessive time pressure, but the situation is worsened by leaps and bounds when the biotic-cum-physical environment is heavily depleted. In Bangladesh, for instance, the widespread shortage of organic materials both for fertilizer, fuel and building leads to people spending much time simply on scavenging. Frequently the least advantageous sources of organic materials have to be exploited, especially by the poor. A case in point is the use of water-lillies that contain some 90 per cent water for fuel. The process of drying them is very demanding both in human time and settlement space-time. Briscoe (1978) aptly characterizes the energy system of most Bangladeshi villages as 'frugal and inefficient', the criterion being the input energy in relation to the output energy. By this token, the energy system of the Raiapu Enga or the Tsembaga is efficient; the productivity of human time is high because of relatively accessible organic inputs (cf also Bayliss-Smith 1977). In Bangladesh, to take one example, inefficient cooking technology imposes both heavy time demands, primarily affecting the women, and all too high energy requirements. Moreover, a skewed distribution of existing organic and land resources increases pressures on poorer households. Since the average regional pressure on land based resources is high, however, the majority of households tend to suffer from low productivity of time and tend to compensate for this not only by time use intensification but also by expanding their joint time resources, notably by investing in children.

The literature on child labour in peasant societies such as those of Bangladesh, Pakistan, India or Java is full of examples of the contributions made by children:

> The farmer's children can be of considerable assistance, even while they are young. A son or daughter can bring grass and water for the cattle before going to school at eight in the morning, can help in the field in the afternoon if necessary, and can graze the cattle in the evening...
>
> If the farmer's wife has no young children, it would mean intolerable hardship. She would have to walk to the fields to deliver two meals and one tea every day. The walk over and back, the wait while everybody eats — so the utensils can be taken back, washed, and cleaned for the next meal — can take as much as four hours /per day in peak season when men have to be served/....
> (Mamdani 1972:99) /For a similar Mexican case, cf Figure 9:12./

Cain when describing Char Gopalpur village (cf Figure 9:9) notes:

> /In spite of/ the small size of the sample ... the most important conclusion ... can be drawn with great confidence. Children of both sexes, regardless of economic class, do a great deal of work in comparison to adults, and, in terms of time worked, reach adult equivalency at relatively young ages. (Cain 1977:219)

The situation is the same in the village of Kali Loro described by White (1976), where the children allocate a good portion of their time to collectively useful activities (Figure 9:10).

One principally very important way in which children can contribute their time and effort to their parental household is to tend other children, viz. their younger siblings. At first this may not seem to be very 'productive', but it does have the effect of releasing the parents to do other activities of benefit to the family. Children thereby impose fewer time demands on their parents, which, of course, reduces the 'costs' of having children. Many academics and planners analysing Third World population growth have tended to take it for granted that children incur demands on parental resources similar to those in urban-industrial societies. This is a misunderstand-

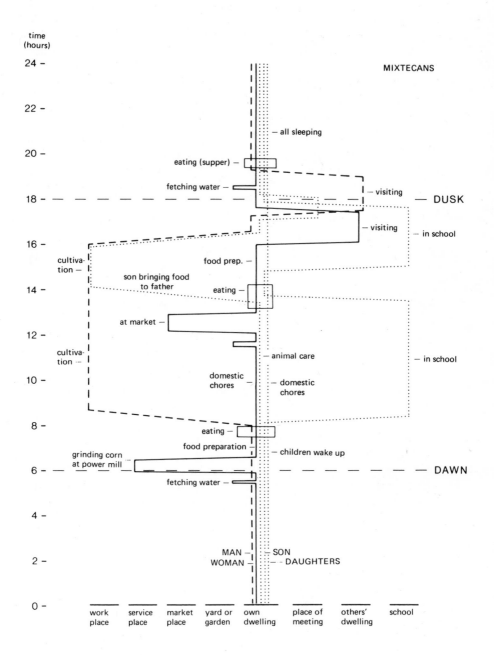

FIGURE 9:12 The daily round of a Mixtecan household in Juxtlahuaca in Mexico. The son takes care of animals and also takes food to the father in the afternoon as well as works with his father in the fields. The eldest daughter has taken on a good deal of the household work and stays home, whereas the youngest daughter is less needed and might as well attend school. The graph is reconstructed on the basis of verbal information in Romney and Romney (1966).

ing, however, since child-tending activities in rural societies are often incidental to other activities taking place within the home range, such as cooking, cleaning or manufacturing. Few special arrangements obtain that make child care costly, and once children are beyond the first two or three years, they are often looked after by any person older than them in the vicinity.

> We should remember also that the biologically-determined subsistence requirements of children are less than those of adults, as are their culturally-determined actual consumption levels... so that the productive value of children, relative to their drain on household income through consumption, is higher than the figures in ... /Figure 9:10/ imply. (White 1976:285-6)

In a study of the time costs of child rearing in the rural Philippines, Ho (1979) found that while the first child took some 3 hours per day of the mother's time resources, as more children were added (up to six or more), there was no further increase in the time spent on children by the mother. After a few years, the older children would step in and take care of the younger ones and take over other activities as well. It was not family size that affected the mother's time allocation but rather family composition.

> Child care costs on mother's time for each additional child decrease as family size increases... These observations help to explain the high fertility levels in rural societies. They indicate that the economic costs of child rearing may indeed be quite low when measured in terms of the mother's forgone market time ... /etc./ Taken together with already existing evidence that children in rural societies make important contributions to household activities even while still young, these observations indicate that childbearing may in fact be a profitable investment... (Ho 1979:659)

In his study of Char Gopalpur, Bangladesh, Cain found that children become *net producers* as early as 12 years, and long before that they started to share household work-loads.

> From the perspective of the parents ... high fertility and large number of surviving children may be economically 'rational' propositions... /in view of/ ... the contributions of children during the period when they are members of their parents' household and their output is controlled by parents ... male children become net producers as early as age 12, compensate for their own cumulative consumption by age 15, and compensate for their own and one sister's cumulative consumption by age 22. (Cain 1977:201)

In Java where sex roles are not so rigid that females cannot work outside the homestead or village, one may suspect that the costs of children are even lower.

Chayanov's assumptions were very far off from these results. He defined children as *consumers only*. They were mainly a burden by contributing nothing until the age of 15, when they all of a sudden attained 70 per cent of the capacity of an able-bodied man. It is no wonder then that Chayanov's consumer/worker (dependency) ratio tended to soar up to such high values, or, when measuring intensity by proxy of output (rather than in input), that there was a positive correlation between family size and output. When the level of technology is not so advanced, skills are acquired relatively early in the life-cycle, and if this is in line with (socio-culturally conditioned) parental intentions, childrens' time resources can be fruitfully mobilized at early ages.

These arguments lead on to the factors of *timing and spacing,* which are of general importance both in the population system and the activity system.* Ho (1979) points out that the allocation of time resources in the activity system has an impact on the timing and spacing of children as elements of fertility behaviour. Looking beyond the conventional arguments that parents invest in children for old age security, children may, in fact, be of service to their parents long before the latter reach old age, since the children's output is controlled (and confiscated) by the parent generation. For how long this may continue depends on how long the children stay with their parents which in turn is often related to the children's age of marriage. These marriages are generally *arranged* by the parents and given that the children in peasant South-asian societies rarely leave their parental family prior to marriage, the mean age of marriage of course affects the reproductive time span of the new generation of women. Mamdani (1972) reports for his northern Indian study area that parents tend to keep their girls a few years longer nowadays. This factor accounted for the minor drop in fertility rates that had taken place rather than the family planning propaganda:

> ... the marriage age among Jat /dominant farmer caste/ girls have risen, regardless of whether they attend school or not. This is for two reasons, the intensification of agriculture has increased the

---

* The reader might at this stage be reminded that the whole idea of individual paths in time space by Hägerstrand (1963) emerged from his spatial generalization of the *life-line* concept developed in demography by Lexis (cf p. 41 above). These developments permitted a link between *infrequent life-cycle events* in the population system and the *very highly frequent activities* in the activity system, mediated by the continuous 'flow' of *time resources* of the population. In this section, we have returned to demographic issues again and how time resource allocation affects the reproduction of the population, including the impact on the timing and spacing of life-cycle (demographic) events.

> work load, not only in the field, but also in the farmer's home. The longer the daughter stays unmarried, the more assistance she can lend her family. Secondly, during the years she spends at home, the grown up daughter can earn her dowry by sewing, spinning and weaving. (Mamdani 1972:102)

The parents thus benefit more by keeping their children in the family as long as possible. The spacing of the marriages of these children in time may also have differential effects on the remaining fertility span of daughters, depending on things like the financing of weddings or other transactions associated with these life-cycle events of such demographic significance. These factors of timing and spacing of demographic events thus have little to do with biology, as Chayanov assumed, except with respect to the broadest parameters of fertility such as the time span of fertility in women and men or the duration of pregnancy (the human gestation period). Timing and spacing of demographic events is largely a question of social interaction.

Factors of timing and spacing in the activity system also affect reproductive strategies. Cain notes that:

> Small landholdings and *seasonal fluctuations* in employment opportunities make it necessary for most households and their individual members to exploit a variety of different sources of income and earnings over the course of a year. In general, a household with a greater number of economically active members will be in a better position to diversify and exploit multiple sources of income, particularly when, as is often the case in Char Gopalpur, *the timing of peak opportunities* of different sources coincide. (Cain 1977:209) /Italics added/

Larger households are able to develop internal specialization which may be economically more efficient. The incorporation of more members also generates divisibilities if tasks are spread in several different places. Furthermore, a large household is less vulnerable to rigid sex role norms, since the larger the number of children born, the less is the probability that they will all be of the same sex.

Households can also increase in size by the inclusion of second generation children. Household heads may try to keep their own children as well as their grand-children, thereby building up an *extended family*. This may or may not contribute to a more intensive use of the members' time resources on the average, but it is likely to lead to *differential time pressures* between ages, generations and sexes, e.g. by the eldest generation taking on more managerial (and less labour-time intensive) activities. *

* Extending a family by a three generation strategy is a much slower strategy than that open to men in polygynous societies. (Cf below on polygyny.)

Population growth and system structuration: An intertemporal contradiction involving action and structure

Although the parent generation may be able to get more than compensation for the costs incurred by children, the long-term costs of increased resource demands following from a growing population are transferred forward in time to the next generation.

> Without sons, there is no living off the land. The more sons you have, the less labour you need to hire and the *more savings* you can have. If I have enough, maybe we will be *able to buy some more land,* and then fragmentation /at the time the sons should have their share/ will not matter.' (Argument presented by a male household head in Mamdani 1972:78) /Italics added/

It is when the new generation has grown up and begins to produce children of their own and when the parent generation dies that the original household splits up and 'fragmentation' or splitting the farm takes place. Then the capacity to earn income in the original family in order to buy more land may have fallen short of expectations, since land is now likely to be scarcer than before and inflation has driven up prices accordingly. This is not surprising, as many other households have in the meantime applied the same strategy of expansion. The net outcome of this process is increased land use intensification.

The negative effects on households of increased competition for environmental resources by the gradual introduction of still more (competitive) actors are *unintended consequences* of what at the household level is rational action, at least in the short-run. These negative effects are *delayed*, however, so this particular contradiction between action and structure is expressed as an *intertemporal* (and intergenerational) allocation conflict. The structuration of the wider system over time is thus far from an outcome of the intentionality of household actors. Furthermore, the overall system is not simply reproduced over time through action/activity; it is *structurally transformed,* so that the action possibilities of the new generation are not equal to those of the old generation.

Just as Boserup, Harris and Wilkinson argued that 'population pressure' leads to intensification and 'time pressure', it is equally likely that 'time pressure' may contribute to household expansion and hence additional 'population pressure'. Although this formulation in 'pressure' terms is merely a journalistic summary of a much more complicated problem (cf p. 292-300), the social and ecological demand mechanisms for human time and settlement space-time are at the core of both the intensification process and the social-cum-environmental structuration process. In these processes, demands and relative 'scarcities' are not necessarily equalized among actors/holders.

When competition for limited resources increases, social polarization and an unequal distribution is a more probable outcome. Presently we shall thus turn to the vertical segmentation in the population time-budget associated with class and sex.

Polygyny as an alternative strategy of household expansion

From the perspective of politically dominant adult males, household expansion may in some societies also take the form of polygynous marriage. Such a marriage generally brings in an additional *adult* at full working capacity. If we thus differentiate between *rates* of household expansion, the time investment by men in alliances that facilitate polygynous unions may be a faster method of household expansion than natural reproduction which takes many years, although one strategy by no means excludes the other.* With or without polygyny, however, women are often a scarce resource to the men. Waddell, for instance, notes that:

> ... newly married Raiapu men feel constantly threatened by the precariousness of their union where there is an effective scarcity of women and where rules concerning the sexual division of labour are strictly adhered to. The presence of females in any production unit is vital. (Waddell 1972:27)

Sahlins (1972) only approaches the theme of household expansion as a *strategy of action* when he discusses the role of 'big-men' in Melanesia as intensifiers of production (cf p. 250, 252 above):

> ... the big-man would quickly surpass the narrow base of auto-exploitation. Deploying his resources /including the time of his wife (wives)/ the emerging leader uses his wealth to place others in debt. Moving beyond the household, he constructs a following whose production may be harnessed to his ambition. The process of intensification is thus coupled to reciprocity in exchange... (Sahlins 1972:136)

This reciprocity of exchange also includes the exchange of women. *Intensification* in the time use of external people mobilized to the ends of a given household through reciprocity relations is not a strategy separate from that of household *expansion* by the long-term inclusion of one more adult woman. According to Chayanov's rule the addition of able-bodied adults would not lead to time use intensification, but in practice it might very well do so, since the 'demand pressure' *for women* is not a separate issue from the 'time pressure' *on women*, our next topic of discourse.

* The effects on population dynamics will differ, however.

TIME DEMAND PRESSURES AND SEX ROLE ASYMMETRIES

The six graphs depicting the population time-budgets of Machiguenga, Ushi, Raiapu Enga, Char Gopalpur, Kali Loro and Panajachel (Figures 9:6-11) show not only that there are substantial variations in the extent time is mobilized for different activities, such as food production and domestic activities, but also that there are great differences in the distribution of activities between the sexes. In two of these agricultural societies, Char Gopalpur (Bangladesh) and Kali Loro (Java), women spent only an hour or less in agriculture, while Machiguenga women spent less than an hour in agriculture plus somewhat less in gathering and fishing. The men were the main agriculturalists. In Ushi and Raiapu Enga society the situation was the reverse. The women allocated much more of their time to agriculture than men. This is in stark contrast to the Muslim women in Char Gopalpur, who did an extraordinary amount of domestic work, but were affected by very strict norms pertaining to activities outside the home. In each and every case, however, we find rather marked asymmetries in time use between men and women, with a far from surprising dominance of women doing domestic chores. Cooking took a great deal of time in all of these societies except Raiapu Enga. Variations by sex in overall work-load in those activities which comprise the definition of 'work' applied in this chapter have been illustrated by shading and hatching (Figures 9:6-11). By this particular definition, women spend more time on 'work' in five out of these six (random) cases, Panajachel in Guatemala being the exception.

A main point emerging from this survey is that only a minor part of the asymmetries found between the sexes is attributable to food production ecotechnology. Household technology, such as the time it takes to husk, etc. a weekly ration of rice, can tell us why certain activities are more or less time-consuming, but it does not tell us why the women have been allocated this kind of activity and why it is not distributed in some other way between the sexes (or age groups for that matter). Hence there is little room for technological determinism in the explanation of the allocation of time to activities between the sexes, even though there is an important technological component.

It is also inadequate to treat the household as an undifferentiated unit, as Chayanov does when speaking of 'auto-exploitation' within the household. There are social and political relations within the family/household, so whatever time pressures there are, these are not necessarily equally distributed. Chayanov's theory of household intensity levels has no sex role dimension to it, nor do the general intensification theories proposed by Boserup (1965), Sahlins (1972), Wilkinson (1973) or Harris (1977). These theories revolve more around the various factors behind the overall increase in work-load than how 'work' is distributed between women and men.* Yet there is little doubt that the allocation of activities and tasks between the sexes is a crucial aspect of the carrying capacity for action and inter-action in a population system.

Sahlins found no pressures on time supply in primitive and to some extent peasant societies, because the *average* work-load was taken to be low. But it is not always averages that count or are most instructive of how systems unfold. There are, for instance, seasonal or temporary variations in time pressure that may have considerable structural importance just as there is uneven distribution of time demand among age-sex categories. Seasonal bottlenecks can have a decisive impact on the total output from a cultivation project. Agronomists in Africa, for instance, frequently point out that harvests could be much better if only there had been enough labour mobilized to sow the crop in time once the rains have come (Ruthen-berg 1968). Similarly, weeding is another activity having a strong effect on total output, but if the task is left for the women only, the productivity of time in agriculture as a whole may suffer.

Among the Ushi, we found that sex role norms were considerably relaxed during the critical wet season and that the men, whose total input into agricultural activity was comparatively low, then com-pleted numerous tasks they would otherwise not have done (cf p. 371). Similarly, shifts in the allocation of tasks between men and women are often induced by the migration of men in search of cash income and the consequent withdrawal of their time resources. As the young men of the Gusii of Kenya migrated, some of their tasks were transferred to children, while women had to do more work in agriculture and take on the milking of cows, a task which had been a typical activity for young men. The introduction of cash crops above subsistence requirements has likewise led to time use intensi-fication in agriculture, not least for the women (LeVine and LeVine 1966:14). Time demand pressure has thus been transferred onto women and adolescents.

As we move towards land use intensive societies such as Char Gopalpur, Kali Loro or Panajachel, there is a tendency for male time to be more intensively mobilized relative to that of women. In part, this perhaps reflects the fact that general time use intensity is higher, so that women cannot take on too disproportionate pressure on their time and energy. On the other hand, it is striking how Char Gopalpur society is willing to accept a high cost for keeping the women active only in the home sphere. Given this spatial range and

---

*   Boserup has, however, completed a study of woman's role in economic development and the topic of activity distribution by sex is certainly broad enough for a separate volume (cf Boserup 1970).

prism constraint, the women fill their time with a good deal of purely domestic activity, although it is probably true that in those environments where there is a relatively great depletion of local organic and energy resources, the domestic work sector compensates for low productivity by increased intensity.

White describes how the sex role structure is much more relaxed in Java and how this facilitates extreme 'occupational multiplicity'. Individual household members switch between activities as external opportunities for income arise. If women for some reason are tied to the house, they engage in additional activities (such as mat-weaving) which yield the best income in that given situation. If they are not tied to small children, for instance, they reduce their domestic activities to earn more. Such sex role flexibility increases the productivity of time.

Larger households are in a better position to diversify and exploit multiple sources of income (cf p. 339), which may be yet one factor stimulating a higher level of fertility. However, if only daughters are born into the household, this may be a self-defeating strategy in a society with rigid sex roles. In some South Asian societies where the system considers sons more valuable than daughters, there is a tendency towards higher mortality rates for female infants by neglect, if not even some practice of female infanticide, especially in the old days.

A closer study of the daily round in terms of paths and prisms will reveal some of the underlying logic in the way activities are allocated between the sexes and how much time each sex spends on what activities. In Figures 5:10 and 5:11, for instance, we could see how the Siane women were normatively obliged to feed and take care of their children in the morning and yet cook for their menfolk in the evening. Given certain time constraints associated with the cooking technology in use as well, these factors determined about how much time these women could maximally spend on agriculture in a day. If they worked more in the gardens, they would be late home and the evening meal would be too late (normatively speaking). The Muslim women in Char Gopalpur were on the other hand tied to their home location, and this excluded many activities to them. Coupling constraints such as women having to cook a meal in the middle of the day may also split their day-prisms in the manner shown in Figure 5:12, and also narrow down the range of action and activities that can be selected in the female role set. For the men, there may be other factors. The way activities can be *combined into role sets* is, consequently, not merely a function of power and norms (regulatory constraints), as is often assumed, but also a matter of *practical feasibility* as influenced by capability and coupling constraints. (Cf Barth 1967b for a discussion on norms and practical allocation.)

A NOTE ON PASTORALISTS

Given the mobility structures associated with pastoral activities, it is impossible to conduct meaningful studies of time allocation in dominantly pastoral societies (esp. nomadic ones) without considering spatial movements. How dominant in quantitative terms are the movement activities in different societies with pastoral ecotechnology? This is difficult to assess due to shortage of data. During the few months when the Basseri, for example, moved from highland to lowland areas or vice versa, they spent an average of 3 hours per day on movement (Barth 1961:150). Yet there seemed to be little time pressure as such, and this moving affected all household members rather evenly. Once in highland or lowland areas, only some household members would move with the animals away from camp. Often the seasonally felt time pressure on pastoralists is during the lean season, i.e. the dry season in the tropics when the animals have to be cared for very selectively to optimize foraging, or when water has to be lifted from wells.

Societies heavily dependent on herd management must have an effective division of activities between age-sex categories. Households must strike a dual balance: there must be enough members with their time input to manage the animals; yet there must also be enough yielding animals to feed their human caretakers. In Chayanovian terms, what then constitutes the relations of intensity between households and herds (just as there was one to land). Do herds expand with family size, for instance? Given that different species and age-sex categories of animals have their given optimal herding requirements and have to be sorted out in space accordingly, households must be *divisible* enough, i.e. have enough members to fit each category of animal (if pastoralism is to be efficient). At the same time, there are clear multiplier effects in that one shepherd might manage one hundred animals as easily as ten, while ten animals obviously produce much less food. So when herds of different categories reach a size beyond individual management, they must be split up and require double the time input. Mixing small herds of animals with different foraging needs to save human time and personnel, on the other hand, lowers the efficiency of herding (cf Dahl and Hjort 1979; Dyson-Hudson 1966:68). In predominantly pastoral societies we find, in fact, that there are numerous social mechanisms for *adjusting household size and composition to herd size and composition, and vice versa,* i.e. exchanges of people and animals among households. This kind of insitutional flexibility has the effect of *keeping overall work-load down* at the same time as it gives better levels of food production, i.e. it *lowers time use intensity and raises productivity.*

SPECIALIZATION, CASTE AND OTHER FORMS OF VERTICAL SEGMEN-
TATION IN A POPULATION TIME-BUDGET

In the previous case studies described in Figures 9:6 to 9:11, the only
form of vertical segmentation shown was based on age and sex
(and to an extent marriage status, as in Figure 9:8). These were the
only vertical dimensions affecting the packing of activities. Any
activities done by a minority of the individuals within an age-sex
category were averaged out, as if they were performed in a smaller
dose by all in that catgory. To depict time use in this manner is a
generalization which is acceptable as long as we are dealing with
relatively undifferentiated and unspecialized societies. However,
even the 'post-traditional' peasant societies of Char Gopalpur, Kali
Loro and Panajachel were characterized more by occupational
multiplicity than by occupational specialization. The application
of preindustrial technology means that skills are acquired at a fairly
young age (compared to modern industrial societies), so the same
individual is technically able to engage in a multitude of different
activities according to opportunities and desires. However, social
structures such as occupational specialization, caste, class, and many
other forms of social differentiation or segmentation leads to the
mobilization of human time for given types of activity along more
specific population categories. This is expressed graphically in the
shape of *vertical cleavages* in the population time-budget.

In Panajachel, for instance, 77 people out of some 780 had their
special occupations, but if this time were averaged out for men,
women and children, these activities would amount to only 45, 12
and 4 minutes respectively, or 3.1, 0.8 and 0.3 per cent of total time
use. If we assume an 8-hour working day (to borrow a yardstick from
modern industrial societies), the specialist activities would be enough
for merely 18 people out of 780. In the real case, of course, hardly
any of these specialists operated on a 'full-time' basis, so for a society
such as Panajachel and the others discussed, no grave violence to
reality is done by ignoring minor occupational specialization in the
generalized representation of the time-budget of the population.

Sometimes, however, we have forms of vertical segmentation
which are strongly instituted even in pre-industrial societies. The
Indian caste system, for instance, comes readily to mind (cf Figure
8:5 and 9:15). Traditionally there was a clear service specialization in
the caste system mainly associated with religious norms of purity and
pollution. Certain activities were regarded as ritually poluting and
were systematically shunned by some caste categories, and this
constitutes a kind of negative specialization. Other activities, such as
work by blacksmiths required specific skills, and caste endogamy
(prohibition on intercaste marriage) generally implied that skills were

transmitted within the caste. Some castes were only part-time specialists and agriculture was their basic livelihood activity.

Leaving aside many of the other structural properties of the caste system, it is interesting to note that in many Indian caste villages, domestic water used to be carried from the wells to the households by people of a caste having this as their main activity (cf Figure 9:14). In an African village, by contrast, water carrying would be a task done mainly by women or children from *all* households (as in Figure 9:13). In the Indian case we have a narrow *vertical* packing of this particular activity in the population time-budget, but in the African case we have an essentially *horizontal* packing of water provisioning. This difference has an interesting time allocation consequence when these respective societies were exposed to technical changes or innovations. While the introduction of metal hand pumps that could be placed in every farmer's own yard would have been perceived as a 'convenient' time and energy saving innovation in an unspecialized African village, but in the Indian caste village it would render the services of one caste category redundant and thereby cause 'unemployment' as well as disrupt a system of service exchange relations. Elder (1970) reports how the caste *(jajmani)* system underwent drastic change due to innovations (i.e. due to *displacement effects* of innovations, cf Carlstein 1978b). The main activity of the Potter caste (no 4 in Figure 9:15) was displaced by new metal vessels, and the services of Blacksmiths (caste 8) were displaced by new metal tools and implements bought in nearby towns. These structural changes affecting caste-specialized Indian villages are in some respects similar to those operating in urban-industrial societies as particular branches of industry decline.*

However, displacement effects coupled to specific 'vertically delimited' population categories can also be found, for instance, in African age-graded or age-class societies. The innovation of *pax* (cf Chapter 5) had similar effects on young males who were formerly engaged in defence and warfare but had to give up these kinds of pursuits when colonial governments imposed their peace. Examples are legion, but let us take one, the Sonjo (described earlier on p. 281):

> In its traditional form, the Sonjo /age-defined/ warrior class tied up a large amount of potential labour /-time/ which might other-wise have been applied to exploitative activities. Setting the average age at which a young man leaves the warrior class at about twenty-seven, this means that seven or eight years of his adult life are devoted to non-economic activities. (Gray 1963:160) †

*   Time resources, interaction and various aspects of specialization will be dealt with at great length in Volume 2 of Time Resources, Society and Ecology.
†   Cf the topic of age pyramid effects and working age on p. 357-59 above.

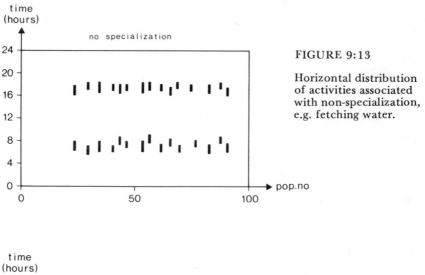

FIGURE 9:13

Horizontal distribution
of activities associated
with non-specialization,
e.g. fetching water.

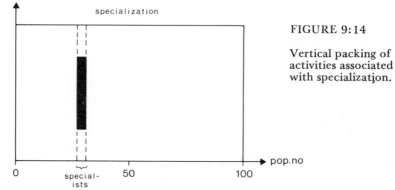

FIGURE 9:14

Vertical packing of
activities associated
with specialization.

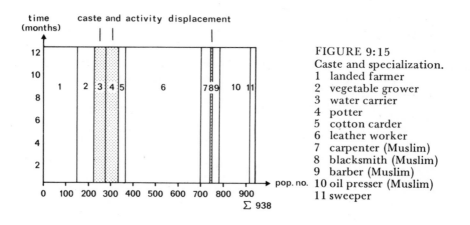

FIGURE 9:15
Caste and specialization.
1 landed farmer
2 vegetable grower
3 water carrier
4 potter
5 cotton carder
6 leather worker
7 carpenter (Muslim)
8 blacksmith (Muslim)
9 barber (Muslim)
10 oil presser (Muslim)
11 sweeper

Curiously, the age and sex role structure that evolved prior to *pax* is often maintained even long after the initial impact of the innovations. Sex roles may thus outlive the material conditions generating them. These long shadows cast by the past could also be seen in the Ushi case (Figure 9:7), both in the population sex ratios (redundant males migrating), and in women still doing most work in agriculture. Similar imprints of 'extinct' sex role structures were also found among the Gusii of Kenya (cf p. 404 and Figure 9:2).

Of course, the discussion of effects of innovations in terms of vertical or horizontal distributions also pertains to changes such as land reforms, consolidation movements or enclosure processes. If, for instance, villages are split up and farms are moved out to their respective consolidated lands, a great deal of time may be saved for the individual farming households. This time can either be used to intensify production or else to reduce the overall work-load, i.e. the time savings are distributed horizontally in the population. If, however a number of landless or near landless households loose their jobs by this process, there is a negative vertical 'saving' of time (cf also Chisholm 1962:13. Cf also below, distribution of land by class.)

CLASS, PRODUCTIVITY AND FACTORS OF INTENSIFICATION

One of the neglected dimensions of intensification theory, particularly the version taking population growth as universal prime mover, has been the aspect of 'pressure of people on people', i.e. the institutional dimension of distribution of resources or 'means of production' among holders in the population. From the perspective of human time resources and the realization of projects, the output resulting from the application of *human time* is very much a function of the coordinated (synchronized and synchorized) input of *other resources* as well, notably space-time (land), energy (measured in Kilowatt-hours), water-time, animal-time, and other inputs, indivisible or divisible. The manner in which these other resources are made accessible as a function of normative, contractual and other institutional-regulatory constraints is another dimension affecting the structuration of socio-environmental added to the practical and logistic dimensions of capacity and coordination (coupling).

In his investigation of Char Gopalpur, Cain compared the time use of different population categories who were defined according to *land ownership*, the latter being a basic feature of *class*. He compared large landowners to small landowners and landless households (Cain 1977:217-9), and some very clear class differences emerged.

The adult males above 13 years in the large landowner category put most of their time into crop production and animal husbandry on their own farmland and engaged very little in work for wages. Small landowners spent about half as much time on own farming and the other half on wage-labour, while the landless men allocated about as much time to wage work as the men in the large landowner category did to own agricultural production. All in all, while men in the large landowner class spent 7.6, 7.5 and 8.2 hours on work (AC, AN, F, W, TD, and D), men in the same three age categories (13-15, 16-21, and 22-59) in the small landowning bracket spent 11, 10.1 and 9 hours on work, which is substantially more. The landless farmers, however, spent still more time, or 10.4, 11.4 and 10.8 hours respectively. These class differences, which should have been represented by *vertical segments* in Figure 9:9 were averaged out, as were, of course, the differences between households. However, the general trend is very clear: The lower the class status, and the less land people have, the lower is the average productivity of their time, as mediated via the monetized sector, and the more this low productivity has to be compensated for by higher intensity of time use.

If we look at the women of corresponding ages, 13-15, 16-21 and 22-59, we similarly find a clear pattern of time allocation by landowning class. The work-load of women is more uniformly high than in the case of men. Only the girls of 13-15 years work more in the landless class, largely because of their greater involvement in wage-work. For age categories 16-21 and 22-59 the range of difference is little more than an hour (from 8.7 to 10.1 hours), but the main tendency is for women in land rich households to have the highest work-load because housework requirements increase with wealth. As was the case with the men, women engage more in wage work the less land that their household owns. Women in richer households do no wage-work at all, while women above 16 in the landless class work 1.2 hours or so for wages. Girls of 13-15 undertake some 3.5 hours of wage work, superceeding by far the other female age categories. This is partly related to reduced role expectations in terms of female seclusion applying to these girls. If we add up what Cain defines as total 'productive' labour (i.e. AC, AN, W, TD and miscellaneous), the class trend is again clear. The less land, the more women engage in these directly income generating activities and the less time they can afford to allocate to purely domestic activities.

As for the children, the class impact on both intensity and kind of activity is that poor boys spend more time in wage work and fishing (the latter being a typical 'time-stretching' activity), while the more land a family has, the more their boys work in animal husbandry and crop production, where the productivity of their time is higher. Young girls in landless households do more domestic work, since they have to take care of the younger children as well as do other tasks which release their mothers for more productive activities.

Class and the short-term exchange of (labour-)time resources

Again it may be of some interest to return to Chayanov's theory (1966), since he made the more than bold assumption that the size of the area cultivated was a function merely of household size and age composition or the consumer/worker ratio. According to Patnaik (1979), a fervent critic of Chayanov, this is the assumption of an homogeneous peasantry with equal access to means of production. What Chayanov forgets to mention is that while family size only trebles or so, the size of a farm in pre-revolutionary Russia could increase by factors as high as a hundred or even much more (Patnaik 1979:380). The consumer/worker ratio of the household explains the size of land holdings as poorly as it does time use or land use intensity in most rural societies. However, *the ratio of household time resources to household land resources* is a factor which within a *certain range* can explain the degree of land use intensity as well as that of time use for cultivation activities. This ratio also incorporates the class dimension, if the *size distributions* for all land holdings of households in the region or local society are taken into account. Class is, needless to say, a matter of distribution.

What we thus find in many agricultural societies with a general limit to land in the region (so that farmers cannot simply increase their holdings by spatial expansion) and where there is an uneven distribution of land between holders, is the previously described tendency for households with relatively *small holdings* to cultivate these *more intensively.* By contrast, households with large holdings of land work these less intensively, *ceteris paribus.* This is, in fact, homologous to how an entire population in a large region would use land less intensively than the same population living in a small region, other things equal.

There are limits, however, to how intensively a household can apply its joint time resources to land. For one thing, there is an upper limit to the time resources of a given size household, regardless of how much land this group holds. If the household owns very large areas of land, its time resources will not suffice to use all this land intensively. In this case it would probably apply its labour-time to the most naturally productive land, albeit tempered by locational considerations so that a time-saving pattern of using the most distant land least intensively would emerge (cf p. 211-20, 228-35). The household with extremely little land, on the other hand, would cultivate this all the more intensively. It would certainly do so in the sense of applying more human time per space-time unit (definition 3, p. 349), but also in the sense of minimizing fallow (definition 1, p. 348), most probably in combination with a more time demanding cropping pattern that gives a higher *output* per space-time unit. Moreover, it is also likely that smallholder households apply more human time per capita, i.e. work at a higher temporal intensity (definition 2, p. 348), since by having less land, the household members can neither choose land quality so freely (i.e. concentrate on the best land), nor choose a cropping pattern which gives the best yield per unit of time put in (cf how the Ladinos in Panajachel could chiefly grow coffee, for instance; p. 384-5). The households poor in land have to choose crops that yield most per unit of land, although this implies that they will have to work relatively more and at a lower productivity of time than larger landholders. But rather than stretching out certain activities with extremely low productivity to the utmost, small landholders or landless households will offer their time to households having more land, thereby expanding the time resources of the latter.

This leads on to Chayanov's thoroughly unrealistic assumption that households produce for own use only. In practically all societies there is exchange of land, human time or products of human time (e.g. food) in one form or another, with or without money as a medium of exchange.* In most peasant societies, households with a shortage of land tend to exchange their time so that they get access to either more land or more produce from land. Households who control more land than they are able to or want to cultivate at a given level of own effort and time use intensity, can either *hire in more labour-time* (and thereby extend their time resources without taking on the broader social responsibilities thereof), or else they can *hire out their land.* (Selling their land to those with less of it, would, of course, alter the situation under discussion.) However, households controlling relatively large land holdings are in the stronger position. If they choose to lease it on a sharecropping basis, they can get produce without doing the necessary work. This obviously raises the productivity of their time, just as it reduces it in the household renting land. The poor farmers thus get less for their labour-time. An alternative strategy of the large landholder is to hire labour instead. The people hired will then have to be supervised which may be rather time-consuming unless the owners supervise only other supervisors (a time-saving hierarchization of relations found in many social systems). This is the model chosen by absentee landlords.

---

* Sahlins (1972:78) says that 'the issue is not the social composition of work' and that 'cooperation remains for the most part a technical fact.' It does not violate his assumptions in the DMP-model, he maintains, since cooperation is not a 'production structure with its own finality'. However, since intensification is a question of input (not output and 'finality'), much labour is *de facto* exchanged between households through these forms of 'technical' cooperation. It is not unlikely that households short in labour are able to compensate this to a fair extent by participating in time exchanges of this kind.

The administrative-cum-supervisory tasks can also be delegated to some relative, perhaps a grown up son. In the latter case, the higher intensity of work is shifted from the parent generation to the successive generation.

Given a very uneven distribution of land among households, small landowning households would want more land, and often invest in children to enlarge their time resources for work as a means to acquire more land. This was the self-defeating policy in the long run (as discussed earlier on p. 401), although it was rational for the parents in the short term of a few years. Households with large land holdings also invest in more children, since they have so much land to work. Hence, both in situations of land abundance and shortage, an expansive household policy is equally attractive for the parent generation (cf p. 395-401). However, 'natural reproduction' is a slow method of expansion; it has its biological limits in monogamous marriages to which are added social limits in societies permitting polygynous marriages. On the other hand, expanding household time supply by hiring people allows even faster expansion and one that is very flexible over time as well.

In many societies, particularly in contemporary Third World countries, those who are rich and hold land of far above average size are in a better position to produce above their subsistence needs and to invest in technical facilities that raise the productivity both of their time and their land. However, they are also in a better position to invest in additional land. This implies that a situation of land shortage can develop at a faster rate than the general rate of population growth, so even when the latter rate is high, the development of land shortage for an increasing number of households can develop still faster. The land shortage felt at the household level is just as much a function of land distribution as overall 'population pressure'. One reason is that when mechanisms of land allocation are released from the *timing* of events in the *population system*, such as birth, initiation, marriage, divorse, death, or similar occasions of land transfer which occur at very *low frequencies* and are instead tied to *highly frequent* events and transactions in the *activity system*, the number of occasions for reallocation increases a great deal. The development of a market for land ties land allocation to the activity system, and tendencies of land concentration may accelerate. So while class at a given point in time reflects land holding and helps to explain land and time use intensities at that particular point in history, the way class structure is generated and reproduced as well as the way the entire system undergoes structuration require us to look into the dynamics of action, interaction and transaction. And this in turn requires an understanding of the *capacity to act, interact and transact.*

INTENSIFICATION THEORY: SOME FINDINGS AND COMPLICATIONS

Chapter 7 concluded that *land use intensification* is a more complex process than a straightforward outcome of population growth, and that there were many other dimensions to this general process. The time-space structure of the settlement system, for instance, had to enter the picture, as did other sources of demand for land, as well as various factors affecting land distribution among holders in the population (cf p. 292-300).

At the end of Chapter 9, we are similarly forced to conclude that the process of *time use intensification* is a much more complex one than a linear increase in work-load as societies evolve towards more land use intensive forms of technology. The six cases presented in Figures 9:6 to 9:11 together with the Bushman example do not, of course, suffice to test the hypothesis of increasing temporal intensity of 'work', as formulated by Boserup (1965), Sahlins (1972), Wilkinson (1973) or Harris (1977), nor is this indeed our main objective, since it is debatable whether this broad hypothesis is testable without a good deal of further qualification, and above all, whether it is fruitfully formulated as it stands.

Essentially the hypothesis of increasing work-load deals with technology only. Each society *as a whole* is classified according to the set of activities in which the dominant *food production technology* is applied, e.g. food foraging, shifting cultivation, short-fallow cultivation, irrigated cultivation, and pastoralism. What is evident from the materials presented in previous chapters, however, is that in the majority of empirical cases, we find a *mixture of different technologies* or a *combination of sub-types,* each having its own possibilities and constraints, and we illustrated their joint articulation and integration by means of time-space structural models. Populations referred to as 'agricultural' may thus practise hunting, gathering or fishing as well as pastoral activities, just as they may combine agricultural technologies operating at different levels of land use intensity. Societies also differ with respect to a crucial technology such as the *mode of transport.* As was demonstrated earlier on by means of prisms, transport technology has an immediate impact on local carrying capacity, as well as the capacity for daily action and interaction in general. So even when rural societies are ordered on a scale of increasing land use intensity, we cannot realistically define societies by their dominant food production technology. Given further that the activities of food production do not even quantitatively dominate in the population time-budget, to base a time allocation theory on the *ideal or pure types* taken from a macro-oriented typology of cultural evolution is not a viable proposition.

The many case studies presented in this book indicate clearly that while there are *very wide variations in land use intensities* for the societies discussed, from the Bushman case in Chapter 3 to Panajachel in Chapter 9, *the range of variation in time use intensity for food production activities is not at all that wide.* If we look at the cases presented roughly according to increasing (average) land use intensity in Figures 9:6 to 9:11, their food production work-loads conform anything but well to the hypothesis of increasing time use intensity (outlined graphically in Figure 9:4). This mounts to the conclusion that time use intensity or work-load is not at all well explained by food production technology alone. If we further take into account that a society generally uses a mix of different technologies and go on to assess their joint effects, thereby defining work much more broadly, we are in a better position, but even then, many factors of time allocation intensity do not follow from technology. The increasing work-load hypothesis must be further qualified in a number of ways relating to three groups of factors:

1) the time-space distribution of activities, settlement structure, prism constraints and movement possibilities;

2) factors affecting the productivity of human time; and

3) factors associated with the distribution and exchange of resource inputs as well as produce between holders.

If we start with the first group of factors related to settlement structure and time-space distributions, these were analysed at depth in Chapters 3 to 7. When we take the time for movement to and from fields, for instance, this factor is usually not considered in intensification theory; travel is simply included in food production activity, although the contribution of travel to the overall work-load depends on the structure of the settlement system. This system, however, is not structured merely by the dominant food production technology but by many other factors, such as whether settlement is clustered or dispersed, the size of individual settlement units, the degree of consolidation of individual holdings, the impact of the defence factor, the location of ritual activities and their relative importance, distances between dwellings and water sources, and so forth. The general point to be made is that if, for instance, the women in one society are found to work an hour more per day in food production than in another society, this need not be attributed to differences in food production technology or land use intensity. It may equally likely be a function of settlement structure unless we have calibrated for this particular factor.

The second group of factors affecting overall work-load relate to the *productivity of human time,* i.e. the relations between time input and output. The thesis of Boserup (1965), Wilkinson (1973) and Harris (1977) is that as technologies become more land use intensive or industrial, there is a *decline in productivity* per person-hour which has to be offset by people working longer to reach the same level of output as before intensification (and the increased level of environmental pressure). However, in comparing different societies-habitats, the many factors of productivity vary a great deal. The amount and distribution of rainfall, quality of soil, yield per hour of input for different crops, the mix of crops grown, productivity of sub-types of technology, or mix of agricultural and pastoral activities, etc.; all these factors influence the productivity of time. Where productivity is higher, the same or higher output can be achieved at lower levels of time use intensity. (Cf, for example, the use of root crops, p. 201.)

The productivity of human time is thus generally a function of the other resources with which time is combined which has implications in terms of synchronized economizing in resources. A greater input of water or auxilliary energy (e.g. by the use of draught animals), may raise the average productivity of human time, for example. Even though animal husbandry also demands human time, it is often the efficiency with which peaks can be cut in the agricultural cycle that determines *total output* of crops, and hence the *average productity* of human time.

If we dare draw some general conclusion from the few cases presented in Chapter 9, it would be that as mean land use intensity increases, it is not the work-load in food production that goes up but rather that in other sectors of activity such as domestic work, trade and wage labour outside agriculture. (Increased wage-labour in agriculture is more a function of distribution between population segments although this also involves differential productivity of time.) The high level of time input in food preparation in Char Gopalpur, Kali Loro and Panajachel may indicate that the choice of relatively high yielding cereals such as rice and maize, leads to higher time demand in food preparation, e.g. due to the tedious grinding of grain prior to daily cooking or baking which reflects an improductive domestic technology, the time costs of which are paid by the women.

By and large, Boserup (1965), Wilkinson (1973) and Harris (1977) tend to overemphasize the decline in productivity of time in their hypotheses of increasing work-load. Harris (1977), when comparing preindustrial societies to present-day industrial societies, argues that it is *higher productivity* which leads to high time use intensities for work. He does not, however, adequately consider the overall *increase in per capita output that follows from industrialization,* although, admittedly, some of this output is diverted from the household sector towards capital-intensification in other sectors, a topic we will return to in Volume 2, where we also examine the extent to which work-load increases with industrialization (cf also Minge-Klevana 1981).

Intensification theory is also compounded by a third group of factors pertaining to the *distribution* of inputs and products *among holders in the population*. Intensification theory must not only consider that a society is segmented into a set of holders who control resources at a given moment in time, but also that these holders engage in various activities of exchange and transaction which reallocate both resource inputs (including human time) and products (products of human time). From the holochronic perspective, distribution is an *ongoing process*, not a state of affairs. Chayanov's kind of assumption that households are isolated and produce only for own use by own means can at best serve as an initial point of departure, but it must soon be abandoned if one wants to tackle the complex forms of interactions that take place in societies-habitats. At the household level assumptions of *homogeneity* also fall short of real world conditions, as we pointed out with respect to differences in time use intensity between women and men, just as these differences cannot be explained solely by reference to technology. At the local or regional level, assumptions that all households are equal in terms of resources controlled are unrealistic, especially so in the analysis of historical as well as contemporary *peasant* societies. Regardless of the absolute availability of resources in a region or locality, demand pressures on human time are mediated by all sorts of social mechanisms, from the exchange systems in which 'big-men' occupy key roles to systems of taxation imposed by the state to extract 'surpluses'. Commercialization, money and market exchange systems also lead to intensification, both in the production of a 'surplus' for sale above own subsistence needs and in the production and sale of specialized commodities that give a higher productivity of time and/or land. Alternatively, own labour-time resources may be sold. In either case, productivity and intensity of time is strongly affected by the terms of trade or price levels.

We may thus conclude that when intensification theory is broadened by the many dimensions discussed above, it emerges as a more *general theory of resource mobilization* in society-habitat, a theory that is of great relevance to the study of structuration, development and evolution.*

---

*   A geographer who has recently done a great deal of work on population and intensification theory and who has generalized the Boserup model is Grigg (1980).

AFTERWORD

A NOTE ON SCARCITY, STRUCTURATION AND THE CAPACITY FOR
ACTION AND INTERACTION

There are many reasons why time resources have been neglected in
the social sciences. In sociology and particularly in social and cultural
anthropology, one reason is that the perspectives of resources and
action have not always been adequately interrelated, another is the
devious notion of scarcity, a concept basic to what I have chosen to
call the 'economistic paradigm'. While economists have conventionally studied *how* people allocate 'scarce resources', many anthropologists have questioned *whether* pre-industrial peoples allocate
'scarce resources' or 'economize' in resources (cf Frankenberg 1967;
LeClair and Schneider, eds. 1968). Initially we rejected many of the
conventional concepts and assumptions inherent in the economistic
paradigm, such as production and consumption; land, labour and
capital; goods and services; and so on (cf Chapter 2 on living and
activity/action possibility boundaries). The rejection thereof is
probably one reason why some readers will find this book awkward
or 'different'. Like most true paradigms in the Kuhnian sense, the
economistic paradigm is taken for granted. Its concepts have so
thoroughly infiltrated social science and every-day language that
these categories are regarded as indispensable and priceless, i.e.
without alternatives and opportunity costs. I do not believe this to
be so and reject many of these concepts at the level of ontology.

The extension of the economistic paradigm as regards 'means,
ends, scarcity and choice' has given rise to its particular obfuscations
illustrated in the debate between formal and substantive schools of
thought on matters economic. As was argued from Chapter 1 and
onwards, resources do not have to be 'scarce' in the conventional
sense to be allocated. All action is perforce limited and selective and
from the time-geographic perspective with its logistic form of logic,

allocation as well as resource utilization is an ontological condition of all action, interaction and transaction. It is the production bias, i.e. that production is a precondition for consumption, and that production is a problem while 'consumption' is not, that is at the root of the obfuscation of human time as a general resource for action and interaction. The insersion of labour — a mere 'part-time' activity — as a substitute concept for human time in general has done much to *prevent integrative approaches in social science and the development of more solid and comprehensive theories of action, interaction and system dynamics and structuration.*

One scholar, however, who has expressed a very broad-minded statement on 'scarcity' which in essence (albeit not in format) is very much in line with the general philosophy of the present study, is Brookfield (1975) in 'Interdependent development':

> ... Resources, even though they are a creation of human perception and technology, exist also in a finite material base. The quantity of human labour power is limited. Space is limited... if scarcity exists then the development problem becomes, at base, a problem of adaptation and allocation at all times and in all places... scarcity is also compounded by the absolute or relative immobility of resources. Social systems whether based on reciprocity, redistribution or market exchange ... *all* exist to mediate this immobility and adapt to absolute scarcity ...
>
> ... scarcity itself is an inherent condition, the effects of which are capable only of amelioration and not of removal... If scarcity did not exist there would be no limits to growth /or the rate of growth, one might add/ ... If the constraints of absolute material scarcity and relative scarcity in the dimensions of time and space were absent, then Utopian solutions would be feasible...
>
> The uneven distribution of scarce resources underlies the whole development problem ... Any redistributive system must have its nodes ... /and the/ ... holders of these nodes have become dependent on the network and its flows, but have compensated this dependence by acquiring control over the allocation of scarce resources and production — that is, power...
>
> ... /socio-politically/ induced scarcity and real /material/ scarcity are not opposites; the former is merely the exploitation and intensification of the latter... It is only the fact of scarcity that makes it possible to exploit scarcity... (Brookfield 1975:205-7)

Symptomatically these words are written in Brookfield's final chapter labelled 'A conclusion that is an introduction'. Although it did not serve as the introduction to the present volume, in all systems in motion and evolution, a conclusion, product or sedimented structure is always an introduction to the next temporal layer to be sedimented in space. It is precisely in the analysis of motion, evolution, development and structuration that the capacity for action of human beings as well as other organisms and agents in nature is of crucial importance.

If we want to understand how components and systems constituting society-habitat *unfold, develop or evolve* over time, we must have clear and profound ideas about the capacity of individuals and populations for *action and interaction* in relation to time and space. Any form of interaction involving human individuals and other populations, living or non-living in the ecotechnological sense defined on p. 8, has to be examined in terms of *process and flow*. The contention made is that this is better accomplished in studies that are *holochronic*. This implies the adoption of a broader *activity systems approach* to society-habitat and a time-space conceptualization in which *structure* and *process* are inseparably fused. Structure then becomes the structure and structuring of process, and structural relations are expressed in terms of flows in space and time, the chosen ontology being that of *matter and systems in motion*. In this way the contradictions or obfuscations inherent in ontologies where time is first abstracted and then reified can be avoided.

In this study, human time has been defined as a resource because all human beings have *limited capacity to act* in relation to a given unit of time. The implication is that there are limits to the generation and articulation of social or environmental phenomena. All kinds of institutions or organizations, for instance, produce and are reproduced within limits, and all structural outcomes from given situations of origin are not equally feasible. Human projects are not realized out of nothing; they thrive on action and action draws on resources, among them human time which we have defined as a *real resource*. In the many models and sub-models presented in this book to interrelate different real-world systems, the problem of coping with process led us to regard resources not only in real terms but also in *real-time* terms. Land was thus defined as *space-time* just as other resources were treated as real-time resources, for example, water-time, animaltime and energy ('power-time', as it were, expressed in, say, Kilowatt-hours). One type of interaction which thereby becomes susceptible to processual study is the *synchronized allocation* and economizing in several different resources put into human projects; an idea that can be generalized to the analysis of animal ecology or processes in nature as well. So when 'time resources' are referred to in the title of this study, we are not only alluding to human time but also to other real-time resources.

However, the condition that some resources sometimes are not 'fully used' does not imply that resources are not limited or subject to allocative mechanisms, nor does it mean that resources are actually and automatically *accessible* in the *specific time-space locations and situations* where they are ingredients in human projects. When resources are tied to their *physical carriers, media or vehicles* and when the latter are expressed as paths in time-space, it can be readily seen that due to constraints on divisibility and coupling, the limits to path allocation make it impossible to use all resources equally and fully due to different displacement effects. One of the main objectives of this first volume has been to illustrate this logic of path allocation as it impinges on human action, resource utilization, the structure of different systems found in society-habitat and the structuration of these systems over time. This is a theme we will return to in Volume 2 of this study, in which a number of important time-space structures can be illustrated better by the use of materials from industrial societies.

BIBLIOGRAPHY

Abler, R., Adams, J. and Gould, P. (1971) *Spatial Organization: The Geographer's View of the World.* Englewood Cliffs: Prentice Hall.

Allan, W. (1949) Studies in African Land Usage in Northern Rhodesia. *Rhodes—Livingstone Institute Paper* 15. Manchester.

Anderson, J. (1971) Space-time budgets and activity studies in urban geography and planning. *Environment and Planning* 3:353-68.

Barth, F. (1956) Ecologic relationships of ethnic groups in Swat, North Pakistan. *American Anthropologist* 58:1079-89.

——(1959a) *Political Leadership among Swat Pathans.* London: Athlone Press.

——(1959b) The land use pattern of migratory tribes in South Persia. *Norsk Geografisk Tidskrift* 17:1-11.

——(1961) *Nomads of South Persia: The Basseri Tribe of the Khamseh Confederacy.* Boston: Little, Brown and Co.

——(1966) Models of social organization. *Royal Anthropological Institute Occasional Paper* 23. London.

——(1967a) 'Economic spheres in Darfur,' in *Themes in Economic Anthropology.* Ed. by R. Firth. London: Tavistock.

——(1967b) On the study of social change. *American Anthropologist* 69:661-68.

Bayliss-Smith, T.P. (1977) 'Energy use and economic development in Pacific communities,' in *Subsistence and Survival: Rural Ecology in the Pacific.* Ed. by T.P. Bayliss-Smith and R.G. Feachem. London: Academic Press.

Becker, G. S. (1965) A theory of the allocation of time. *Economic Journal* 75:493-517.

Belshaw, C.S. (1954) The cultural milieu and the entrepreneur: A critical essay. *Explorations in Entrepreneurial History* 7:144-146.

Bharadwaj, K. (1974) *Production conditions in Indian agriculture: A study based on farm management surveys.* (University of Cambridge Department of Applied Economics, Occasional Paper 33.) Cambridge: Cambridge University Press.

Blaikie, P.M. (1971) Spatial organization of agriculture in some north Indian villages. *Institute of British Geographers, Transactions.*

Bohannan, P. and Bohannan, L. (1968)   *Tiv Economy*. London: Longmans.

Boserup, E. (1965) *The Conditions of Agricultural Growth: The Economics of Agrarian Change under Population Pressure.* London: George Allen and Unwin.

—— (1970) *Woman's Role in Economic Development.* London: Allen and Unwin.

Bottrall, A. (1978) 'The management and operation of irrigation schemes in less developed countries,' in *The Social and Ecological Effects of Water Development in Developing Countries.* Ed. by G. Widstrand. Oxford: Pergamon Press.

Briscoe, J. (1979) Energy use and social structure in a Bangladesh village. *Population and Development Review* 5:615-639.

Brookfield, H. C. (1962) Local study and comparative method: an example from central New Guinea. *Annals of the Association of American Geographers* 52:242-54.

—— (1969) The environment as perceived. *Progress in Geography* 1: 52-80.

——, ed. (1973) *The Pacific in Transition: Geographical Perspectives on Adaptation and Change.* London: Edward Arnold.

—— with Hart, D. (1971) *Melanesia: A Geographical Interpretation of an Island World.* London: Methuen.

—— (1975) *Interdependent Development.* London: Methuen.

—— and Brown, P. (1963) *Struggle for Land: Agriculture and Group Territories among the Chimbu of the New Guinea Highlands.* Melbourne: Oxford University Press.

Brown, P. and Brookfield, H.C. (1967) Chimbu settlement and residence: a study of patterns, trends and idiosyncrasy. *Pacific Viewpoint* 8:119-51.

Brown, P (1973) *The Chimbu.*

Buck, J. L. (1930) *Chinese Farm Economy: A Study of 2866 Farms in Seventeen Localities and Seven Provinces in China.*   (Doctoral Dissertation.) Chicago.

—— (1937) *Land Utilization in China.*

Bylund, E. (1956) Koloniseringen av Pite Lappmark t.o.m. år 1867. *Geographica* 30. Uppsala.

Cain, M.T. (1977) The economic activities of children in a village in Bangladesh. *Population and Development Review* 3:201-227.

Carlstein, T. (1966) *Födoaktivitetens och bosättningsmönstrets återspegling i individernas tidrumsbudget på Raroia i föreuropeisk tid.* (The impact of food activities and residential pattern on the time-space budget of individuals on Raroia island in pre-European times.) Mimeo. Department of Geography, University of Lund.

Carlstein, T. (1972) Regional or Spatial Sociology? Mimeo. Department of Geography, University of Lund.
Also in *Social Issues in Regional Policy and Regional Planning*. Ed. by A. R. Kuklinski. The Hague and Paris: Mouton. 1977.

—— (1973) *Population, Activities and Settlement as a System: The Case of Shifting Cultivation*. Mimeo. Department of Geography, University of Lund.

—— (1974) *Time Allocation*. Mimeo. Department of Geography, University of Lund.

—— (1978a) *Time Allocation, Innovation and Agrarian Change: On the Capacity for Human Interaction in Space and Time*. Mimeo. Department of Geography, University of Lund. 365 p.

—— (1978b) 'Innovation, Time Allocation and Time-Space Packing.' in *Human Activity and Time Geography*. Ed. by Carlstein, T. Parkes, D. and Thrift, N. London: Edward Arnold.

—— (1978c) *Steel Axes as a Time-Saving Innovation*. Mimeo. Department of Geography, University of Lund. 48 p.

Carlstein, T., Lenntorp, B. and Mårtensson, S. (1968) Individers dygnsbanor i några hushållstyper. *Urbaniseringsprocessen* 17. Department of Geography, University of Lund.

Carlstein, T., Parkes, D. and Thrift, N. eds. (1978) *Making Sense of Time*. Timing Space and Spacing Time I. London: Edward Arnold.

—— (1978) *Human Activity and Time Geography*. Timing Space and Spacing Time II. London: Edward Arnold.

—— (1978) *Time and Regional Dynamics*. Timing Space and Spacing Time III. London: Edward Arnold.

Carlstein, T. and Thrift, N. (1978) 'Afterword: Towards a time-space structured approach to society and environment,' in *Human Activity and Time Geography*. Ed. by T. Carlstein, D. Parkes and N. Thrift. London: Edward Arnold.

Carneiro, R. L. (1960) 'Slash-and-Burn Agriculture: A Closer Look at its Implications for Settlement Patterns,' in *Men and Cultures*. Ed. by A. F. C. Wallace. Philadelphia: University of Pennsylvania Press.

—— (1968) 'Slash-and-Burn Cultivation among the Kuikuru and its Implications for Cultural Development in the Amazon Basin,' in *Man in Adaptation: The Cultural Present*. Ed. by Y. Cohen. Chicago: Aldine.

Chapin, F. S. Jr. (1965) *Urban Land Use Planning*. Urbana: University of Illinois Press.

—— (1974) *Human Activity Patterns in the City: Things People Do in Time and Space*. New York: John Wiley and Sons.

Chapman, M. (1970) *Population Movement in Tribal Society: The Case of Duidui and Pichahila, British Solomon Islands*. Doctoral Dissertation, University of Washington.

Chayanov, A. V. (1966) *The Theory of Peasant Economy.* Homewood, Ill.: Richard D. Irwin for the American Economic Association. /Originally published in Russian in 1925./

Chisholm, M. (1962, 1972) *Rural Settlement and Land Use.* London: Hutchinson University Library.

Christaller, W. (1933) *Die zentralen Orte in Süddeutschland.* Jena. (Transl. by C. W. Baskin as *Central Places in Southern Germany.* Englewood Cliffs, N.J.: Prentice Hall.

Christiansen, S. (1975) *Subsistence on Bellona Island (Mungiki): A Study of the Cultural Ecology of a Polynesian Outlier in the British Solomon Island Protectorate.* Copenhagen: C.A. Reitzels.

— — (1978) Infield-outfield systems — Characteristics and developments in different climatic environments. *Geografisk Tidskrift* 77:1-5. Copenhagen.

Clark, C. and Haswell, M. R. (1964) *The Economics of Subsistence Agriculture.* London: Macmillan.

Clarke, W. C. (1966) From extensive to intensive shifting cultivation: A succession from New Guinea. *Ethnology* 5:347-59.

Clarkson, J.D. (1968) Ecologic and spatial analysis — Towards adaptive research in the Developing Countries. *Social Science Research Institute Working Papers* 7. University of Hawaii, Honolulu.

Cleave, J.H. (1974) *African Farmers: Labour Use in the Development of Smallholder Agriculture.* New York: Columbia University Press.

Conklin, H. C. (1957) Hanunoo agriculture: A report on an integral system of shifting cultivation in the Philippines. *FAO Forestry Development Papers* 12. Rome.

— — (1961) The study of shifting cultivation. *Current Anthropology* 2:27-61.

Connell, J. and Lipton, M. (1973) Assessing village labour situations in developing countries. *Institute of Development Studies Discussion Paper* 35. Brighton. Also publ. 1977 by Oxford University Press, Delhi.

Coon, C. S. (1976) *The Hunting Peoples.* Harmondsworth: Penguin.

Cressey, G. B. (1958) Qanats, Karez and Foggaras. *The Geographical Review* 48:27-44.

Dahl, G. and Hjort, A. (1976) *Having Herds: Pastoral Herd Growth and Household Economy.* University of Stockholm, Stockholm Studies in Social Anthropology.

— — (1979) Pastoral Change and the Role of Drought. *SAREC Report* R2:1979. Stockholm: Swedish Agency for Research Cooperation with Developing Countries.

Djurfeldt, G. and Lindberg, S. (1975) *Behind Poverty — The Social Formation in a Tamil Village.* (Scandinavian Institute of Asian Studies Monograph Series 22.) Lund: Studentlitteratur/London: Curzon Press.

Douglas, M. (1967) 'Primitive rationing: A study in controlled exchange,' in *Themes in Economic Anthropology*. Ed by R. Firth. London: Tavistock.

—— (1962) 'Lele economy as compared with the Bushong,' in *Markets in Africa*. Ed. by G. Dalton and P. Bohannan. Evanston: Northwestern University Press.

—— (1963) *The Lele of the Kasai*. London: Oxford University Press.

Douglas, M. and Isherwood, B. (1979) *The World of Goods: Towards an Anthropology of Consumption*. London: Allen Lane.

Dyson-Hudson, N. (1966) *Karimojong Politics*. Oxford: Clarendon Press.

Dyson-Hudson, N. and R. (1970) 'The food-production system of a semi-nomadic society: The Karimojong, Uganda,' in *African Food Production Systems: Cases and Theory*. Ed. by P. F. M. McLoughlin. Baltimore and London: The John Hopkins Press.

Eggan, F. (1966) *The American Indian: Perspectives for the Study of Social Change*. Chicago: Aldine.

Eldblom, L. (1968) *Structure foncière — organisation et structure sociale. Une étude comparative sur la vie socio-économique dans les trois oasis libyennes de Ghat, Mourzouk et particulièrement Ghadamès*. (Meddelanden från Lunds Universitets Geografiska Institution, Avhandlingar LV.) Lund: Uniskol.

Elder, J. W. (1970) 'Rajpur: Change in the Jaimani system of an Uttar Pradesh Village.' in *Change and Continuity in India:s Villages*. Ed. by K. Ishwaran. New York: Columbia University Press.

Ellegård, K. (1977) *Utveckling av transportmönster vid förändrad teknik - En tidsgeografisk studie*. Department of Geography, Lund University. (Mimeo.)

Ellegård, K., Hägerstrand, T. and Lenntorp, B. (1977) Activity organization and the generation of daily travel: Two future alternatives. *Economic Geography* 53:126-52.

English, P. W. (1968) The origin and spread of qanats in the Old World. *Proceedings of the American Philosophical Society* 3.

Epstein, T. S. (1962) *Economic Development and Social Change in South India*. Manchester: Manchester University Press.

—— (1968) *Capitalism, Primitive and Modern: Some Aspects of Tolai Economic Growth*. Canberra: Australian National University Press.

Erixon, S. (1938) Regional European Ethnology II, functional analysis — time studies, *Folkliv* 3:263-94.

Fernea, R. A. (1970) *Shaykh and Effendi: Changing Patterns of Authority Among the El Shabana of Southern Iraq*. Cambridge, Mass.: Harvard University Press.

Firth, R. ed. (1967) *Themes in Economic Anthropology*. London: Tavistock.

Forde, C. D. (1934) *Habitat, Economy and Society: A Geographical Introduction to Ethnology.* London: Methuen.

Foster, G. (1948) *Empire's Children: The People of Tzintzuntzan.* Mexico, D.F./Washington, D.C.: Smithsonian Institution, Institute of Social Anthropology, Publication 6.

Freeman, D. (1955) *Iban Agriculture.* (Colonial Research Studies 18.) London: Her Majesty's Stationery Office.

——(1970) *Report on the Iban.* (London School of Economics, Monographs on Social Anthropology.) London: Athlone Press.

Friedman, Jonathan (1972, 1979) *System, Structure and Contradiction in the Evolution of »Asiatic» Social Formations.* (Social Studies in Oceania and South East Asia 2.) Copenhagen: The National Museum of Denmark.

——(1974) Marxism, structuralism and vulgar materialism. *Man* 9: 444-469.

Frödin, J. (1930) 'Svenska fäbodar: Övergångsformer inom vårt fäbodväsen,' in *Svenska kulturbilder* 3. Ed. by S. Erixon and S. Wallin. Stockholm: Skoglunds.

Geertz, C. (1963) *Agricultural Involution: The Process of Ecological Change in Indonesia.* Berkeley and Los Angeles: University of California Press.

Ghez, G. R. and Becker, G. S. (1975) *The Allocation of Time and Goods over the Life-Cycle.* National Bureau of Economic Research. New York: Colombia University Press.

Giddens, A. (1979) *Central Problems in Social Theory: Action, Structure and Contradiction in Social Analysis.* London: Macmillan.

Godelier, M. (1969)   La monnaie de sel des Baruya de Nouvelle-Guincé. *L'Homme* 9:5-37.

——(1972) *Rationality and Irrationality in Economics.* London: Monthly Review Press.

——(1975) 'Modes of Production, Kinship and Demographic Structures,' in *Marxist Analyses and Social Anthropology.* Ed. by. M. Bloch. London: Malaby Press.

Goody, J. ed. (1958) *The Developmental Cycle in Domestic Groups.* Cambridge: Cambridge University Press.

Gray, R. F. (1963) *The Sonjo of Tanganyika: An Anthropological Study of an Irrigation-Based Society.* London: Oxford University Press.

Gregory, D. (1978a) 'Social Change and Spatial Structures,' in *Making Sense of Time.* Ed. by T. Carlstein, D. Parkes, and N. Thrift. London: Edward Arnold.

——(1978b)   *Ideology, Science and Human Geography.* London: Hutchinson.

Grigg, D. B. (1976)   Population Pressure and Agricultural Change. *Progress in Geography* 8:133-176.

—— (1980) 'Migration and overpopulation,' in *The Geographical Impact of Migration*. Ed. by P. White and R. Woods. London and New York: Longmans.

Grove, A. T. (1961) 'Population densities and agriculture in Northern Nigeria,' in *Essays on African Population*. Ed. by K. M. Barbour and M. Prothero. London.

Gullstrand, R. (1972) *La localisation des lieux d:habitat et de la population rurale, Changement 1890-1970*. Unpublished seminar paper delivered at conference in Lund, Sweden 1972.

Hägerstrand, T. (1947) En landsbygdsbefolknings förflyttningsrörelser. Studier över migrationen på grund av Asby sockens flyttningslängder 1840–1944. *Svensk Geografisk Årsbok* 23:114-42.

—— (1953) *Innovationsförloppet ur korologisk synpunkt*. (Meddelanden från Lunds Universitets Geografiska Institution, Avhandlingar, 25.) Lund. Transl. 1967 by A. Pred. *Innovation Diffusion as a Spatial Process*. Chicago: Chicago University Press.

—— (1957) 'Migration and area,' in *Migration in Sweden*. Ed. by D. Hannerberg, T. Hägerstrand and B. Odeving. (Lund Studies in Geography, Series B.) Lund: C. W. K. Gleerup.

—— (1963) 'Geographic Measurements of Migration: Swedish Data,' in *Human Displacements: Measurement Methodological Aspects*. Ed. by J. Sutter. Monaco.

—— (1970) What about people in Regional Science? *Papers of the Regional Science Association* 24:7-21.

—— (1972) The impact of social organization and the environment upon the time use of individuals and households. *Plan (International)* Stockholm.

—— (1973a) *The Impact of Transport on the Quality of Life*. Fifth International Symposium on Theory and Practice in Transport Economics. European Council of Ministers of Transport. Athens.

—— (1973b) 'The domain of Human Geography,' in *Directions in Geography*. Ed. by R. Chorley. London: Methuen.

—— (1974a) On socio-technical ecology and the study of innovations. *Ethnologica Europaea* 7:17-34.

—— (1974b) 'Ecology under one perspective,' in *Ecological Problems of the Circumpolar Area*. Ed. by E. Bylund, H. Linderholm, and O. Rune. Luleå: Norrbottens Museum.

—— (1975b) 'Space, Time and Human Conditions,' in *Dynamic Allocation of Urban Space*. Farnborough: Saxon House.

—— (1978a) 'Survival and arena: On the life-history of individuals in relation to their geographical environment,' in *Human Activity and Time Geography*. Ed. by T. Carlstein, D. Parkes and N. Thrift. London: Edward Arnold.

—— (1978b) 'A note on the quality of life-times,' in *Human Activity and Time Geography*. Ed. by. T. Carlstein, D. Parkes and N.

Thrift. London: Edward Arnold.

Haggett, P. (1965).   *Locational Analysis in Human Geography.* London: Edward Arnold.

—— (1972) *Geography – A Modern Synthesis.* New York: Harper and Row.

Haggett, P., Cliff, A. D. and Frey, A. (1977) *Locational Analysis in Human Geography.* (2nd rev. ed.) London: Edward Arnold.

Harris, M. (1971) *Culture, Man, and Nature: An Introduction to General Anthropology.* New York: Thomas Y. Crowell Co.

—— (1977) *Cannibals and Kings: The Origins of Cultures.* New York: Random House.

Hawley, A. (1950) *Human Ecology.* New York: Ronald Press.

Herskovits, M. J. (1952) *Economic Anthropology: The Economic Life of Primitive Peoples.* New York: Norton.

Highsmith, R. M. (1961) *Case Studies in World Geography: Occupance and Economy Types.* Englewood-Cliffs: Prentice Hall.

Hindess, B. and Hurst, P. Q. (1975) *Pre-capitalist Modes of Production.* London: Routledge and Kegan Paul.

Ho, T.J. (1979) Time costs of child rearing in the Rural Philippines. *Population and Development Review* 5:643-62.

Hultblad, F. (1968) *Övergång från nomadism till agrar bosättning i Jokkmokks socken.* (Meddelanden från Uppsala Universitets Geografiska Institutioner, Ser. A 230.) Stockholm: Almqvist och Wiksell.

Hunter, G. (1976) 'Organizations and Institutions,' in *Policy and Practice in Rural Development.* Ed. by G. Hunter, A. H. Bunting, and A. Bottrall. London: Croom Helm.

Hyden, G. (1972) *Socialism och samhällsutveckling i Tanzania.* Bo Cavefors.

Ingold, T. (1978) The rationalization of reindeer management among Finnish Lapps. *Development and Change* 9:103-32.

Ingold, T. (1980) *The significance of storage in hunting societies.* Conference paper: Troisieme reunion 'conservation des grains'. Levroux, 24-29 novembre 1980. 25 p.

Izikowitz, K. G. (1951) *Lamet – Hill Peasants in French Indochina.* (Etnologiska studier 17.) Göteborg: Etnografiska Museet.

Jahnke, H. E. and Ruthenberg, H. (1976) 'Organizational aspects of livestock development in the dry areas of Africa,' in *Policy and Practice in Rural Development.* Ed. by G. Hunter, A.H. Bunting and A. Bottrall. London: Croom Helm.

Johnson, A. (1975) Time allocation in a Machiguenga community. *Ethnology* 14:301-310.

Johnson, D. (1969) *The Nature of Nomadism: A Comparative Study of Pastoral Migrations in Southwestern Asia and Northern Africa.* University of Chicago, Dept. of Geography Research Paper 18.

Jones, W. O. (1957) Manioc: An example of innovation in African economics. *Economic Development and Cultural Change* 5: 97:117.

—— (1959) *Manioc in Africa.* Stanford: Stanford University Press.

Kamuzora, C.L. (1980) Constraints to labour time availability in African smallholder agriculture: The case of Bukoba district, Tanzania. *Development and Change* 11:123-135.

Kay, G. (1964a) Chief Kalaba's village. *The Rhodes-Livingstone Papers* 35. Manchester: Manchester University Press.

—— (1964b) 'Aspects of Ushi settlement history: Fort Rosebery District, Northern Rhodesia,' in *Geographers in the Tropics Liverpool Essays.* Ed. by R. W. Steel and R. M. Prothero. London: Longmans.

—— (1967) *Social Aspects of Village Regrouping in Zambia.* Institute for Social Research, University of Zambia. (Also in Department of Geography Miscellaneous Series, 7, University of Hull.)

Kemp, W. B. (1971) The flow of energy in a hunting society. *Scientific American* 224:104-15.

Knight, C. G. (1974) *Ecology and Change: Rural Modernization in an African Community.* New York: Academic Press.

Larsson, G. (1947) *Inflytande på avståndet från brukningscentrum till inägojorden på arbetsbehov, driftsformer och driftsresultat.* Stockholm.

Lea, D. (1973) 'Stress and adaptation to change: an example from the East Sepik District, New Guinea,' in *The Pacific in Transition.* Ed. by H. C. Brookfield. London: Edward Arnold.

Leach, E. R. (1954) *Political Systems of Highland Burma: A Study in Kachin Social Structure.* London: G. Bell and Sons.

—— (1959) Hydraulic society in Ceylon. *Past and Present* 15:2-25.

—— (1961a) *Pul Eliya — A Village in Ceylon: A Study of Land Tenure and Kinship.* Cambridge: Cambridge University Press.

—— (1961b) *Rethinking Anthropology.* (London School of Economics Monographs on Social Anthropology 22.) London: Athlone.

Le Clair, E., Jr. and Schneider, H. (1968) *Economic Anthropology: Readings in Theory and Analysis.* New York: Holt, Rinehart and Winston.

Lee, R. B. (1968) 'What hunters do for a living, or, how to make out on scarce resources,' in *Man the Hunter.* Ed. by R. B. Lee and I. Devore. Chicago: Aldine.

—— (1969) '!Kung Bushman subsistence: An input-output analysis,' in *Environment and Cultural Behaviour.* Ed. by A. P. Vayda. New York: Natural History Press.

—— (1979) *The !Kung San: Men, Women and Work in a Foraging Society.* London: Cambridge University Press.

Lee, R. B. and Devore, I. eds. (1968) *Man the Hunter.* Chicago.

Lenntorp, B. (1976) *Paths in Space-Time Environments: A Time-Geographic Study of Movement Possibilities of Individuals.* Lund: Gleerups. (Lund Studies in Geography, Series B. 44.)

—— (1977) A time-geographic approach to individuals' daily movements. *Rapporter och Notiser* 33. Department of Social and Economic Geography, Lund University.

—— (1978) 'A time-geographic simulation model of individual activity programmes,' in *Human Activity and Time Geography.* Ed. by T. Carlstein, D. Parkes and N. Thrift. London: Edward Arnold.

LeVine, R.A. and LeVine, B.B. (1966) *Nyansongo: A Gusii Community in Kenya.* New York: John Wiley.

Lewis, I. M. (1961) *A Pastoral Democracy: A Study of Pastoralism and Politics among the Northern Somali of the Horn of Africa.* London: Oxford University Press.

Lewis, O. (1951) *Life in a Mexican Village: Tepoztlan Restudied.* Urbana: University of Illinois Press.

Lexis, W. (1875) *Einleitung in die Theorie der Bevölkerungsstatistik.* Strassbourg: Trubner.

Linder, S. B. (1970) *The Harried Leisure Class.* New York: Columbia University Press.

Linton, R. (1933) *The Tanala: A Hill Tribe of Madagascar.* (Field Museum of Natural History, Anthropological Series, 22.) Chicago.

—— (1940) A neglected aspect of social organization. *American Journal of Sociology* 45:870-86.

Lipton, M. (1961) Working paper for Myrdal's Asian Drama.

Loficie, M. F. (1978) Agrarian crisis and economic liberalization in Tanzania. *The Journal of Modern African Studies* 16:451-75.

Long, N. (1968) *Social Change and the Individual: A Study of the Social and Religious Responses to Innovation in a Zambian Rural Community.* Manchester: Manchester University Press.

Ludwig, H.-D. (1968) 'Permanent farming on Ukara: the impact of land shortage on husbandry practices,' in *Smallholder Farming and smallholder development in Tanzania.* Ed. by H. Ruthenberg. (IFO-Institut fur Wirtschaftsforschung, Munchen, Afrika-Studien 24.) Munchen: Weltforum.

Lösch, A. (1954) *The Economic of Location.* New Haven: Yale University Press.

Mabogunje, A. (1972) *Regional Mobility and Resource Development in West Africa.* Montreal and London: McGill-Queens University Press.

Macmillan, W. (1978) 'Mathematical Programming Models and the Introduction of Time into Spatial Economic Theory,' in *Time and Regional Dynamics.* Ed. by T. Carlstein, D. Parkes and N. Thrift. London: Edward Arnold.

Mamdani, M. (1972) *The Myth of Population Control: Family, Caste and Class in an Indian Village*. New York: Monthly Review Press.

Manker, E. (1953) *The Nomadism of the Swedish Mountain Lapps: The Siidas and their Migratory Routes in 1945*. Stockholm: Hugo Gebers.

— (1963) *De Åtta Årstidernas Folk*. Göteborg: Tre Tryckare.

Manshard, W. (1974) *Tropical Agriculture: A Geographical Introduction and Appraisal*. London and New York: Longman.

Mårtensson, S. (1977) Childhood interaction and temporal organization. *Economic Geography* 53:99-125.

— (1980) *On the Formation of Biographies in Space-Time Environments*. (Lund Studies in Geography, Series B 47.) Lund: Gleerup.

Mascarenas, A. C. (1977) After villagization — What? *Bureau of Resource Assessment and Land Use Planning, Service Paper* 77/1. University of Dar es Salam. Also in *Towards Socialism in Tanzania*. Ed. by B. Mwansasu and C. Pratt.

Maude, A. (1973) 'Land shortage and population pressure in Tonga,' in *The Pacific in Transition*. Ed. by H. C. Brookfield. London: Edward Arnold.

McCarthy, F.D. and McArthur, M. (1960) 'The food quest and the time factor in aboriginal economic life,' in *Records of the Australian-American Scientific Expedition to Arnhem Land, Vol. 2: Anthropology and Nutrition*. Ed. by C.P. Mountford. Melbourne: Melbourne University Press.

Meier, R. L. (1959) Human time allocation: A basis for social accounts. *Journal of the American Institute of Planners* 15:27-33.

— (1962) *A Communications Theory of Urban Growth*. Cambridge, Mass.: The MIT Press.

Minge-Klevana, W. (1980) Does labour time decrease with industrialization? A survey of time-allocation studies. *Current Anthropology* 21:279-298. /Includes comment by T. Carlstein./

Moore, W. E. (1963) *Man, Time and Society*. New York: Wiley.

Müller-Wille, W. (1936) *Die Ackerfluren im Landesteil Birkenfeld*. (Dissertation.) Bonn.

Mumford, L. (1966) *Technics and Civilization*. New York: Hartcourt, Brace and Co.

Murdock, G. P. (1968) 'The current status of the world's hunting and gathering peoples,' in *Man the Hunter*. Ed. by R. B. Lee and I. Devore. Chicago: Aldine.

Myrdal, G. (1968) *Asian Drama: An Inquiry into the Poverty of Nations*. Vol. 2. New York: Pantheon.

Netting, R. McC. (1968) *Hill Farmers of Nigeria: The Cultural Ecology of the Kofyar of the Jos Plateau*. (The American Ethnological Society, Monograph 46.) Seattle and London: University of Washington Press.

Nilsson, M. P. (1920) *Primitive Time Reckoning*. Lund.

Nulty, L. (1972) *The Green Revolution in Pakistan: Implications of Technological Change.* New York: Praeger.

Odum, H. T. (1971) *Environment, Power and Society.* New York and London: Wiley.

Olsson, G. (1965) *Distance and Human Interaction: A Review and Bibliography.* Philadelphia: Regional Science Research Institute.

Park, R. E. (1934) Human ecology. *American Journal of Sociology* 42:1-15.

Parrack, D. W. (1969) 'An approach to the bioenergetics of rural West Bengal,' in *Environment and Cultural Behaviour.* Ed. by A. P. Vayda. New York: The Natural History Press.

Patnaik, U. (1979) Neo-Populism and Marxism: The Chayanovian view of the agrarian question and its fundamental fallacy. *Journal of Peasant Studies* 6:375-420.

Pehrson, R. N. (1966) *The Social Organization of the Marri Baluch.* Compiled by F. Barth. (Viking Fund Publications in Anthropology 43.) New York: Wenner-Gren Foundation for Anthropological Research.

Pelto, P. J. (1973) *The Snowmobile Revolution.* Menlo Park: Cummings.

Pelzer, K. J. (1945) *Pioneer Settlement in the Asiatic Tropics.* (American Geographic Society, Special Publication 29.) New York: Institute of Pacific Relations.

Peters, D. U. (1950) Land use in Serenje District: A survey of land usage and the agricultural system of the Lala of the Serenje plateau. *The Rhodes-Livingstone Papers* 19. London: Oxford University Press.

Piementel, D. and Piementel, M. (1979) *Food, Energy and Society.* London: Edward Arnold.

Pred, A. (1973) 'Urbanization, domestic planning problems and Swedish geographic research,' in *Progress in Geography.* Ed. by C. Board et al. London: Edward Arnold.

—— (1977) The choreography of existence: Comments on Hägerstrand's time-geography and its usefullness. *Economic Geography* 53:207-221.

Polanyi, K. (1968) *Primitive, Archaic and Modern Economies.* Ed. by G. Dalton. New York: Doubleday.

Pospisil, L. (1963) *Kapauku Papuan Economy.* (Yale University Publications in Anthropology 67.) New Haven.

Punjab Board of Economic Inquiry (1938) *An Economic Survey of Durrana Langana: A Village in the Multan District of the Punjab.* Publication 54. General editor J. W. Thomas.

Rahman (1981, forthcoming) Ecology of Karez Irrigation: A Case of Pakistan. *GeoJournal.*

Rald, J. and Rald, K. (1975) *Rural Organization in Bukoba District.*

Uppsala: Scandinavian Institute of African Studies.

Rapp, A. and Hellden, U. (1979) *Research on Environmental Monitoring Methods for Land-Use Planning in African Drylands.* (Lunds Universitets Naturgeografiska Institution, Rapporter och Notiser 42). Lund.

Rappaport, R. A. (1968) *Pigs for the Ancestors: Ritual in the Ecology of a New Guinea People.* New Haven and London: Yale University Press.

—— (1969) 'Ritual regulation of environmental relations among a New Guinea people,' in *Environment and Cultural Behaviour.* Ed. by A. P. Vayda. Garden City; The Natural History Press.

Redfield, R. (1960) *The Little Community* and *Peasant Society and Culture.* Chicago: University of Chicago Press.

Reining, P. (1970) 'Social factors and food production in an East African peasant society: The Haya,' in *African Food Production Systems: Cases and Theory.* Ed. by P.F.M. McLoughlin. Baltimore: John Hopkins Press.

Richards, A. I. (1939) *Land, Labour and Diet in Northern Rhodesia.* London: Oxford University Press.

Romney, K. and Romney, R. (1966) *The Mixtecans of Juxtlahuaca, Mexico.* New York: John Wiley.

Ruthenberg, H. ed. (1968) *Smallholder Farming and Smallholder Development in Tanzania.* Munchen: Weltforum.

Sahlins, M. (1957) Land use and the extended family in Moala, Fiji. *American Anthropologist* 59:449-62.

—— (1960) 'Political power and the economy in primitive society,' in *Essays in the Science of Culture.* Ed. by G.E. Dole and R.L. Carneiro. New York: Crowell.

—— (1968) 'A note on the original affluent society,' in *Man the Hunter.* Ed. by R. Lee and I. DeVore. Chicago: Aldine.

—— (1972) *Stone Age Economics.* London: Tavistock.

—— (1978) Culture as proteins and profit. (A review of M. Harris' book Cannibals and Kings.)

Salisbury, R. F. (1962) *From Stone to Steel: Economic Consequences of Technological Change in New Guinea.* London and Melbourne: Cambridge University Press.

De Schlippe, P. (1956) *Shifting Cultivation in Africa: The Zande System of Agriculture.* London: Routledge and Kegan Paul.

Scholz, F. (1974) Belutschistan: Eine sozialgeographische Studie des Wandels in einem Nomadenland seit Beginn der Kolonialzeit. *Göttinger Geographische Abhandlungen* 63. Göttingen: Verlag Erich Goltze.

Schultz, T. W. (1964) *Transforming Traditional Agriculture.* New Haven: Yale University Press.

Scudder, T. (1962) *The Ecology of the Gwembe Tonga.* Manchester:

Manchester University Press.

Skum, N. N. (1938) *Same Sita – Lappbyn.* Stockholm: Thule.

Smith, Carol ed. (1976) *Regional Analysis.* New York: Academic Press. (2 vols.)

Sorokin, P. (1943) *Sociocultural Causality, Space, Time.* Durham, N. C.

Sorokin, P. and Berger, C. Q. (1939) *Time Budgets of Human Behavior.* Cambridge.

Soule, G. (1955) *Time for Living.* New York: Viking Press.

Spate, O.H.K. and Learmonth, T.A. (1967) *India and Pakistan: A General and Regional Geography.* London: Methuen.

Spencer, J. E. (1966) *Shifting Cultivation in Southeastern Asia.* Berkeley and Los Angeles: University of California Press.

Stauder, J. (1971) *The Majangir: Ecology and Society of a Southwest Ethiopian People.* Cambridge: Cambridge University Press.

Stenning, D.J. (1958) 'Household viability among the Pastoral Fulani,' in *The Developmental Cycle in Domestic Groups.* Ed. by J. Goody. Cambridge: Cambridge University Press.

Steward, J. (1955) *Theory of Culture Change: The Methodology of Multilinear Evolution.* Urbana: University of Illinois Press.

Stoddart, D. R. (1967) 'Organism and ecosystem as geographical models,' in *Models in Geography.* Ed. by R. J. Chorley and P. Haggett. London: Methuen.

Suttles, W. (1968) 'Coping with abundance: Subsistence on the northwest coast,' in *Man the Hunter.* Ed. by R. B. Lee and I. Devore. Chicago: Aldine.

Sweet, L. (1969) 'Camel pastoralism in North Arabia and the Minimal Camping Unit,' in *Environment and Cultural Behaviour.* Ed. by A. Vayda. Garden City: The Natural History Press.

Szalai, A., Converse, P. E., Feldheim, P., Scheuch, E. K. and Stone, P. J. (1972) eds. *The Use of Time: Daily Activities of Urban and Suburban Populations in Twelve Countries.* The Hague: Mouton.

Tax, S. (1953, 1963) *Penny Capitalism: A Guatemalan Indian Economy.* Chicago: The University of Chicago Press.

von Thünen, J. H. (1826) *Der isolierte Staat in Beziehung auf Landwirtschaft und Nationaloekonomie.* (Transl. *Von Thunen's Isolated State.* Ed. by P. Hall. London: Oxford University Press.

Turner, B.L., Hanham, R.Q. and Portararo, A.V. (1977) Population Pressure and Agricultural Intensity. *Annals of the Association of American Geographers* 67:384-396.

Vayda, A. (1961) Expansion and warfare among swidden agriculturalists. *American Anthropologist* 63:346-58.

— ed. (1969) *Environment and Cultural Behavior: Ecological Studies in Cultural Anthropology.* Garden City: The Natural History Press.

Waddell, E. (1972) *The Mound Builders: Agricultural Practices, Environment, and Society in the Central Highlands of New Guinea.* Seattle and London: University of Washington Press.

—— (1973) 'Raiapu Enga adaptive strategies: Structure and general implications,' in *The Pacific in Transition.* Ed. by H.C. Brookfield. London: Edward Arnold.

Wagner, P. L. (1960) *The Human Use of the Earth: An examination of the interaction between man and his physical environment.* Glencoe: The Free Press.

Wallin, E. (1974) *Rätten till framtida tillgångar.* Mimeo. Department of Geography, University of Lund.

—— (1980) *Vardagslivets Generativa Grammatik: Vid Gränsen Mellan Natur och Kul-tur.* (Meddelanden från Lunds Universitets Geografiska Institution, Avhandlingar 85.) Lund: Gleerup.

Watson, J. B. (1965a) From hunting to horticulture in the New Guinea highlands. *Ethnology* 4:295-309.

—— (1965b) The significance of a recent ecological change in the Central Highlands of New Guinea. *Journal of the Polynesian Society* 74:438-450.

—— (1967) Horticultural traditions of the Eastern New Guinea Highlands. *Oceania* 38:81-98.

—— (1977) Pigs, fodder and the Jones Effect in postipomoean New Guinea. *Ethnology* 16:57-70.

Wester, E. (1960) Några skånska byar enligt lantmäterikartorna. *Svensk Geografisk Årsbok* 36.

White, B. (1976) Population, involution and employment in rural Java. *Development and Change* 7:267-290.

White, C. M. N. (1959) A preliminary survey of Luvale rural economy. *The Rhodes-Livingstone Papers* 29. Manchester.

Wilkinson, J. C. (1977) *Water and Tribal Settlement in South-East Arabia: A Study of the Aflaj of Oman.* Oxford: Clarendon.

Wilkinson, R. G. (1973) *Poverty and Progress: An Ecological Model of Economic Development.* London: Methuen.

Wilson, M. (1951) *Good Company: A Study of Nyakyusa Age-Villages.* Boston: Beacon Press.

—— (1963) 'Effects on the Xhosa and Nyakyusa of Scarcity of Land,' in *African Agrarian Systems.* Ed. by D. Biebuyck. London: Oxford University Press.

Wittfogel, K. A. (1956) 'The hydraulic civilizations,' in *Man's Role in Changing the Face of the Earth.* Ed. by W. L. Thomas. Chicago.

—— (1957) *Oriental Despotism.* New Haven: Yale University Press.

Zipf, G. K. (1949) *Human Behaviour and the Principle of Least Effort.* Cambridge, Mass.: Addison-Wesley Press.

INDEX

AUTHOR INDEX

Abler, R. 216n
Adams, J. 216n
Allan, W. 17, 148

Barth, F. 30–1, 51, 110–2, 114–5, 229, 321, 395, 406
Bayliss-Smith, T. 346
Becker, G. S. 28, 41, 303
Belshaw, C. 328
Berger, C. Q. 25, 254, 333
Bharadwaj, K. 232, 233–4
Binford 90
Blaikie, P. M. 228
Bohannan, P. & L. 149, 178, 192–3, 202, 223
Boserup, E. 10–13, 18–9, 142n, 148, 156, 157, 160, 187, 189, 190–1, 195–6, –202, 211, 213–4, 220, 222, 225, 233, 237–40, 242–4, 289–97, 344, 348, 352–4, 357–8, 393, 401, 403, 404, 415–17
Briscoe, J. 395
Brookfield, H. C. 17–9, 24–5, 149, 180–1, 190, 196, 199, 221, 247, 292, 350, 372, 420
Brown, P. 17, 18, 149, 180, 221, 247
Buck, J. L. 229, 233
Bylund, E. 143

Cain, M. T. 375, 377–8, 394–6, 400, 411
Carlstein, T. 4, 24, 27, 28n, 36, 45, 51, 53, 57, 63, 154, 168n, 201n, 206n, 296, 303n, 332, 333n, 334, 343, 381, 408
Carneiro, R. L. 14, 241
Chapin, F. S. Jr. 24
Chapman, M. 182–3
Chayanov, A. V. 234, 235, 237–9, 242–3, 299, 370, 386–91, 399, 400, 402–3,

412, 413, 418
Chisholm, M. 4, 149, 166, 190, 211, 213, 216, 220, 226, 228, 232, 240, 410
Christaller, W. 4, 218
Christiansen, S. 17, 149, 184, 213, 221
Clark, C. 10, 294n
Clarke, W. C. 18
Clarkson, J. D. 3
Cleave, J. H. 335, 342
Cliff, A. D. 216n
Conklin, H. C. 147, 148
Connell, J. 24, 25, 36, 341, 344, 347–8, 386
Coon, C. S. 97
Cressey, G. B. 264n

Dahl, G. 406
Dahl, S. 119, 120n
Devore, I. 65, 90, 362
Djurfelt, G. 267
Douglas, M. 28n, 30, 357–9
Dyson-Hudson, N. & R. 127–8, 130, 291n, 406

Eggan, F. 98, 99
Eldblom, L. 277
Elder, J. W. 30, 307, 408
Ellegård, K. 57
Epstein, T. S. 178, 266
Erixon, S. 333

Fernea, R. A. 260, 274, 287–8, 290
Firth, R. 321
Forde, D. 37, 98, 101–2, 104n, 117
Foster, G. 333
Frankenberg, R. 419
Freeman, D. 148, 164, 165–6, 168, 174–5, 181
Frey, A. 216n

Friedman, Jonathan 3, 13, 58—9, 149n, 248, 271
Frödin, J. 131

Geertz, C. 2—3, 11—13, 19, 272—3, 291—2, 299, 381, 393
Ghez, G. R. 303
Giddens, A. 392
Godelier, M. 13—4, 22—3, 28, 52, 28—9, 61, 248, 271, 315—6, 391—3
Goody, J. 235
Gould, P. 216n
Gray, R. F. 259, 261—4, 281—2, 408
de Grazia, S. 341
Gregory, D. 60, 62—3
Grigg, D. B. 189, 293—4, 418
Grove, A. T. 18, 191, 214, 225, 226
Gullstrand, R. 204, 205

Hägerstrand, T. 4, 8, 24—6, 39—49, 56—7, 216n, 217, 235, 236, 246, 249n, 320, 304, 311, 315, 317, 327, 399
Haggett, P. 92, 212, 216n
Hanham, R. Q. 372
Harris, M. 12—3, 19, 66, 141—2, 147, 176, 189, 233, 257, 291, 293, 348, 352, 354, 357—8, 393, 401, 403, 415—7
Hart, D. 18, 190, 196, 199, 221, 372
Haswell, M. R. 10, 294n
Hawley, A. 2, 24
Herskovits, M. 321
HIghsmith, R. M. 264
Hindess, B. 259
Hjort, A. 119, 120n, 406
Ho, T. J. 398—9
Hultblad, F. 138
Hurst, P. Q. 259
Hunter, G. 252
Hyden, G. 206

Ingold, T. 121, 140, 363
Isherwood, B. 28n
Izikowitz, K. G. 148, 172

Jahnke, H. E. 121
Johnson, D. 103—7, 114, 122—3, 126, 130, 143—4, 146
Johnson, A. 341—2, 364—6, 368
Jones, W. O. 199

Kamuzora, C. L. 335
Kay, G. 24, 164, 178, 364, 369—71
Kemp, W. B. 5
Keynes, J. M. 39, 53

Larsson, G. 228, 232
Lea, D. 17
Leach, E. R. 30, 31, 148, 196, 259, 266—7
Learmonth, T. A. 286

LeClair, E. Jr. 55, 419
Lee, R. B. 5, 65—6, 78—86, 90, 360—2
Lenntorp, B. 24, 57, 63, 76
Levi-Strauss, C. 55, 58, 60, 64
LeVine, R. A. & B. B. 336, 339, 404
Lewis, I. M. 116
Lewis, O. 148
Lexis, W. 41, 399
Lindberg, S. 267
Linder, S. B. 28
Linton, R. 148, 162, 164, 166, 307, 359
Lipton, M. 24, 25, 36, 341, 344, 347, 348n, 386
Loficie, M. F. 206, 209
Long, N. 31, 164, 177—9, 182, 193, 199
Lösch, A. 226
Ludwig, H.-D. 224, 225

Mabogunje, A. 199
Macmillan, W. 271n
Malthus, T. 10, 393
Mamdani, M. 394—6, 399, 400—401
Manker, E. 136, 137, 140
Manshard, W. 260
Mårtensson, S. 24, 57
Marx, K. 24, 28, 39, 53, 60n
Mascarenhas, A. C. 206
Maude, A. 18, 19
Maurer 299
McArthur, M. 343
McCarthy, F. D. 343
Meier, R. L. 24
Minge-Klevana, W. 359, 417
Moore, W. E. 23
Müller-Wille, W. 226
Mumford, L. 9
Murdock, G. P. 94, 97
Myrdal, G. preface, 344, 345

Netting, R. McC. 31, 148, 180, 196, 214, 224
Nilsson, M. P 27
Nulty, L. 261

Odum, H. T. 1, 2, 7
Olsson, G. 92, 216n

Park, R. E. 2
Parkes, D. 4, 24, 27, 36, 45, 57, 63, 154, 333n
Parrack, D. W. 4
Patnaik, U. 412
Pehrson, R. N. 117
Pelto, P. J. 140
Pelzer, K. J. 148
Peters, D. U. 17, 148, 178
Piementel, D. & M. 361
Polanyi, K. 7
Portararo, A. V. 372

Pospisil, L. 19, 25, 148, 195, 221–2, 343, 372
Pred, A. 57
Punjab Board of Economic Inquiry 268

Rahman, 265
Rald, J. & K. 335, 336, 341–2
Rapp, A. 143
Rappaport, R. A. 3, 5, 17, 149, 181, 221
Redfield, R. 2
Reining, P. 335–6, 338
Ricardo, D. 28, 211
Richards, A. I. 148, 172, 180, 219, 333
Romney, K. & R. 397
Ruthenberg, H. 121, 149, 404

Sahlins, M. 12–3, 25, 31, 35, 65–6, 90–91, 190, 233, 235, 238–44, 248, 250–54, 292, 308, 327–8, 333, 342–3, 352, 354, 358, 362, 387–91, 402–3, 413n, 415
Salisbury, R. F. 25, 31, 168, 219, 221–2, 306, 321, 328, 337, 374
de Saussure, F. 55
de Schlippe, P. 148, 178
Schneider, H. 55, 419
Scholz, F. 107, 109
Schultz, T. W. 294n
Scudder, T. 28–9, 148–9, 178, 196–7, 230, 238
Skum, N. N. 137
Smith, A. 28, 39, 53
Smith, C. 34
Sorokin, P. 25, 73, 154, 333
Soule, G. 28
Spate, O. H. K. 286

Spencer, J. E. 148
Stauder, J. 149, 177, 179–81, 185, 248n
Stenning, D. J. 115, 121
Steward, J. 2, 3
Stoddart, D. R. 15
Suttles, W. 96–7
Sweet, L. 116, 119
Szalai, A. 25, 36, 333, 335

Tax, S. 333, 343, 383–5
Thrift, N. 4, 24, 27, 28n, 36, 45, 53, 57, 63, 154, 303, 333n, 334, 343
von Thünen, J. H. 4, 211–2, 242, 387
Turner, B. L. 372

Vayda, A. 3, 176
Vidal de la Blache, P. 61n

Waddell, E. 18–9, 181, 196, 199, 201, 221, 222, 247–8, 372–6, 379, 402
Wallin, E. 57
Watson, J. B. 199
Wester, E. 231
White, B. 299, 300, 334, 336, 340, 361, 379–82, 384, 394–6, 398
White, C. M. N. 199, 202
Wilkinson, R. G. 2, 11–13, 15, 19, 66, 143, 189, 254, 348, 352–4, 357–8, 393, 401, 403, 415–17
Wilkinson, J. C. 265–7, 273–6, 280, 284, 285
Wilson, M. 180, 229, 296
Wittfogel, K. A. 12, 257–9, 263, 291, 379

Zipf, G. K. 30, 218, 219

SUBJECT INDEX

accessibility 38, 74, 165 (see also prisms)

activities 5, 22, 48, collective 324

activity possibility /boundaries/ 33, 52-55, 105

adaptation 2, 3, 13

activity system 48, 245—49, 317, 334, 346, 399, 419—22

affluent society, original 65

age-sex categories 88, 100, 135, 168, 306, 307, 346—8, 356, 408

agro-pastoralism 126—36

animals Ch. 4, 224

animal-time 228, 268—9

annual cycle of activities 88, 105, 108, 111—13, 122—25, 127—30, 133, 136—9, 155, 275, 280—82, 286—88, 388—89, 400

Anthropology 18, 34, 185, 321, 327—30, 333

apiculture 181

Baluchi 107, 109, 126, 143

Bangladesh 243, 377—9, 394

Baruya 315—6

Basseri 110—14, 145—6, 406

Bedouin 119

Bellona 184

Bemba 172, 177, 180, 191

Big-men 250, 252, 402

Blackfoot 99, 102

Boserup's intensification model 293—4

bundles 42, 332

Bushmen (San) 66, 78—89, 360—62

Bushong 357—59

capability constraints 25—6, 46—7, 50, 105, 286, 405

carrying capacity 5, 10, 14—18, 19—21, 81—2, 92, 111—15, 151, 162—4, 240, 292

capitalism 50

caste system 307—8, 408—9

Ceylon 196, 260, 274

Char Gopalpur (Bangladesh) 375, 377—9

Chayanov's theory 234—9, 242, 386—91

children, economic role of 395—399

Chimbu 18—9, 180, 221—2, 239, 350

China 229, 257, 261

Chinook 96

choice 26, 47—8, 55, 75 (see also scarcity)

class 410—414

consolidation of holdings 229—32

consumption 28, 52, 54

convenience 28—32

coupling constraints 26, 46, 105, 121, 282—88, 325—6, 405

cropping cycles 155—58

cropping frequency 11, 156—8, 191

culture 3, 5, 33

cultural evolution 9, 12, 13, 22, 65, 66, 103

cultural landscape 5, 241

daily round 44, 168, 170—2, 174, 334—9, 346, 397, 405

decisions (see choice)

deintensification 140, 195—98, 224

delegation of activities 313

demand pressure 192, 293—300

demography 41, 185, 235—6, 242—4, 305, 342, 391—402 (see also: population)

diachronic study 22 (see also: holochronic)

dialectics 63

displacement effects 327, 408

distance and interaction 92—4, 117—9, 168, 215, 216—20

divisibility (cf indivisibility)

domains 45, 101, 228—35
domestication of animals 104
domestic groups, developmental cycle of
    235—8 (see also: family cycle)
domestic mode of production 12, 239—44,
    250—55, 308, 390—91
dynamic analysis 22

ecological adaptation 13
ecological balance 2
ecology, biological 1, 5—7
ecology, cultural 3
ecology, human 2—3, 5—7, 15, 38
ecosystem 3, 15
ecotechnology 1, 5, 8—9 (definition)
elbowroom 152—53
effort minimization 30 (see also: Zipf, move-
    ment minimization)
energy 15, 17, 20, 105, 226, 260, 346, 360-
    61
energy crisis 15, 39
energy time-budgets (see: time-energy bud-
    gets)
employment (see: labour, work)
England 297
expansion (in space) 135, 141—3, 237

factors of production 10
fallows 152, 156—8, 289—90
family (see: domestic units, households)
family, extended 400
farm size 234, 237—8 (see also: class)
farmstead 228—38
fertility, demographic 125, 395—400
fertility maintenance (of land) 19, 135, 147,
    160, 210, 213, 214, 221—7, 258, 290
fishermen, sedentary 94—7
fodder 131—35
food-chains and webs 4, 39
food preparation 97, Figures 9:6—11
foraging (cf hunting-gathering)
forces of production 59
forest fallow 156—9
fusion—fission 90, 91, 121—25, 128—9, 135,
    156—8, 175—85, 227

genre-de-vie 61
Geography 18—19, 32, 34—7, 60—64, 216
geometric approaches 61
gestation periods 314
Chadames 277
goals 47—8 (see also: projects)
gravity model 217 (see also: distance and
    interaction)
Guadalcanal 182—84
Guatemala 383—85 (see also: Panajachel)
Gwembe Tonga 196—7, 230, 238

habitat 5
Haida 96

holdings, size of 232—8
holism 35
holochronic analysis 23, 63, 149, 421
horses 98—102, 117
horticulture 148
households 42—3, 228—44, 334—39, 386—
    400
household size 237—8, 386—400, 406
hunting-gathering 9, Ch. 3, 360—63

Iban 164—8, 174—5, 180, 182, 239, 241
ILO 36
impaction 196—8, 224
India 232—4, 243, 266—7, 305
indivisibility 40, 49, 56, 68—9, 121, 126,
    251, 271—2
Indonesia 2 (see also: Java, Kali Loro)
Indus 286—88
industrial revolution 12, 15
industrial societies 19
industrial technology 261
infrastructure—superstructure   model   60,
    249
infield-outfield systems 213
inheritance 228
innovations 11, 15, 19, 101, 209
institutions 33, 48, 242
integrative theory 34—6, 419—22
intensification 9, 18—9, 135, 141—3, 161,
    187, 190, 193—5, 199, 210—16, 259,
    289—300, 348—59, 386—418
intensification theory 10—18, 292—300,
    415—17
intention(ality) 48, 249 (see also: projects)
interaction theory 3, 14, 33—4, 244—9, 269,
    295, 327
involution 11
Ipomoean Revolution 199
irrigation agriculture Ch. 7, 379—85
irrigation cycles 276—80
Islam 264, 277, 312

jaimani system 408
Java 291, 299, 379—82, 394
Juxtlahuatl (Mexico) 397

Kababish 117, 120, 122—5, 127
Kali Loro (Java) 379—82
Kapauku 19, 221, 239, 343, 350
Karimojong 127—30, 145, 261
Kazaks 117
Kofyar 180, 196, 214, 223—4, 240, 262
Kurdshuli 111
Kwakiutl 96

labour 28—9, 340—47, 386—9, 412—14
labour theory of value 60
labour utilization identity 344—5
Lala 178, 193, 201, 202, 239
Lamet 165, 172, 239, 241

land distribution 223—5 (see also: class)
land reforms 227, 229—32
land use (see: settlement, space-time)
Lapps 136—40, 143
Lele 357—9
Libya 277
life-line 41, 399
life-paths 38—47, 399
life-span 49
lingua franca 101
living and activity possibility boundaries
    (see: activity possibility boundaries)
local prism habitat 81—2, 144, 162—3, 207
location theory 4, 5
Luvale 202

Machiguenga 364—8
Majangir 177, 179, 180—81, 185
manuring (see: fertility maintenance)
marginal utility 10
marxist approaches 52
marxist-structuralist approaches 58, 62 (see
    also: Godelier, Friedman)
Masai 261
Mesopotamia 286—8, 290
Mexico 397
migration 125, 217, 297
mixed farming 223—7
Moala 242
mobility (spatial) 73, 85, 94—102, 107,
    144—6, 172
mode of production 391—2 (see also: dom-
    estic mode of production)
money (and time) 327—30
mortality 125
movability 69 (see also: portability)
mounted hunters 97—102
movement minimization 78, 218—9
multicropping 156, 259, 290
Mutair 117

Neolithic Revolution 66, 105
New Guinea 17, 221—3
Ngoni 177
Nigeria 191, 214, 225
Nile 287
nomads and nomadic pastoralism Ch. 4
Nootka 96
norms 48, 88 (see also: sex roles)
Northwest Coast Indians 94—7
Nyakyusa 180, 229, 296

Oman 264—6, 272, 274, 276, 280, 284—6
ontology 419
oriental despotism 257—8

Pakhtuns (Pathans) 229
Pakistan 109, 265
Panajachel (Guatemala) 343, 383—5
paradigm, economistic 419—20

parallel cultivation systems 19
parcel 151 (definition)
pastoralism 9, Ch. 4, 406
pasturage 108, 110, 114
path allocation 46, 125, 244
pax/Pax Britannica 178, 196—7, 266 (see
    also: warfare)
Philippines 261
physics 27
pig cycle 17, 181
pig husbandry 222—3 (see also: Raiapu Enga)
Plains Indians 97—102, 181
plough 195, 227
polygyny 402
political dominance 13, 299—300 (see also:
    class)
population of artifacts 8, 43, 297
population density 14—18, 94—5, 203, 240
population dynamics and growth 3, 10—14,
    19, 91, 125, 292—300, 393—402
population pressure 12, 141, 242—4, 292—
    300
population size 90, 160 (see also: village size)
population system 244—5, 317—21, 399n
population time-budget 20, 32—3, Ch. 8,
    407—10 (see also: time-budget)
portability 69, 96, 226—7
porterage 134
predatory pastoralism 138—40
principle of least effort 218 (see also: Zipf)
principle of return 70, 240
principle of reunion 70—73, 240
prism 74—8, 83—4, 87—8, 92—6, 100, 111,
    117—19, 132, 135, 144—6, 163, 167,
    174, 177, 203, 210, 214
production 52, 258
production possibility boundaries 52, 59—
    60, 240, 302
productivity factors 147, 189, 190, 227,
    290, 297, 345, 389, 395, 410—14
pseudo-intensification 199—206
Pul Eliya 266—7, 282
Punjab 230, 267—9, 286—8

quanat 264—6

Raiapu Enga 195, 221, 222, 247, 372—75,
    379
regulatory constraints 48—9, 242, 288
relativity theory 154
residence 144—6, 163, 164, 209
residential shifts 89, 158—64
Russia 238

Salish 96
satellite settlements 130—33, 146, 164—6
Saudi Arabia 272
Scandinavia 130—40
scarcity 26—8, 321—2, 419—20
schooling 43, 312

seter system 130–33, 145–6
settlement schemes 206–11
settlement size (see: village size)
settlement system 44, 68, 92–4, 246–7
sex roles 403–5 (see also: age-sex categories)
shifting cultivation 5, 9, Ch. 5, 364–71
short fallow cultivation Ch. 6, 372–79
Siane 165, 168–71, 179, 221–22, 241, 306
Sicily 216
situations (in time-space) 55–6
sleep 313
snow-mobile 140
socialism 50
social structure 13, 61
society-habitat 62
sociology 2
soil fertility (cf fertility maintenance)
Somali 116
Sonjo 261–4, 281–3, 408
space-time (as land) 19–20, 150–53, 97, 348
space-time budget 19–21, 150–53, 215, 223–4, 270, 275, 285
Spain 216
spatial analysis 4, 15
spatial density 152
spatial mobility (see: mobility)
spatial organization 15, 24, 64
spatial structure 60–64
specialization 407–10
stall-feeding 223–5
stations 43, 67–70
storage 70
storage versus mobility 67–9, 91, 96–7, 99, 127–35
structural causality 14, 248–9
structuralism 55–64
structure 52, 58
Sweden 131–3, 227–8, 230–31, 235–6, 297, 305
swidden agriculture Ch. 5
synchorization 270
synchronic study 22
synchronization 73, 270

Tanala 162
Tanzania 149, 206–11, 226, 229, 250–55, 261
technology 5–9, 114–5, 188, 395
temporal allocation of water 269–88
temporal intensity 151–2, 237
territories/territoriality 101, 121, 158, 160, 198, 205–6
terroirs 206–6
time allocation 153–4, Ch. 8, Ch. 9
time-budget 19, 20–21, 22–25, 77, 154, 233–4
time demand Ch. 8
time distance 76
time-energy budgets 5, 17

Time Geography 34, Ch. 2, 153
time indication 27
time reckoning 27
time resources 22–33, 166–8, Ch. 8, Ch. 9
time-saving innovations 101
time sharing of water 274–79
time supply Ch. 8
Tiv 191–2, 193, 202, 223
Tlingit 96
Tonga 19
traction animals 223, 226–7
trajectories (see: life-paths)
transhumance 130
transport costs 211–2
transport possibility boundaries 53 (see also: prisms)
travel time 32, 172–4, 184, 209, 218–9, 229–32, 253, 340, 345
trophic exchanges 3, 7
Tsembaga 221–2, 239, 350
tubers 199–206, 221–3
Turkey 142–3

Ujamaa villages 207–11, 250–5
Ushi 178, 369–71

viability 121
village size 170, 175–85, 201, 206–11, 224, 241
villagization 206–11, 250–55

Wakara 224–5
warfare (see also: pax) 176–8
water 17, 79–80, 82, 115–17, 120–21, 228, Ch. 7 (see also: irrigation)
water-time 270–88
watering intensity 276–80
wells 267–9
work 7, 331–2, 340, 344, 386–89
work-load 349–56
World Employment Programme 36

Xhosa 296

Zaire 204–5
Zambia 164